Birgit Schneider, Thomas Nocke (eds.)
Image Politics of Climate Change

Image | Volume 55

Birgit Schneider, Thomas Nocke (eds.)

Image Politics of Climate Change

Visualizations, Imaginations, Documentations

[transcript]

Supported by the Fritz Thyssen Foundation and
the Potsdam Institute for Climate Impact Research

Bibliographic information published by the Deutsche Nationalbibliothek
The Deutsche Nationalbibliothek lists this publication in the Deutsche Nationalbibliografie; detailed bibliographic data are available in the Internet at http://dnb.d-nb.de

Cover layout: Kordula Röckenhaus, Bielefeld
Cover illustration: Collage by Bijan Dawallu, aromaberlin, 2011.
(The graphs are taken from Michael Mann, Micheal E. and Lee R. Kump.
Dire Predictions. Understanding Global Warming.
New York: Dorling Kindersley, 2008.)
Proofreading: Helen Carter
Printed and bound in Great Britain by Marston Book Services Ltd, Oxfordshire
Print-ISBN 978-3-8376-2610-0
PDF-ISBN 978-3-8394-2610-4

Table of Contents

CHAPTER 3
IMAGES OF CLIMATE CHANGE IN THE PRESS AND ON THE WEB

CHAPTER 4
FROM VISION TO ACTION? MAKING THE INVISIBLE IMAGINABLE THROUGH ART AND PHOTOGRAPHY

Chapter 5
Images of Climate Control

Image Politics of Climate Change: Introduction

BIRGIT SCHNEIDER, THOMAS NOCKE

In September 2013, the *Intergovernmental Panel on Climate Change* (IPCC) presented its fifth report about climate change during a press conference in Stockholm. On this occasion, the audience was also presented with some of the latest scientific images produced for the new Summary for Policy Makers. In the manner of the previous reports colored global maps, barren graphs and curves served as most simplified condensates for visually communicating the findings of the highly complex research field of climate change.

For the fifth report, the graphs seemed to have been shaped even more directly, woodcut-like, in order to put across the urgency of the worrying messages delivered to the decision makers even more clearly this time. Thomas Stocker, Professor of Climate and Environmental Physics and head of the IPCC Working Group I ("the physical science basis of climate change") directly addressed the imaginative power of scientific imagery. While pointing to a global map marking mean temperatures in bright red he declared: "This is the face of the surface of our planet if you look at the atmosphere. It is red. The world has been warming. The trend that you see is clearly given as colors of red."[1] His co-speaker, Michel Jarreaud, Secretary-General of the World Meteorological Organization, pointed to another highly condensed graph, showing the comparison of temperature scenarios with two curves in blue and red climbing more or less steeply into

1 The press conference was held in Stockholm on 27 September 2013 to present the Summary for Policymakers of the Working Group I contribution to the Fifth Assessment Report. The quotes were taken from the webcast which in the meantime has been deleted.

possible climate futures. He commented on the graph saying: "The children born now will *see* the last part of the graphic." [our italics][2]

Future is invisible; likewise, global temperatures can neither be experienced nor seen. When, during the press conference, the IPCC scientists pointed to the highly abstracted and conventionalized graphs, they communicated the essence of climate scientific findings with the help of visual media. By speaking *colloquially* about the images, the scientists *metaphorically* related the data abstractions back to their concrete meaning for mankind.

Alongside scientific graphs, many alternative and even opposing image strategies exist today that also strain to testify to climate change. The online picture library "Global Warming Images," compiled by the photographer Ashley Cooper during the last ten years, might serve as an example. In contrast to the scientific extractions like global graphs and maps, the online library photographically documents the impact of climate change on landscapes, people and wildlife by capturing visually the consequences of weather extremes in specific locations that are related to climate change. Differently to scientific graphs, but with a similar outcome, the photographs make evident climate change as an inevitable process which deeply transforms the world as we know it. Cooper's photographs can be taken as examples of the vast amount of images which pinpoint tangible and local hot spots where possible impacts of climate change can be experienced and seen directly. Cooper traveled around the globe to witness, with the help of his camera, droughts in Australia, floods in Bangladesh, coastal erosions in the UK, glacial retreats in France, or rising sea-levels in Amsterdam, to name but a few.

Both categories of images—the abstract condensates of global climate science and the pictures of concrete climate events—have entered the collective "cultural memory" (Jan Assmann). Images have started to shape the imagination of a world under the conditions of climate change.

VISUALIZING THE UNIMAGINABLE—MOTIVATION TO WRITE THIS BOOK

The present book seeks to analyze the full diversity of pictures originating from different fields which all claim to make climate change meaningful on a visual level. With this publication, we want to add to contributions made by Julie Doyle, Mike Hulme, Libby Lester, James Painter or Stephen Sheppard who have

2 Ibid.

all published on certain aspects of images in the climate change discourse in recent years.[3] Since there is no single discipline appointed to deal with the subject, we decided to compile this anthology in order to make the many approaches to climate pictures accessible by combining them across the disciplines.

A starting point for putting together this anthology has been the observation that since scientists began to make their findings public, the number of images picturing climate change has been growing immensely: the reports of the IPCC are filled with a plethora of colorful graphs; every week, another popular book on climate change broadcasting scientific data graphics is published, but also plenty of expensive photobooks have been compiled depicting landscapes under the impact of climate change; magazines print high quality photo series of a changing world and the latest natural disasters caused by climate impacts; art galleries have put together a great number of shows in the last few years addressing climate change issues at an artistic level; numerous movies have been produced focusing on the issue of a final and global catastrophe or addressing climate change from a documentary perspective; innumerable websites and blogs with scientific, journalistic or political backgrounds exist which make use of climate change pictures; pictures of climate change have even become a permanent backdrop for the visual language of advertising.

To sum up, scientific research on climate change has given rise to a variety of images picturing climate change. Alongside the diversity of scientific pictures, many other fields have started to translate climate change findings into languages other than science. In this case, the highly conventionalized and regulated structures of scientific graphs, tables, bar charts and map structures undergo a different framing. Although originating in climate science, in the cultural spheres alternative images are produced in order to communicate and comprehend the different meanings and impacts of climate change.

3 Lester and Hutchins, *Environmental Conflict and the Media*, 2013; Painter, *Climate Change in the Media*, 2013; Sheppard, *Visualizing Climate Change*, 2012; Doyle, *Mediating Climate Change*, 2011; Hulme, *Why We Disagree About Climate Change*, 2009. The following should also be mentioned regarding individual articles: Himmelsbach and Oldenburg, *Ecomedia,* 2007; Dobrin and Morey, *Ecosee*, 2009. Articles focussing on the issue: Knebusch, "Art and Climate (Change) Perception," 2008; Rosentrater, "Representing and Using Scenarios for Responding to Climate Change," 2010; Schneider, "Climate Model Simulation Visualization from a Visual Studies Perspective," 2012, Schneider, "Image Politics," 2011; Walsh, "Visual Strategies to Integrate Ethos across the "is/ought" Divide in the IPCC's Climate Change 2007," 2009; Yussof, "Mapping the Disaster," 2006.

Due to their quantity, it is necessary to critically analyze climate pictures, because a reverse movement can be observed, as is the case for images in general. It is a special characteristic of images that once in use, they become normalized;[4] they acquire the characteristic of unquestioned naturalness, which makes us forget their constructiveness. The images gain their "own reality."[5]

Climate Change Knowledge is Mediated Knowledge

The sheer observation of quantity is contrasted by another observation that is also causally connected to climate science: most phenomena studied in climate science are invisible.[6] Climate change as a long term process cannot be seen. It needs to be constructed on the basis of physics, chemistry and big data: measurements, simulations and statistics. Because climate is a scientifically constructed object, there is no way to learn about it other than through media devices. We *need* media to learn about climate change. Visualization methods represent an important means of mediating climate change besides using language. Therefore one needs to ask how (Western) culture obtained this view.

Pictures have played a significant role in climatology ever since its beginnings around 1800. The history of climatology began with the transferal of increasing amounts of weather data into graphs and maps, such as the first climate map based on information visualization which Alexander von Humboldt realized on the basis of only 58 short term measurement series from weather stations (see Figure 6 on page 42).[7] Synopsis (literally 'seeing together'), the method of creating images based on statistical thinking, gave form to the data collections of otherwise shapeless and spatially detached weather events.

It was only through these new strategies of visualizing numbers of observation and measurement that the climate became evident, and as such, an epistemic object of research. By the same token, it is thanks to the existence of charts that

4 Gugerli and Orland, *Ganz normale Bilder*, 2002; Blumenberg, "Lebenswelt und Technisierung," 1993.

5 Heintz, Huber, "Der verführerische Blick," 2001, 25 (translated by the auhtors).

6 Knebusch, ibid.

7 This is what German naturalist Alexander von Humboldt (1769-1859) did in 1817, when he outlined the first set of isotherms—statistical lines of average temperature—on a map structure. By introducing isotherms he was able to give an idea of climatic zones. Cf. Schneider, "Reisepfade der Erkenntnis," 2012; Monmonnier, *Air Apparent*, 1999 and the article by Grevsmühl in this volume.

climate change became detected. Without such graphical methods, climate warming would have been much more difficult to interpret or might even have gone unnoticed. The discovery of climate change is a cultural achievement originating in the scientific history of knowledge production and the globalization processes accompanying this development.

Therefore we propose to examine in this book the hypothesis that today, the media and the climate as scientific object are inextricably linked.[8] Knowledge about climate change on a global level cannot be acquired through direct experience but only via translations and abstractions accomplished by media. Abstract scientific graphs are where climate change first became visible, like the curve plotting first begun by Charles David Keeling in 1958 to document the ongoing change in concentration of carbon dioxide in Earth's atmosphere. The constructed abstractions of climate science have been the primary catalyst for images of climate change.

Climate Change can be Successfully Visualized but not Imagined

Although a full range of pictures successfully visualizing climate as a scientific entity already exists, scientific graphs have been repeatedly blamed, from very different sides, for being highly limited when it comes to communicating climate change to those not involved in climate research. Here, we need to differentiate between questions of visualization and imagination. Some findings are so vast and huge—they contradict all experiences we have about the world today; in fact they are *unimaginable*. We can look at the IPCC curves as long as we want—the meaning of climate futures stays unimaginable although the colored lines follow a clear rationality. Here, two distinct forms of expression clash. In the case of climate change, the vast metaphor of nature's revenge, according to Lorraine Daston, evokes a feeling of "terror," because passions cannot respond to the global violation of nature—our senses fail.[9]

---

8 Climate change in this global totality might even be called an "absolute metaphor" (Hans Blumenberg), a thing we are able to name but we cannot fully relate to (like "the world," "history" etc.).

9 Talk given on the conference *The Artwork between Technology and Nature*, January, 22 2010, National Art Gallery Denmark, Kopenhagen. For Daston's concept of terror, horror and wonder as reactions to the violation of natural orders see Daston, *Wonders and the Order of Nature*, 1998.

In response to this, various actors have urged the need for images distinct from science to fill this gap. Climatologist Hans Joachim Schellnhuber, for example, during a conference called "Tipping Point" organized by the British Council in Potsdam/Germany in 2008, put forward his idea for an "exhibition of the worst nightmares of climate scientists." The conference brought artists from different fields into dialogue with climate scientists from the Potsdam Institute for Climate Impact Research. Schellnhuber explained that he felt the urge to use a format like this because in the highly regulated framework of scientific practice there was no space for scientists' emotional reactions, such as fear, to the findings. Such emotions had to be assiduously suppressed when making scientific statements.

Alongside climate scientists, humanities scholars like cultural scientists or philosophers, environmentalists, curators and artists have pointed out the alternative potential of artistic approaches to imagining climate changes and to communicating the matters concerning the climate crisis.

It has been claimed that contemporary art in particular has the potential "for discovering new ways of seeing the world" and for "rethink[ing] categories and phenomena that we usually take for granted."[10] In the case of climate change, it has been suggested that art might be able to "help focus attention on the cultural dimension of climate change while also outlining the contours of a range of wider social, political, and aesthetic changes that point towards a new cultural awareness."[11] Given these expectations, art is usually placed on the communicating and the imaginative side of the problem. The artists do not offer solutions to the problems, "but images which can be employed as tools for reflection, discussion, insight—and possibly, action."[12]

Therefore we suggest that the problem is one of imagination rather than of representation. Because we *do* have pictures that successfully communicate the concept of climate change especially in terms of science.

Climate Images are Political Images

Although it can be claimed that images of climate change make visible the unimaginable, at the same time actors in all fields repeatedly state that the current images of climate change wouldn't be strong enough to deliver the urgent information effectively enough. For this reason, more emotive images are sought,

10 Witzke, "Introduction," 2009, 9.

11 Ibid.

12 Hansen, "Rethink Kakotopia," 2009, 12.

ones that are more "expressive" and clear, and provide better metaphors and more tangible future scenarios. The argument goes like this: We *know* enough about climate change, but still so little is done on all levels—ergo we need stronger, more striking and engaging images. But the claim—or hope—that images have the power to produce action and influence decisions, and in so doing, to change reality, is only rarely brought into question.

When such statements are made, the political dimension of images is addressed—the practice and theory of influencing people's opinions and decisions. Climate pictures are often produced for normative purposes; they are fabricated to change what they show: possible futures are blueprinted in order to prevent the futures shown with the help of the curves from coming true. The power of evidence and the power of persuasion are two main features which are especially valid for climate pictures. Climate pictures are produced to testify, prove and explain the concept, impacts and social and cultural meanings of climate change.

Because of their normative character, what is special about images depicting climate change is that they have implicitly also become *political images*. This is likewise true for scientific images of climate change. Climate science is facing considerable pressure from all sides: when scientific graphics produced by climatologists started to gain currency in the field of policy, as climate change became a key issue within risk society, they encountered different values and expectations. Climate change is not only a science subject, it is also at the core of socio-political interests.[13] Climate science is the paradigmatic field in which images assume a role as political agents. Discussions about the trustworthiness of the latest findings are therefore often conducted based on key images, since they offer evidence that attracts the most careful attention. Scientific graphics about climate change are used as the basis on which political decisions are discussed—the heated discussions of climate change skeptics and mainstream climate science researchers might serve as the most obvious example of where the political role (and the blurring of roles) of climate science pictures is revealed.

Is there a general link from knowledge to action, from image to action? It is claimed that using pictures which are as effective as possible helps to vividly put across to the recipients what they are unable to visualize and become conscious of by means of text alone. It is this implicit basic assumption of the power of images to influence reality that we address in this book. How are images supposed to influence reality? What different assumptions underlie image usage in the

13 Welzer and Leggewie, *Das Ende der Welt wie wir sie kannten*, 2009; Hulme, *Why We Disagree About Climate Change*, 2009.

broad field of climate change communication? What problem-solving approaches are taken in the different disciplines?

THE GENERAL ROLE OF THE VISUAL

Media society thrives on a predominance of the visual. Messages are to be accompanied by pictures. In the case of climate change, pictures have at least the same importance as text in shaping our knowledge of the world. Therefore, we want to point out the most important functions and roles ascribed to images in this field.

It is generally assumed that pictures have a pedagogical ability to show complex connections in an easy, comprehensible and instant manner. Successful pictures are meant to reveal knowledge and meanings at a glance. Because of their greater vividness, images are thought to be "a more easily accessible medium of communication than (conceptual) language."[14]

This is especially true for scientific images.[15] In the sciences, successful climate pictures "condense" knowledge; they are the extraction in a cascading process of simplification.[16] Many climate pictures more or less directly rely on the complex research of climate change findings (big data, statistics, global perspectives). But in the process of making images, complex knowledge is radically reduced to core statements and concise synopses. The process of knowledge acquisition remains invisible—in the background of the pictures. The picture extracted might be thought of as an iceberg where only the very top can be seen; still the outstanding peak signifies for the whole iceberg.

The unique benefit of condensing knowledge into a single picture frame is that images, in contrast to texts alone, have the ability to very directly address and concentrate human perception.

Parallel to these productive epistemic functions images can fulfill there are, however, many strong reservations opposing these positive views on the power of images. These reservations are likewise contrasted with the understanding of texts. From a negative perspective, pictures are believed to provide merely a blurred version of reality. They are said to generally appeal more to emotions than intellect. They are considered to belong to the realm of emotions and sub-

14 Hüppauf and Weingart, "Wissenschaftsbilder—Bilder der Wissenschaft," 2009, 14.

15 Elkins, *The Domain of Images*, 1999; Bredekamp, Dünkel and Schneider, *The Technical Image,* 2014.

16 Lenoir, *Inscribing Science*, 1998; Latour, "Drawing Things Together," 1990.

jectivity, whereas texts are thought to belong to the field of rationality. Accordingly, images leave the door open for deception, propaganda, illusion and seduction, and can attract a great deal of mistrust.

However, the ease with which fabricated images can be distributed indicates their potential power. The result of the simplification and extraction process referred to is an eidetic and concise gestalt. This gestalt can be easily reproduced, distributed, recombined and scaled. This is what Bruno Latour summed up with the term "immutable mobile."[17] In the context of scientific pictures of climate this means when the figures are communicated and distributed beyond the field of science, the accompanying texts might be cut out in the process and the images might start to travel independently, detached from their original background. Images start to migrate from one sphere to another. They might keep their basic gestalt (immutable), but as picture agents they are able to serve various interest groups, trigger different associations and offer new perspectives.

Images can start to travel and migrate in different ways. Initially, they are reproduced, copied or reworked when a scientific paper is brought to a news magazine and an image might be redrawn by a graphic designer. Or a photograph is printed in a magazine. But when images get taken up by the media they gain collective visibility and something else can happen. Pictures can imprint their gestalt onto the memory of the observers: some images become *iconic*: in other words, they start to influence or even dominate the concept of a world under the threat of climate change. This happens when pictures start to enter the collective memory by acquiring "collective symbolism" (Jürgen Link).[18] W. J. Thomas Mitchell has described images metaphorically as quasi-living organisms (spreading even like viruses) that populate the world, and have even received the status of subjects in our culture.[19] By ensnaring people within a structure of desire and power, images are even able to develop omnipotence. The Hockey Stick graph and the polar bear would be the most obvious examples for iconic images of climate change today; but also photographs picturing droughts, floods and retreating glaciers would fall into this category, or even old religious images like Judgment Day or biblical plagues which become linked with images of a threatening climate catastrophe. Such images today influence how climate change is imagined.

This broad impact of images is, we believe, due to the fact that images contain a surplus value that cannot be fully controlled. This aligns with a unifying

17 Latour, ibid.

18 Link, Hall, "The Normalistic Subject and Its Curves," 2004.

19 Mitchell, *What Do Pictures Want?*, 2007.

factor present in all these articles, which is the idea that pictures in general, even the most straightforward charts, do more than merely illustrate data information.[20] Rather, they inevitably imbue the findings with an additional quality beyond the objective and the rational. This quality is often emotional, sometimes turning contemporary climate pictures into ambivalent agents, depending on who makes use of them.

To sum up: when images are widely distributed, they are indeed able to shape how the world is thought about and seen. In this respect, they are able to also shape futures—by opening up thoughts about futures but also by narrowing the perspective, as in the case of geo-engineering phantasies. They might be visions with only fictional status but they can become the catalyst for future actions. Today's images might become the blueprint for tomorrow's realities. Therefore climate pictures have a key role in making the future imaginable.

DESCRIPTION OF THE BOOK

This book takes as its subject the visual aspect of the climate discourse, treated here in sixteen interdisciplinary case studies. It takes a critical look at the very varying strategies that are used by the different contexts. For this purpose, it combines a wide interdisciplinary range of perspectives and questions, in order to discuss the different strategies and conceptions that lie behind pictures concerning climate. By analyzing very disparate kinds of image-making and image interpretation in the field, one might even say the articles compiled in this volume portrait distinct cultures of image-making. But by looking at the images through the lens of "images of climate change" we consequently needed to bring together very different disciplines in order to give an outline of a fragmented field.

A particular feature of this book—the authors' specialist backgrounds—is reflected in the interdisciplinary nature of the subject. Their specializations include both visual practice and theory: in the fields of climate sciences, computer graphics, art, curating, art history and visual studies, communication and cultural science, environmental and science and technology studies. The close interlinking of these viewpoints promotes in-depth insights into issues of production and analysis of climate visualization, some of which have the character of a work-in-progress.

20 Bredekamp, Dünkel and Schneider, "The Image," 2014, 12.

All articles in this book share a common basic feature: the lines of inquiry they pursue remain close to the material, in other words, they use case studies as examples. At its heart are images which depict the climate and climate change in very different ways—landscape photography, press photos, maps and diagrams used in climate science, and artists' treatments of the subject and the visual strategies of environmental activism. In the process, the visual part of the climate discourse is critically analyzed in the context of politics, technology, science, media and society. The visual examples originate mainly from the current climate discourse—most concern climate *change*. As the volume also includes historical case studies, it will be possible to relate the history of climate graphics to current practice.

The volume is divided into five thematic sections, or chapters, each of which highlights the subject from a different perspective. The sections are organized not according to disciplines, but themes. This enables them to be viewed simultaneously from different disciplinary perspectives. It becomes apparent on reading the individual articles which concepts and pictorial logics underlie the particular forms of visualization, also at which points frictions emerge due to different expectations and conceptions when specialist pictures in particular circulate amongst the politics, science, press and publicity scenes.

The first section—The Epistemic Value of Visualization in Climate Sciences—examines the issue of the epistemological status pictures have in climate science, both historically and currently; in other words, how climate as a research object is defined using visualization. Sebastian Grevsmühl discusses the importance of visual metaphors for climate science for the example of the "ozone hole" to explain how the visual creation of geophysical invisible phenomena takes place. Thomas Nocke provides an overview of how visualization is used within the heterogeneous climate science field in the form of an analysis method, and illustrates recent developments and new data visualizations.

The second section—Communicating Results: The Status of Climate Expert Graphs in IPCC Reports—centers on scientific expert pictures and how they have acquired great significance in the wake of the IPCC (Intergovernmental Panel on Climate Change) reports. There is an apparent juxtaposition of the scientific ideal of sober objectivity and a contrasting sense of anxiety and warning which is particularly evident from the pictures. In this context, Lynda Walsh takes up the issue of graphics in the IPCC Summary for Policy Makers of 2007 in a rhetorical analysis and reveals popular misconceptions about scientific graphs in public perception. Mike Hulme and Martin Mahony perform an analysis of the "burning embers" diagram which visualizes abstract concepts such as risk and danger, investigating the social organization of knowledge production and the role of expert judgments.

The third section—Pictures of Climate Change in the Press and on the Web—looks at the role of different media formats such as print and internet. The main issue here concerns what intentions, conventions and forms of logic are behind the use of pictures in these areas, and aims to provoke discussion around the subject of climate and climate change. Elke Grittmann analyzes the highly conventionalized image language used by the press during the United Nations Climate Change Conference in Copenhagen in 2009. Birgit Schneider, Thomas Nocke and Georg Feulner investigate the images employed in skeptical climate media and the argumentation strategy used to visually cast doubt on climate research findings. Wrobel and Reusser review challenges and approaches to visually communicate climate research knowledge on web portals interactively.

The first, second and third chapters deal mostly with maps and diagrams, whereas in chapter four—From Vision to Action? Making the Invisible Imaginable through Art and Photography—the special role of photography and art, and the visual strategies used by climate activists are analyzed. The images discussed here all aim to make visible the invisible in climate change, involving terms such as "shock" and "sublimity," and show the most difficult task of picturing climate change using images in the first place. Julie Doyle identifies a dominance of the photograph within Greenpeace campaigns, and analyzes Greenpeace climate change campaign literature since 1994. Vera Tollmann investigates visual strategies of climate activists, reviewing the example of the polar bear and its visual deconstruction. Ulrike Heine presents a systematic analysis of several elaborate photobooks which have been published on the subject of climate change within the last decade. Edward Morris and Susannah Sayler—starting from a critical perspective on the power of images to depict climate change—question artistic practices such as pensive photographs as alternative approaches to generate public sentiment and climate change awareness.

In contrast, the fifth section—Images of Climate Control—looks at the use of pictures in terms of how they are used for control and surveillance of climate. Gisela Parak investigates the early collections of aerial survey photography in the United States from the 1940s to the 1970s. She shows how the remote bird's-eye view was used to demonstrate climatic conditions of territory and to indicate manageability by focusing on technologically formed urban and rural landscape patterns. The last two articles focus on the phantasms of geo-engineering: James Rodger Fleming contrasts the ideas of vast geo-engineering concepts with science fiction, art, cartoons and Greek mythology to expose underlying myths of controllability. Isabell Schrickel performs a critical iconic analysis of typical geo-engineering illustrations, which support a normalization of the discourse and lead to a familiarization of technical and natural processes in a geo-engineering context.

HOW TO READ THIS BOOK

This book seeks to provide an overview of a wide spectrum of languages, styles and strategies, all of which aim to make climate phenomena, climate change or climate impacts visible and explicit. In assembling this diversity of 'voices' we hope to provide material which will be useful to a wide range of people or even contemporary decision makers concerned with the topic of climate change and climate change communication. It can be read as a whole by those interested in the entire topic of visual studies regarding climate images, or by disciplinary experts with a more focused view. Some specific suggestions follow:

Climate scientists might first focus on the chapters related to climate data visualization and the climate media discourse (1, 2 and 3), then move to the more cultural framings of the topic within the chapters 4 and 5.
Visualization experts from the fields of scientific and information visualization are encouraged to start reading the articles in chapter 1 as well as the articles "Twist and Shout: Deceptive Graphs in the Skeptical Climate Blogosphere" by Feulner, Nocke and Schneider and "Towards an Interactive Visual Understanding of Climate Change Findings on the Net: Promises and Challenges" by Markus Wrobel and Dominik Reusser in the third chapter.
Art historians and Visual Studies researchers might start with chapter 4, where the role of images in making the invisible imaginable through art and photography is discussed.
Media and communication scholars might begin with chapter 3 and read the article by Elke Grittmann "Between Risk, Beauty and the Sublime: The Visualization of Climate Change in Media Coverage during COP 15 in Copenhagen 2009" or the article by Julie Doyle "Picturing the Clima(c)tic: Greenpeace and the Representational Politics of Climate Change Communication" in chapter 4.
Science and technology study scholars might start with chapter 1, reading "The Creation of Global Imaginaries: The Antarctic Ozone Hole and the Isoline Tradition in the Atmospheric Sciences" by Sebastian Grevsmühl or chapter 5 "Picturing the State of the Nation's Environment: Early Aerial

Photography in the United States from the 1930s to the late 1960s" by Gisela Parak or turn to chapters 2 and 5 where the discussions of today's scientific imagery are covered.

Political scientists might first focus on section 2 and move on to the article in section 5 "Images of Feasibility: On the Viscourse of Climate Engineering" by Isabell Schrickel.

Policy Makers might start at chapter 2, with Lynda Walsh's article ""Tricks," Hockey Sticks, and the Myth of Natural Inscription: How the Visual Rhetoric of Climategate Conflated Climate with Character" or Julie Doyle's article in chapter 4, "Picturing the Clima(c)tic: Greenpeace and the Representational Politics of Climate Change Communication" and before turning to chapters 2, 3 and 5.

CONCLUSION

The approach of this book is to take a critical look at images showing climate change, as pictures in this context are widely used by all concerned, and are connected with very high expectations; we look at these images critically because in their daily use, especially when they are duplicated over and over again, they end up becoming 'transparent.' The more normal these images become in their daily usage, the more they are overlooked and taken for granted.

If we take a close look at the visual elements of the climate discourse, we can demonstrate the complex ways in which images have the power to reinforce and accentuate the threatening results presented by climatologists, but also the ways in which the power of images sometimes becomes precarious. The impacts climate pictures create, which are discussed in this book, are manifold. To name only a few: they might be expressive, influential, documentary, global or regional, fictional, sublime, prognostic, religious, scientific, rational, historical, emotional, engaging, alarming, artistic, activist, misleading, enlightening or convincing.

We take the view that climate change today should not be reduced to a discussion of it as a scientific object of knowledge, but rather be investigated as a cultural event with its impact and conceptions as well. Societies need cultural images in the broadest sense, transferring global scientific knowledge into regional knowledge and imagined scenarios. For this purpose, alternative languages to scientific language have to be created to make the unimaginable imaginable: what kind of Earth it will be that our children will experience? What does it mean locally when the face of the Earth turns red? How will the total transformation of our landscapes, cities, factories and transportation systems change society and culture?[21]

Images might be a means by which visions, phantasies and imaginations could be developed that make climate change more tangible, and hence shape how we see this world and behave in it. Images in the Anthropocene become a crucial way of detecting climate change, hence it is necessary to pose questions regarding their reality and construction.

ACKNOWLEDGEMENTS

We thank all authors of the book for sharing their ideas with us, and for their co-operation and patience. We owe special gratitude to student assistants at Potsdam University: Pia Nawrazala and Simon Hirsbrunner. The present book could not have been completed without their active support and tireless work in putting together this volume. We also owe great thanks to Helen Carter, who scrupulously took care of the copyediting. We thank Katharina Wierichs and the team of the publisher Transcript and Bijan Dawallu who created the cover image for us. Finally, we thank the Fritz Thyssen Foundation, Cologne and the Potsdam Institute for Climate Impact Research, Potsdam for providing us with the financial support that enabled us to realize this book.

BIBLIOGRAPHY

Bredekamp, Horst, Vera Dünkel and Birgit Schneider. "The Image: A Cultural Technology: A Research Program for a Critical Analysis of Images." In *The Technical Image. A History of Styles in Scientific Imagery*, edited by Horst Bredekamp, Vera Dünkel and Birgit Schneider, 11–14. East Lansing: Michigan State University Press, 2014.

—. *The Technical Image. A History of Styles in Scientific Imagery.* East Lansing: Michigan State University Press, 2014.

Blumenberg, Hans. "Lebenswelt und Technisierung unter Aspekten der Phänomenologie." In *Wirklichkeiten, in denen wir leben,* edited by Hans Blumberg, 7–54. Stuttgart: Reclam, 1993.

Daston, Lorraine and Katharine Park. *Wonders and the Order of Nature: 1150-1750.* New York: Zone Books, 1998.

21 Latour, "A Plea for Earthly Sciences," 2009.

Dobrin, Sidney I. and Sean Morey. *Ecosee: Image, Rhetoric, Nature*. New York: Suny Press, 2009.

Doyle, Julie. *Mediating Climate Change*. Farnham: Ashgate, 2011.

Elkins, James. *The Domain of Images*. Ithaca, NY: Cornell University Press, 1999.

Gugerli, David and Barbara Orland. *Ganz normale Bilder. Historische Beiträge zur visuellen Herstellung von Selbstverständlichkeit.* Zürich: Chronos, 2002.

Hansen, Elisabeth Delin. "Rethink Kakotopia." In *Rethink—Contemporary Art & Climate*, Arhus: Alexandra Institute, 2009

Heintz, Bettina and Jörg Huber. "Der verführerische Blick. Formen und Folgen wissenschaftlicher Visualisierungsstrategien." In *Mit dem Auge denken, Strategien der Sichtbarmachung in wissenschaftlichen und virtuellen Welten,* edited by Bettina Heintz and Arnold Benz, 9–40. Zürich: Ed. Voldemeer, Wien; New York: Springer, 2001.

Himmelsbach, Sabine and Yvonne Oldenburg. *Ecomedia. Ecological Strategies in Today's Art.* Ostfildern Ruit: Cantz, 2007.

Hüppauf, Bernd and Peter Weingart. "Wissenschaftsbilder—Bilder der Wissenschaft." In *Frosch und Frankenstein. Bilder als Medium der Popularisierung von Wissenschaft,* edited by Bernd Hüppauf and Peter Weingart, 11–44. Bielefeld: Transcript, 2009.

Hulme, Mike. *Why We Disagree About Climate Change: Understanding Controversy, Inaction and Opportunity.* Cambridge: Cambridge University Press, 2009.

Knebusch, Julien. "Art and Climate (Change) Perception: Outline of a Phenomenology of Climate." In *Sustainability: A New Frontier for the Arts and Cultures*, edited by Sasha Kagan and Volker Kirchberg, 242–61. Frankfurt a.M.: Verlag für Akademische Schriften, 2008.

Latour, Bruno. "A Plea for Earthly Sciences." In *Social Connections: Sociology's Subjects and Objects,* edited by Judith Burnett, Syd Jeffers and Graham Thomas, 72–84. London: Palgrave Macmillan, 2009.

—. "Drawing Things Together." In *Representation in Scientific Practice*, edited by Michael Lynch and Steve Woolgar, 19–68. Cambridge, Mass.: MIT Press, 1990.

Lenoir, Timothy. *Inscribing Science: Scientific Texts and the Materiality of Communication.* Stanford, Calif.: Stanford University Press, 1998.

Lester, Libby and Brett Hutchins. *Environmental Conflict and the Media.* New York: Lang, 2013.

Link, Jürgen and Mirko M. Hall. "The Normalistic Subject and Its Curves: On the Symbolic Visualization of Orienteering Data." *Cultural Critique* 57, (Spring, 2004): 47–67.

Mitchell, W.J. Thomas. *What Do Pictures Want? The Lives and Loves of Images.* Chicago: University of Chicago Press, 2007.

Monmonnier, Mark. *Air Apparent: How Meteorologists Learned to Map, Predict, and Dramatize Weather.* Chicago, Ill.: University of Chicago Press, 1999.

Painter, James. *Climate Change in the Media. Reporting Risk and Uncertainty.* London: Tauris, 2013.

Rosentrater, Lynn D. "Representing and Using Scenarios for Responding to Climate Change." *WIREs Climate Change* 1, no. 2 (2010): 253–59.

Schneider, Birgit. "Linien als Reisepfade der Erkenntnis." In *KartenWissen: Territoriale Räume zwischen Bild und Diagramm*, edited by Stephan Günzel and Lars Nowak, 175–99. Wiesbaden: Reichert, 2012.

—. "Climate Model Simulation Visualization from a Visual Studies Perspective." *WIREs Climate Change* 3, no. 2 (March/April 2012): 185–93.

—. "Image Politics: Picturing Uncertainty. The Role of Images in Climatology and Climate Policy." In *Climate Change and Policy. The Calculability of Climate Change and the Challenge of Uncertainty*, edited by Gabriele Gramelsberger and Johann Feichter, 191–209. Berlin/Heidelberg: Springer, 2011.

Sheppard, Stephen. *Visualizing Climate Change.* London: Routledge, 2012.

Walsh, Lynda. "Visual Strategies to Integrate Ethos Across the "is/ought" Divide in the IPCC's Climate Change 2007: Summary for Policy Makers." *POROI* 6, no. 2 (2009): 33–61.

Welzer, Harald and Claus Leggewie. *Das Ende der Welt wie wir sie kannten. Klima, Zukunft und die Chancen der Demokratie.* Frankfurt a. M.: Fischer, 2009.

Witzke, Anne Sophie. Introduction. In *Rethink—Contemporary Art & Climate*, Arhus: Alexandra Institute, 2009.

Yusoff, Kathryn. "Mapping the Disaster: Global Prediction and the Medium of 'Digital Earth'." *Wild Nature and the Digital Life Special Issue, Leonardo Electronic Almanac* 14, no. 07-08 (2006): 1–11.

Chapter 1
The Epistemic Value of Visualization in Climate Sciences

The Creation of Global Imaginaries: The Antarctic Ozone Hole and the Isoline Tradition in the Atmospheric Sciences

SEBASTIAN VINCENT GREVSMÜHL

This historical essay retraces from the perspective of visual and material culture how ways of analyzing and visualizing atmospheric data dramatically affect how scientific phenomena are perceived. The article first describes how the idea of an "ozone hole" moved in the 1930s from a hypothetical, theoretical construct within the astronomical community to a description of one of the most important global environmental problems of the late twentieth century. The chapter then explains in detail how NASA scientists reframed the local phenomenon of ozone depletion as a global environmental risk through their use of contour maps, displaying large quantities of global satellite data in synoptic form, coupled with the introduction of a new powerful metaphor: the "ozone hole."

INTRODUCTION

In 1990, Michael Lynch observed that the analysis of metaphors had received far more attention than had visual representations.[1] Since then, research conducted in history of science and science studies has changed the scholarly landscape considerably. Beginning already in the 1980s, visual studies have received major attention from a broad variety of disciplines and today it seems that in the history of science, there is almost no major visualization technique which has attracted closer scrutiny in the form of a case study. Despite their diversity, all case studies share the claim that scientific visualizations shape scientific

1 Lynch, "The Externalized Retina," 1988, 201.

knowledge in a broad variety of ways.[2] Indeed, as Norton Wise has pointed out, most of the history of science could be written as a visual history in which scientific visualization would have to be integrated as "materialized arguments" instead of mere illustrations accompanying scientific discourse.[3] His observation implies that scientific visualizations may indeed be useful in indicating how scientists think and interpret their findings. His call for a "materialized epistemology"[4] is largely informed by two major influences which shaped the field of the history of science during the last thirty years: laboratory studies and the anthropological approach to lab-work on the one hand, and cultural history and the so-called "practical turn"[5] on the other. Both fields have shown that close analysis of the actual gestures involved in scientific production, the material settings and the actual workspace play a major role in the shaping of scientific facts. Material culture is therefore one of the most promising methodologies in the analysis of scientific images and their cultural, social and historical meaning.[6]

This chapter will focus on a detailed study of the instrumental setting and of the gestures associated with the construction of a macroscopic geophysical phenomenon and major environmental threat widely known as the "ozone hole." Although omnipresent today, "ozone hole" visualization helped to increase public awareness of false-color imagery in combination with mappings of atmospheric processes on a global scale. I will argue that an analysis of the interaction between the visual, material and metaphorical elements may show that the metaphor of the "ozone hole" is a direct consequence of the practical

---

2 Good places to get an idea of the growing field of visual studies in science studies and history of science are the following reviews: Dommann, "Vom Bild zum Wissen," 2004; Borck, "Bild der Wissenschaft," 2009; Hentschel, "Bildpraxis in historischer Perspektive," 2011.

3 Wise, "Making Visible," 2006, 75–82.

4 Ibid., 81.

5 The practical turn is intimately linked to the work of Ian Hacking, Nancy Cartwright, Peter Galison, David Gooding and Bruno Latour. An excellent introduction is: Lenoir, ed., Inscribing Science, 1998.

6 Mitchell proposed in 1992—as a corollary to Richard Rorty's "linguistic turn"—the "pictorial term," an encouragement to analyze sign systems beyond purely linguistic sign systems. An expanded version of his article was published in: Mitchell, Picture Theory, 1994, 11–34. The "iconic turn" was proposed two years later by German art historian Gottfried Boehm, reviving the idea of images in philosophical argumentation. See: Boehm, "Die Wiederkehr der Bilder," 1996, 11–38.

and material settings of NASA's research. Although NASA scientists missed out on the opportunity to announce ozone depletion in the first place (researchers of the British Antarctic Survey (BAS) published the actual discovery of ozone depletion in 1985), the powerful metaphorical framing of the phenomenon as a "hole" in the ozone layer was nonetheless an American invention. Visual and material culture provide essential clues as to why British researchers were referring "merely" to the phenomenon as ozone depletion, whereas the US colleagues saw a veritable hole in the sky. This case study is therefore an attempt to understand why different research communities see different things while apparently observing the same geophysical phenomenon. Finally, I would like to argue that strict micro-historical analysis has to be complemented by a broader historical inquiry into the longue durée of the "ozone hole" metaphor as well as the isoline tradition both of which proved essential in the creation of this first major global environmental threat.

THE BRITISH DISCOVERY OF OZONE DEPLETION

In 1985, when Joseph Farman, Brian Gardiner and Jonathan Shanklin of the British Antarctic Survey (BAS) published their findings that total stratospheric ozone values were declining considerably for the third consecutive time during the austral winter in Antarctica, their discovery was received with shock by the community of atmospheric scientists. Indeed, neither computer models, nor field or satellite observations managed to infer such a dramatic decline and none of the published chemical depletion processes could explain the alarming development.[7]

For NASA, since 1975 equipped by legislation with a substantial budget "to conduct a comprehensive program of research, technology and monitoring of the phenomena of the upper atmosphere,"[8] the British primacy was an important backlash. Since then, several explanations have been put forward to account for

7 See: Stolarski, "The Antarctic Ozone Hole," 1988, 20. The discovery seemed even more spectacular in the light of institutional background and funding: the British research team was at the time a rather minor player in the field of atmospheric sciences with modest financial resources. Their findings relied on an instrument invented back in the 1920s which New Scientist columnist Fred Pearce rightly labeled "Joe Farman's old-tech triumph." See: Pearce, "Joe Farman's Old-Tech Triumph," 2008, 46–47.

8 Lambright, *NASA and the Environment*, 8.

the embarrassing episode in NASA's history.[9] It seems clear that Antarctic ozone research relied at the time on two distinct research traditions with expensive American *big science* on the one hand, and British fundamental research on the other, conducted in a conservative way with a comparatively modest budget and rather unspectacular, almost old-fashioned instruments. However, within the Antarctic context, the conservative British approach ultimately proved highly efficient in transforming total ozone measurements into hard scientific facts.

One reason for the British success was the robustness of their data. A BAS report published in 1975 stands like no other document for the continuous British effort in documenting total ozone values observed at two research stations within the Antarctic region.[10] It contains a table showing, on just a few pages, the mean total ozone values for each month of the year between 1957 and 1972. Although instrumental "friction" is not an issue in ground measurements, the values of the BAS report are nonetheless the result of an intensive effort of purification, since numerous presuppositions are attached to the measurements, such as one which states that the maximum concentration of ozone is situated at an altitude of twenty-two kilometers. Nevertheless, the British measurements certainly defined a kind of "standard" for Antarctic stratospheric ozone.

On closer observation, however, the established "standard" proved to be extremely local. Firstly, measurements in Antarctica were taken at only two stations: Halley and Argentine Islands. Secondly, those stations were not part of the global ozone network, put into place during IGY by the World Meteorological Organization (WMO) and formally institutionalized by the International Ozone Commission (IOC).[11] The BAS research group profoundly mistrusted the network and the quality of the measurements, which was the main reason why they did not communicate their measurements. Even the instruments of the British Antarctic Survey were calibrated not by the international standard of the British Met Office but according to their own criteria.[12] In a sense, one might say that the "standard" relied on highly local procedures and a maverick

9 I would like to mention two commentaries in particular which contest the rumor that NASA deleted all the low ozone values below 180 Dobson units: Pukelsheim, "Robustness of Statistical Gossip and the Antarctic Ozone Hole," 1990, 540–542; Christie, "Data Collection and the Ozone Hole," 2004, 99–105.

10 See: Farman and Hamilton, *Measurements of Atmospheric Ozone*, 1975.

11 For a historical survey of the International Ozone Commission, see: Bojkov, The International Ozone Commission, 2010.

12 Grevsmühl, "Interview with Jonathan Shanklin," 29 April 2008.

institutional culture. Yet paradoxically, the *refusal* to participate in the (to some eyes compromised) international network was precisely the reason for BAS' credibility. By building their own institutional and material culture, their own standards and techniques, BAS could successfully establish the credibility of the only long-term dataset of atmospheric ozone measurements on the Antarctic continent. In other words, the British research group could, unlike any other, speak with legitimate authority on Antarctic ozone values.

When, at the beginning of the 1980s, total ozone values started to drop significantly during the austral spring, Joseph Farman decided to send Jonathan Shanklin to Halley with a newly calibrated instrument in order to assure that no calibration error or instrumental malfunction was biasing the measurements. Confidence could only be reestablished when both instruments, the new Dobson spectrophotometer number 123 and the old instrument number 31, worked side-by-side for two weeks, producing more or less coherent measurements, allowing the measurements of the last three years to be connected to all the other total ozone values established beforehand.[13] Even though several unsuccessful attempts were made to communicate with the proper satellite research group at NASA in order to obtain some sort of confirmation, the BAS group had produced, they considered, enough reliable data to submit the results to the journal Nature in late 1984. The results seemed sufficiently robust to signify a general and significant tendency.[14]

13 British Antarctic Survey, *Halley Bay Geophysics Report*, 1982.

14 Jonathan Shanklin's private archives contain two letters written in 1983 to two NASA groups asking for confirmation, but in both cases the wrong people were addressed. See also: Grevsmühl, "Interview with Jonathan Shanklin, 29 April 2008.". This also explains, at least in part, why other measurements of a Japanese research group—yet also alarming and published in 1984—were ignored by the community. Their measurements contained a series of only eleven months and by choosing to publish in a rather unknown journal, the Japanese researchers were not able to attract sufficient attention. Their results clearly did not have the same legitimacy as those of their British colleagues. The first article mentioning stratospheric ozone depletion over Antarctica was published in 1984: Chubachi, "Preliminary Result of Ozone Observations," 1984, 13–19.

THE AMERICAN CONTEXT OF SPACE-BASED ENVIRONMENTAL MONITORING

The American space agency was confronted with completely different problems. In a brief historical reflection on the "ozone hole" discovery, Pawan Barthia, at the time a member of the Ozone Processing Team, explains very well why NASA missed out on the opportunity to announce annual ozone depletion above the Antarctic continent.[15] As one of the scientists involved in writing the actual code for data reduction, Barthia explains that—as with any method of (environmental) remote sensing—a great number of presuppositions had to be introduced, a necessity inherent to the method employed. Indeed, from a purely mathematical perspective, it is impossible to resolve the "problem of inverse observation" without introducing *a priori* information.[16] For the satellite Nimbus 7, Carlton L. Mateer proposed in 1971 the use of a set of standard atmospheric ozone profiles allowing calculation of the total column ozone.[17] Yet none of these standard profiles allowed values below 180 Dobson units, since such values had never been observed before. For NASA, that (at the time) reasonable presupposition justified the introduction of data flags which allowed, in particular, the highlighting of data which might be corrupt, namely values below 180 or above 650 Dobson units.

That decision turned out to be a crucial one, since it initiated, at the end of 1984, a detailed spatial analysis of ozone values below 180 Dobson units, which were becoming more and more numerous in the 1983 data. The analysis showed that spatial distribution of the "errors" was not arbitrary, but that all values were well within the confines of the Antarctic continent. Based on the visualization of the "doubtful" data, NASA researchers started to suspect in late 1984 that ozone values above the Antarctic continent were indeed decreasing. The result of that analysis was presented in the following summer, in 1985, by Pawan Barthia at the joint IAGA/IAMA meeting in Prague (Figure 1) only a few months after the paper of the British Antarctic Survey ozone group was published in Nature.

---

15 See: Bhartia, "Role of Satellite Measurements," 2009, 183–189.

16 See the mathematical explanation of the statistician Rodgers: Rodgers, Inverse Methods for Atmospheric Sounding; book cited in: Bhartia, "Role of Satellite Measurements," 2009, 184. The TOMS instrument manual also mentions that the algorithm relies on the a priori introduction of atmospheric ozone profiles: Fleig et al., *User's Guide for the Total-Ozone Mapping Spectrometer*, 1982.

17 Mateer et al., "Estimation of Total Ozone From Satellites Measurements," 1971, 1307–1311.

Figure 1: Satellite map of total ozone above the Antarctic region for October 1, 1983. In order to produce this map, the standard ozone profiles were arbitrarily extrapolated to include values below 180 Dobson units. (→ color plate 1)

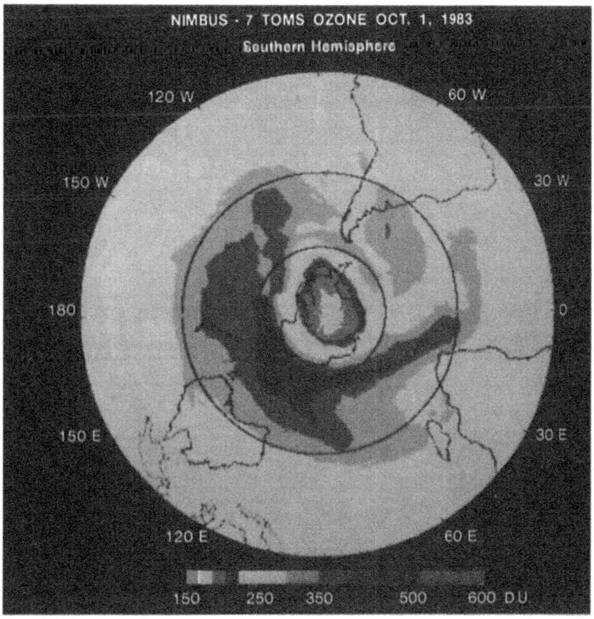

LESSONS WHICH MAY BE LEARNT FROM THE OZONE DEPLETION DISCOVERY

At least three lessons can be learnt from this first part of the story. First of all, the case of ozone depletion reveals a problem which is, actually, generally well known to the field of scientific observation. The Polish philosopher and physician Ludwik Fleck was one of the first scholars to describe, during the 1930s and 40s, a similar problem based on his first-hand experience in medical research.[18] Fleck argues that "in order to see, one has to know what is essential and what is not. [...] Otherwise one looks, but one does not see, one focuses in vain on too many details but one does not perceive the totality of the figure."[19]

18 See Fleck's article, first published in 1947: Fleck, "Schauen, sehen, wissen," 1983, 147–174.

19 Ibid., 148 (translated by the author).

On another occasion, he points out that "one has to learn how to look, in order to see properly."[20]

If we transpose this insight to the context of remote sensing, the question of "seeing properly" indeed becomes an instrumental question which is bound, by definition, to important presuppositions which guide the researcher's gaze. Yet it is precisely at the margin of what may count as credible and what has to be discarded as an instrumental artifact that new knowledge is created. Only by allowing sufficient space for the materialization of the improbable and unpredictable within the process of scientific research can new phenomena emerge. Indeed, in his influential study on experimental research, Hans-Jörg Rheinberger has identified precisely the openness of experimental systems as the principal motor for the production of "epistemic things."[21]

Secondly, one may observe that both groups were confronted in very different ways by the problem of data reduction. The British team members did not have enough measurements which they could consider "reliable" to enable them to speak of a significant statistical tendency. The space community was facing almost the opposite problem: there was a vast amount of data of which an increasing part was considered "doubtful." For one group, the production of new knowledge and confidence in the data had to be established by the *multiplication* of measurements, whereas the other group had first to *reduce* and then to *reclassify* the data to produce new knowledge.

Thirdly, both cases reveal a complex dialectic between the local and the global as an important characteristic of environmental remote sensing. The robustness of the British data and the legitimacy of their research relied on a very particular culture of isolation: measurements were only realized at two research stations, with only one situated on the Antarctic continent, and results were neither communicated to the global ozone network, nor to the appropriate research group within the space community. Quite paradoxically, it was the isolation that accounted for the whole *coherence* of the British research, ultimately enabling the "doubtful" measurements to connect to an extremely solid data base. The isolation allowed, so to speak, the transformation of doubtful data into ozone values sufficiently robust to be published. However, the British discovery was at the time a highly local phenomenon.

In order to produce the global imaginary of the "ozone hole," a truly *global view* had to be introduced which only the satellite could deliver. The TOMS

20 See his 1935 article: Fleck, "Über die wissenschaftliche Beobachtung," 1983, 60 (translated by the author).

21 See: Rheinberger, Experimentalsysteme und epistemische Dinge, 2001, 84.

(Total Ozone Mapping Spectrometer) instrument that was used for those measurements relied fundamentally, as any satellite remote sensing device, on what earth scientists call "ground truthing." The notion calls for information gathered in the field, in order to validate the satellite image interpretation with measurements obtained from ground-based methods. In other words, one may observe an important dialectic between the local and the global: the British research team was in many ways too local, too isolated to be able to speak of a truly global phenomenon and the Americans were too global, too far away from the local context, too attached to the synoptic view in order to effectively relate the satellite data to ground-based measurements.[22] However, once both cultures were united, once a recombination of important elements from both cultures had found a common ground, a veritable "hole" in the sky emerged as one of the most powerful metaphors of a new global environmental risk.

To enable the whole complexity of the "ozone hole" history to be grasped, I propose, in the following, a detailed visual analysis of the actual images mobilized in the transformation of the phenomenon of stratospheric ozone depletion into a global environmental risk.

THE "OZONE HOLE" VISUALIZATIONS

Figure 2 shows a typical example of an ozone depletion visualization based on satellite data obtained from NASA's TOMS instrument on Nimbus 7. This type of image largely helped, as I will show in detail, to transform ozone depletion into the powerful metaphor of the "ozone hole." Produced in 1986 at Goddard Space Flight Center (GSFC) by the groups responsible for the scientific analysis of the satellite data (the Ozone Processing Team and the main contractor, the Systems and Applied Sciences Corporation), this type of image was widely circulated in national and international media during the second half of the 1980s.[23] The first "ozone hole" animations, based on the same kind of image,

22 However, one must mention the ground-based observations which were conducted at US South Pole station, although the measurements which fell into the critical time frame were unfortunately biased by an instrumental error, allowing no confirmation of the observed depletion; see: Bhartia, "Role of Satellite Measurements," 2009, 185.

23 The earliest published "ozone hole" visualization was produced in late 1984 and presented by Pawan Bhartia as member of the Ozone Processing Team at the IAGA symposium in Prague in August 1985. It was included in the following article: Callis

Figure 2: Satellite map showing total ozone in the Southern hemisphere in polar orthographic projection for October 10, 1986. At the center, in a darker shade of blue, the so-called "ozone hole" is depicted (defined as values below 220 Dobson units). Measurements were taken with the TOMS instrument aboard NASA's Nimbus 7 satellite.

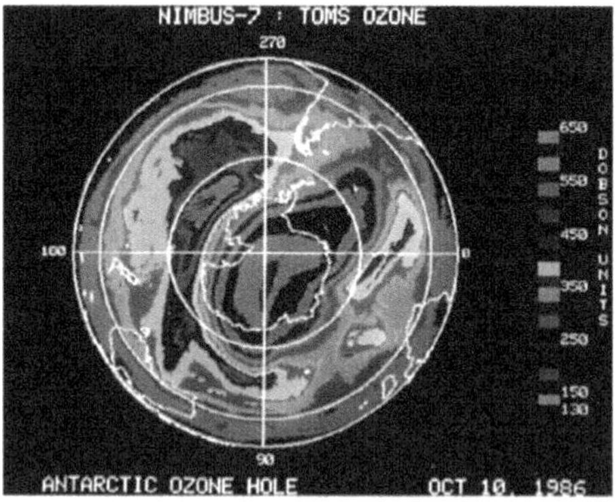

Figure 3: Still image of the animated ozone atlas produced by STX, showing a total ozone map for October 10, 1986 as in the previous figure, but with an altered color code. The "ozone hole" shows up as a violet patch at the center of the image.

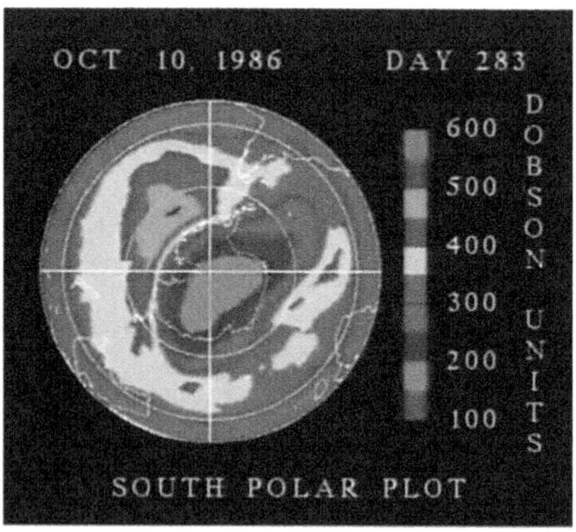

and Natarajan, "The Antarctic ozone minimum," 1986, 10774. The historical context of the first ozone hole visualization was described by Bhartia, "Role of Satellite Measurements," 2009, 183–189.

were produced by former NASA employee Mark Schoeberl who coupled an automatic camera to the mapping routines which were developed under the supervision of James A. Gatlin at Goddard by the Sensor Evaluation Branch.[24] The group automated what their colleague Richard Stolarski had indeed tried before by hand: the coloring-in of total ozone contour plots (see Figure 3 for a still image and Figure 4 for an example of an ordinary contour plot).[25]

Figure 4: For comparison, ozone contour plot of total ozone over Antarctica one year later, on October 10, 1987 with values dropping as low as 150 Dobson units.

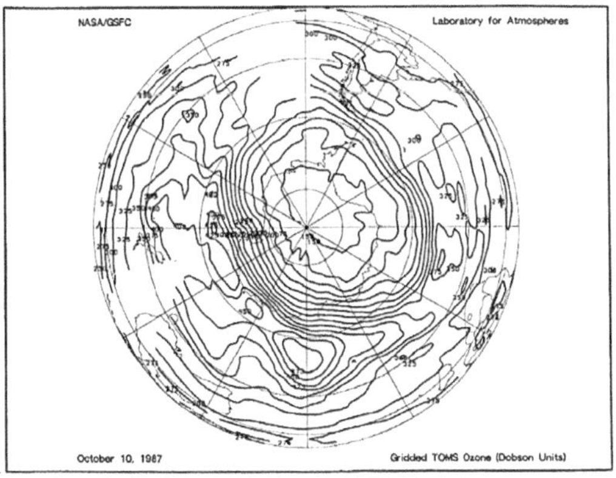

The examples mentioned clearly show that NASA's "ozone hole" visualizations are highly complex data products. They are a bird's eye view, allowing the

24 Personal communication of Richard McPeters (13 September 2010) and Richard Stolarski (30 August 2010). The engineer James Gatlin joined NASA in 1962. He was director of the Sensor Evaluation Branch at GSFC between 1974 and 1984. In 1984, he became director of the Engineering Directorate; see: Reisin, ed., *IEEE Geoscience and Remote Sensing Newsletter 121*, 2001, 7.

25 Personal communication of Richard Stolarski (30 August 2010). See also the statement of Richard McPeters (member of the ozone research team at NASA during the 1980) in: Christie, The Ozone Layer, 2001, 51. The first "ozone hole" animations are also available as an animated atlas, see: Goddard Space Flight Center, Animated Atlas of TOMS Ozone Data, 1989; for a discussion of this important resource of 3440 color visualizations, see: Chesters and Krueger, "A Video Atlas of TOMS Ozone Data," 1989, 1564–1569.

integration of over 140,000 discrete daily total ozone measurements into a single image. In their most common rendition, ozone visualizations unite cartography (namely the polar orthographic projection) with geophysical measurements presented in the form of contour lines. Once received at Goddard Space Flight Center, the TOMS observations gathered on board Nimbus 7 are transformed and reduced to discrete ozone values and a computer algorithm allows the visualization of the data by connecting points of equal value by a line—a line which is generally referred to not only as a contour line but also an isoline. Finally, the image is color-coded, allowing more convenient identification of a specific value range. The whole process may be rendered even more complex by introducing the variable of time: several images can be combined with an animation of the spatial and temporal dynamics of the "ozone hole."

That visual and material culture do indeed not only interfere in the description of the world but also have a performative function may easily be grasped when taking a closer look at the polar orthographic projection, mobilized in most cases to depict polar phenomena. For different reasons, it is considered the most convenient way to move from the sphere to the map, at least in the case of the polar regions. However, as with any other projection type, the polar orthographic projection is only one mode of representation, chosen from amongst a large panoply of cartographic products. One should certainly not consider it "a transparent opening to the world" but rather, as any form of art, "a particular human way [...] of looking at the world."[26] Most importantly, the consequences of the choice of the projection type may be far-reaching, especially in the political domain. Indeed, when considering the orthographic projection, it is easy to understand why territorial claims in Antarctica always take the form of a wedge. The general adoption of this particular projection type imposes in this case its own visual logic on the domain of geopolitics.[27]

THE ISOLINE TRADITION IN HISTORY OF CARTOGRAPHY

One of the most striking elements of the "ozone hole" visualizations is the contour line or the isoline. Just as the projection type, it imposes a certain *pictorial*

26 Brian Harley uses a citation of Blocker, see: Harley, "Deconstructing the Map," 2001, 153.

27 Another quite striking example is the famous Swiss Dufour map and the numerous controversies which it attracted during the first half of the 19th century, see: Gugerli and Speich, *Topografien der Nation*, 2002.

logic which may easily be grasped by looking at some of its historical uses. The isoline tradition first developed within the maritime context towards the end of the 18th century, when cartographers started to include terrestrial features in contour maps. Even though some early examples date back to the 16th century, historians generally consider Edmond Halley to be the author of the first "scientific" contour map, a map which is based on measurements taken on two maritime voyages aboard the Paramour.[28] His 1701 map of the "variations of the compass" (Figure 5), which has been called an "isogonic" map since the 19th century, is indeed the first surviving example of its kind, based on geophysical measurements at some 150 locations and showing contour lines of magnetic declination (angular difference between geographic and geomagnetic north) at a very important scale.[29] Since the beginning of the 18th century, nautical charts have included contour lines indicating navigational dangers and elements of submarine topography. Terrestrial topographical maps as we know them today were introduced around 1760, when Marcellin Du Carla proposed the use of contour lines on terrestrial charts.[30] However, the most emblematic figure in the isoline history is without doubt Alexander von Humboldt, who invented a whole panoply of different isolines, most notably the "isotherms" in 1817, showing mean temperature at a global scale (Figure 6). Within the history of climatology, the isoline may therefore be identified as a key instrument of what Matthias Heymann has called the period of "classical climatology" which relied mainly on the calculation of *mean* values of a great number of meteorological variables for their spatial representation.[31] In a quite masterful way, Alexander von Humboldt achieved with this simple gesture the creation of an illusion of continuous measurement which, as Birgit Schneider has shown more generally within the context of expert visualization, only partially reveals the complexity

---

28 Robinson, *Early Thematic Mapping*, 1982, 211.

29 See: Thrower, "Edmond Halley as a Thematic Geo-Cartographer," 1969, 652–676.

30 See: Du Carla-Boniface, *Expression des nivellemens*, 1762, in particular the map added at the end of the book. On the historical use of the contour line by Du Carla, see: Graber, "Le nivellement, " 2006, 51.

31 See: Heymann, "Klimakonstruktionen," 2009, 171–197. Although the isoline still plays a key role in climatology today, the context has radically change. According to Heymann it has moved from "classical climatology" through "dynamic" or "modern climatology" (mainly influenced by the Bergen school and the inclusion of meteorological knowledge) to "modern climate research" in which computers and models play the leading role.

Figure 5: Detail of a so-called "isogonic" map produced in 1701 by Edmond Halley and believed to be first data-based contour map showing lines of equal magnetic declination.

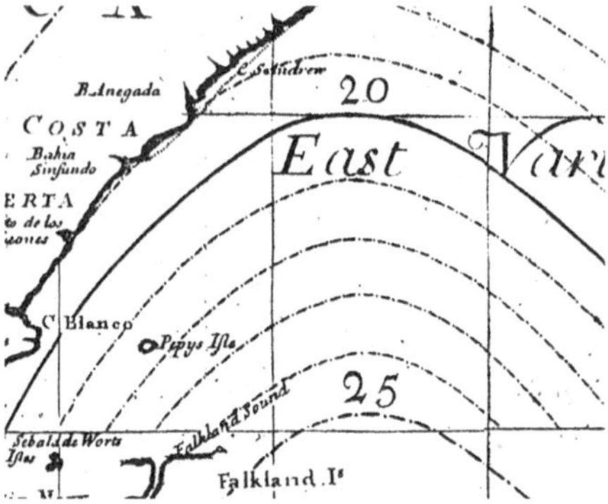

Figure 6: First graph of "isotherms" (upper part of image) showing mean temperature for the Northern hemisphere spanning from America to China. Alexander von Humboldt, "Carte des lignes Isothermes." Paris: V.H. Perronneau, 1817.

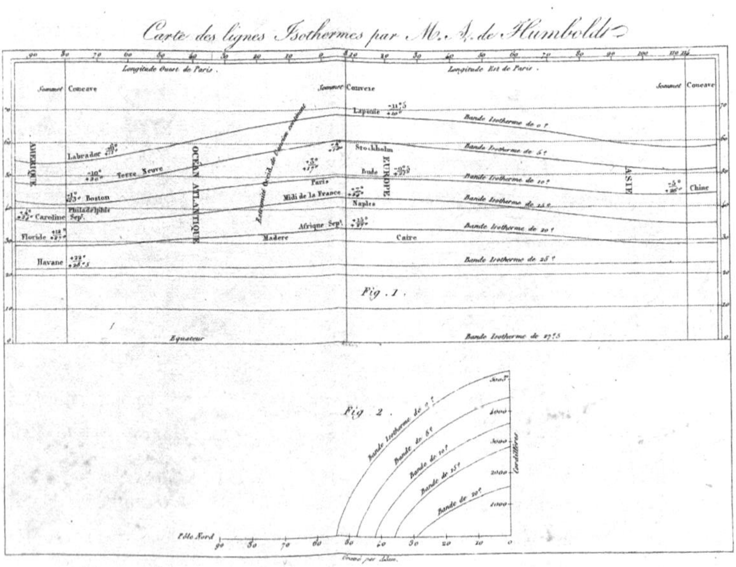

method, and others followed his lead and adopted it in quite diverse contexts, ranging from meteorology to statistics. As Thomas Hankins has shown, from of the construction process.[32] He helped enormously to popularize the 1840 onwards, one can observe a veritable isoline "craze" with the publication of "atlases that described everything imaginable by means of isomaps."[33]

Probably one of the most important aspects of NASA's "ozone hole" animations is its capacity to produce "global imaginaries," like Halley, Humboldt and Gauss masterfully did. They all knew how to create visually, and on a very important scale, geophysical phenomena that nobody would or could ever see. In other words, the history of science shows that the problem which is inherently linked nowadays to climate change, namely its very "imaginability," is by no means a new question. The isoline as a graphic tool plays a crucial role in this history. Its visual power resides in its capacity to create in a very efficient way (and probably like no other visual tool) the *illusion of continuous measurement.* For example, British astronomer John Herschel praised in a critical note on Gauss' latest book (which had as its main subject another global imaginary: terrestrial magnetism), the capacity of the isoline to overcome visually, thanks to graphical interpolation, the restrictions of the need for "observations and real stations" by the introduction of "imaginary stations so distributed over the globe as to afford the greatest possible facility to the calculations."[34]

And indeed, the history of cartography teaches us that the contour line can either largely extrapolate on the basis of comparatively few, or even invented, data points or it can integrate an enormous amount of data into very few lines. Both, and any intermediate settings, are of course possible. By definition, the contour line does not allow for differences in a relatively small data basis, such

32 See: Schneider, "Climate Model Simulation Visualization from a Visual Studies Perspective," 2012, 191–192. For a more detailed discussion of Humboldt's isotherm map, see: Schneider, "Linien als Reisepfade der Erkenntnis," 2011. David Aubin also speaks, in the case of Humboldt's isotherm map, of the erasure of singularity, see: Aubin, "The Hotel that Became an Observatory," 2009, 373.

33 Hankins, "A 'Large and Graceful Sinuosity'," 2006, 624. Fitzhugh Talman, director of the Weather Bureau library, published in 1915 a list of ninety different isolines used in meteorology and their first known citation. Even though his list contains several factual errors (for example the pioneer work of Heinrich W. Brandes is not mentioned), it gives an excellent idea of how the use of isolines in the geosciences expanded rapidly during the 19th century, see: Talman, "A List of Meteorological Isograms," 1915, 195–198.

34 John Herschel quoted in: Hankins, "A 'Large and Graceful Sinuosity'," 2006, 624.

as the 150 or so observations in Edmond Halley's case (all taken at different locations all along the tracks of his two Atlantic voyages), or a truly rich dataset of 140,000 daily ozone readings as in the case of NASA's total ozone visualizations. However, this means that the contour not only creates the illusion of continuous measurement, but it also hides very effectively its own construction process, making it impossible to judge the data basis behind the visualizations. Contour lines are therefore always both an extremely powerful and an extremely doubtful iconographical tool. The decision of NASA scientists to mobilize contour lines on an orthographic map changed how the world viewed an important atmospheric phenomenon. As I will show in more detail in the following analysis, the isoline, as very effective graphical tool, helped to transform a highly local phenomenon into a vast, dynamic feature of continental size. Finally, and maybe most importantly, the use of the isoline also shaped the ways in which public discourse reframed ozone depletion as global environmental risk. In short, to reformulate concisely my main thesis which I will now explore in more detail: British scientists "only" saw *ozone depletion* whereas NASA revealed a veritable *hole in the sky*.

THE VISUAL CONSTRUCTION OF THE "OZONE HOLE"

The decision of NASA researchers to use isolines in combination with orthographic maps modified considerably the way a relatively local scientific phenomenon—located far from the traditional contexts of interest of Northern hemisphere States—was communicated to a large public as a global environmental risk. In adopting distinct instrumental and visual practices, British researchers were "only" able to reveal depletion where their American colleagues saw a veritable "hole" in the sky. I argue that the metaphor of the "hole" can only exist in an image sense, in other words, its ocular metaphoricity relies on an iconographical construction. Holes owe, so to speak, their very existence to a clear delimitation of space, a sharp visual distinction between an inside and an outside. Using this distinction, contour lines can create holes, whereas simple plots cannot. Contour lines create shapes and delimitate space, allowing them—especially in conjunction with color—to create holes where other visualization technologies allow "only" the depiction of tendencies such as decline or depletion. This becomes obvious when one looks at the very first plot Jon Shanklin of the British Antarctic Survey drew for the famous 1985 *Nature* paper on Antarctic ozone depletion (Figure 7). Effective as the graph was, showing monthly mean total ozone values (including its error bars) measured with the help of a Dobson spectrophotometer at Halley station, Shanklin's depiction did not suggest any such thing as an "ozone hole." To put it in another way, the actual visualization technology employed formatted the formulation of the scientific phenomenon. From Shanklin's ozone plot, the reader quickly grasps the sharp local *decline* (which was made even more effective in its final printed version in the *Nature* paper with the use of a considerably shorter ordinate) in total ozone values. The metaphor of the "hole" however does not seem appropriate for this type of visual rendition.

Figure 7: Jon Shanklin's original "Minimum 11-day Spring Total Ozone" plot for the famous 1985 Nature paper, covering 1957-1984, revealing dramatic ozone decline above Halley, Antarctica (author's collection).

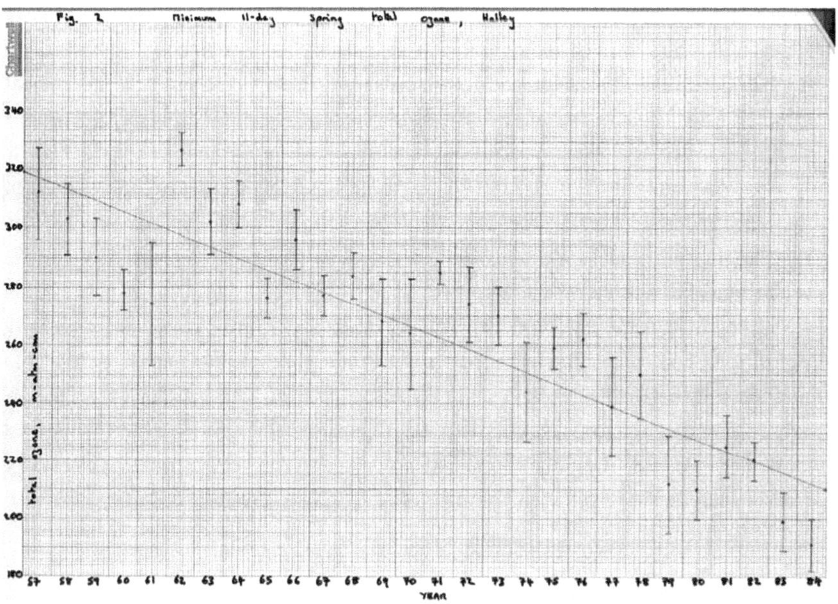

The use of color, or rather the correlation of a certain data-interval with a specific color, helped to add the effect of homogenization to the illusion of continuous measurement. The areas which are sharply delimited by the contour lines appear to be homogenous spaces of measurement, clear and distinct objects of knowledge. The rather chaotic effects of atmospheric diffusion and circulation are in this case clearly domesticated by a powerful false color visualization technology, entertaining, as Jean-Michel Besnier has rightly pointed out, a surprising metaphysical relationship with modern medical imagery, most notably with fMRI and PET scan technologies.[35]

In consequence, it is not surprising that the first appearance of the "ozone hole" metaphor (one definitely has to speak of a metaphor since stratospheric ozone is never removed completely) may be observed in the printed press and in conjunction with the creation of the first total ozone visualizations (based on TOMS data) by NASA scientists. Don Heath, the designer of the satellite instrument that allowed these global ozone measurements (the TOMS instrument on the Nimbus 7 satellite), gave some early color-coded ozone slides to Sherwood Rowland for a speech at the University of Maryland. In preparation for his speech, Rowland suggested the "hole" metaphor to famous New York Times science writer Walter Sullivan, who had already been the chronicler of the International Geophysical Year in the Antarctic region and who ran the first story on the "hole" in the ozone layer on November 7, 1985, using an early black-and-white total ozone contour visualization.[36] As seen above, material and visual culture largely influenced Rowland's, and finally Sullivan's, metaphorical choice. The "hole" metaphor only made sense in conjunction with contour visualizations.

In scientific publications, the "ozone hole" metaphor emerged in 1986 from the very group of NASA scientists which was more or less directly involved in the imaging process: Mark Schoeberl, Richard Stolarski and Richard McPeters.[37] Although one referee of the Stolarski paper objected to the use of the "hole" metaphor in the title of the paper, Paul Crutzen and Frank Arnold were already permitted, less than four months later, to employ in the same journal the metaphor (albeit with quotation marks) in the title of their article.[38] By 1988, a large scientific consensus had formed regarding the physical and chemical causes of ozone depletion and one can observe that the "ozone hole" was finally firmly established as an adequate metaphor of global environmental risk in both

35 Besnier, "De Galilée à Google Earth," 2009, 186.

36 Personal communication of Pawan Bhartia (10 September 2010). The first press article appeared in November 1985: Sullivan, "Low Ozone Level Found Above Antarctica," 1985, B-21. For a brief historical account, see: Grundmann, *Transnationale Umweltpolitik zum Schutz der Ozonschicht*, 1995, 174–177.

37 "The deep minimum, or hole […]," citation on p.810 in: Stolarski et al., "Nimbus 7 Satellite Measurements of the Springtime Antarctic Ozone Decrease," 1986, 808–811.

38 Paul Crutzen and Frank Arnold, "Nitric Acid Cloud Formation in the Cold Antarctic Stratosphere," 1986, 651–655.

spheres, in scientific and in public discourse.[39] At the time of its creation, the metaphorical force of the "ozone hole" lay probably in its heuristic and dramatic value, allowing the powerful designation of an otherwise invisible environmental risk, long before scientific consensus on the ozone destruction mechanism was achieved. Despite considerable scientific uncertainty, the metaphor helped to create an alarming discourse with significant effects within the political sphere, leading finally to the signature of the Montreal Protocol in 1987. From the point of view of metaphor theory, this suggests that although it is indeed impossible—as Max Black's interactionist theory holds—to paraphrase a metaphor without destroying its cognitive core, an active metaphor nevertheless works so effectively because the metaphorical signifier may be read and understood practically in a literal way.[40] A veritable hole in the sky which lets dangerous ultraviolet rays pass almost unimpeded the atmosphere was indeed a very strong image for the environmental movement and legislators alike, pointing finally at the very limits of the biosphere where local actions can provoke effects on a truly global scale.

CONCLUSION

The combined visual and material approach I employed in the above analysis indicates in detail how a historical cartographic tool could be put effectively to use in the construction of an influential global imaginary, showing considerable effects well beyond scientific discourse, most notably in the political sphere. In particular, I argued that the isoline shaped the metaphorical choice of the "ozone hole," transforming thereby a local depletion phenomenon into a global environmental risk. With the help of TOMS ozone visualizations the scientific community, policy makers, and the broader public received both: synoptic visualizations and a powerful metaphor. Together, they could constitute and frame the "ozone hole" as an emblem of a fragile planet, a symbol of global

39 See also quantitative analysis in Grundmann, *Transnationale Umweltpolitik zum Schutz der Ozonschicht*, 1995, 177.

40 American philosopher Donald Davidson defended the rather radical thesis that "metaphors mean what the words, in their most literal interpretation, mean, and nothing more": Davidson, "What Metaphors Mean," 1978, 31–47, here 32. This important shift from semantics to pragmatics can however be useful in explaining the actual use of the "ozone hole" metaphor. For Max Black's interactionist theory, see: Black, "More about Metaphor," 1993, 19–41.

environmental catastrophe as well as an icon of the global environment. It is precisely in this tradition in which contemporary icons of climate change, as for example the so-called red planet, try to inscribe themselves.[41]

And only against this specific historical backdrop is it possible to understand the political responses which led to the signature of the Montreal Protocol and its successive amendments. Scientific consensus was only established *after* the signature of the Montreal Protocol in 1987, most notably following *in situ* NASA measurements conducted in the Antarctic vortex during the austral spring of 1987. That mission, known as the Airborne Antarctic Ozone Experiment (AAOE), allowed for the discovery of the "smoking gun," the decisive visual proof of the "culpability" of chlorofluorocarbons (CFC) in ozone destruction, by showing a strong anti-correlation between the chemically active parts of the CFCs and the overall ozone concentration.[42]

Several explanations have been put forward to account for this at first glance unusual situation. For example, within the political sciences, Peter Haas argued that in the case of ozone depletion, political action, largely informed by the precautionary principle, is a direct consequence of the formation and action of a strong "epistemic community."[43] His notion refers to a group of professionals who share a certain number of values and practices concerning the ways in which to treat environmental problems, as well as an interpretive framework to judge them by. According to Haas, shortly after Farman's article appeared in press in 1985, ozone experts, politicians, and representatives of NGO's and scientific institutions came together to form an "epistemic community," leading to the signature of the Montreal Protocol, even in the absence of a general scientific consensus on the actual causes of annual ozone depletion.

My approach showed that it is indeed possible and legitimate to extend that explanation by accentuating the dynamics of visual and material culture which constituted the foundation of scientific practice, of the knowledge of the "epistemic community" and of the discourses mobilized. As I have shown,

41 On the making and function of climate change icons such as the red planet, see: Schneider, "Climate Model Simulation Visualization From a Visual Studies Perspective," 2012, 185–193.

42 AAOE and the "smoking gun" are discussed in all the important books on ozone depletion, most notably in: Parson, Protecting the Ozone Layer, 2003, 150–153, as well as in: Christie, *The Ozone Layer*, 2001, 60–65.

43 See: Haas, *Saving the Mediterranean*, 1990. See also his articles which discuss the ozone case in detail: Haas, "Obtaining International Environmental Protection," 347–364; Haas, "Banning Chlorofluorocarbons," 1992, 187–224.

ozone research within the Antarctic context goes well back to the mid-20th century and the International Geophysical Year, preceding largely contemporary environmental preoccupations. In the case of stratospheric ozone, the British Antarctic Survey initiated its own values, practices, norms, and standards, assuring thereby a certain degree of coherence which gave their ozone data set an important legitimacy and robustness. Combined with the synoptic satellite views of NASA, this "infrastructural" (Paul Edwards) knowledge formed an important basis for all the main political actions that followed. Hence, material and visual culture contributed together, metaphorically and visually, to the creation of a powerful global imaginary, sending to a vast public the message of urgency and planetary crisis.

Today, in the aftermath of Climategate, the common contesting of expert graphics and climate models calls for a reproduction of the ozone "success story" within the climate change regime are clearly understandable. Most explanations which try to account for the fact that a similar adoption process has not worked for climate change, are to my mind, gross oversimplifications as they systematically underestimate the complexity of the ozone issue.[44] However, and this essay hopefully did not fail to make this point, engaging in longer historical dynamics may help reestablish the historical richness as well as the historical complexity of our objects of study. Indeed, in science studies, in environmental history, and in the history of science we are only just starting to understand the complexity of global environmental issues.

BIBLIOGRAPHY

Aubin, David. "The Hotel that Became an Observatory: Mount Faulhorn as Singularity, Microcosm, and Macro-Tool." *Science in Context* 22 (2009): 365–386.

Besnier, Jean-Michel. "De Galilée à Google Earth—Ce que cela a changé." In *Objectifs Terre. La Révolution des Satellites*, edited by Catherine Cornu, 175–191. Paris: Le Pommier, Cité des Sciences et de l'Industrie, 2009.

Bhartia, Pawan K. "Role of Satellite Measurements in the Discovery of Stratospheric Ozone Depletion." In *Twenty Years of Ozone Decline*, edited by Christos Zerefos, 183–189. Dordrecht: Springer, 2009.

44 Although I agree with Ungar on many points, his paper on climate change and the ozone hole is a typical example of an underestimation of the complexity of the ozone case, see: Ungar, "Knowledge, Ignorance, and the Popular Culture," 2000, 297–312.

Black, Max. "More about Metaphor." In *Metaphor and Thought*, 2nd ed., edited by Andrew Ortony, 19–41. Cambridge: Cambridge University Press, 1993.

Boehm, Gottfried. "Die Wiederkehr der Bilder." In *Was ist ein Bild?*, edited by Gottfried Boehm, 11–38. München: Wilhelm Fink, 1996.

Bojkov, Rumen D. *The International Ozone Commission (IO3C): Its History and Activities Related to Atmospheric Ozone.* Athens: Academy of Athens, 2010.

Borck, Cornelius. "Bild der Wissenschaft." *NTM* 17 (2009): 317–27.

British Antarctic Survey. *Halley Bay Geophysics Report 1982*, AD6/2Z/1982/01, Archives of the British Antarctic Survey, Cambridge, UK.

Callis, Linwood, and Murali Natarajan. "The Antarctic Ozone Minimum: Relationship to Odd Nitrogen, Odd Chlorine, the Final Warming, and the 11-year Solar Cycle." *Journal of Geophysical Research* 91 (1986): 10771–96.

Chesters, Dennis, and Arlin J. Krueger. "A Video Atlas of TOMS Ozone Data, 1978–88." *Bulletin of the American Meteorological Society* 70 (1989): 1564–69.

Christie, Maureen. "Data Collection and the Ozone Hole: Too Much of a Good Thing?" *History of Meteorology* 1 (2004): 99–105.

—. *The Ozone Layer: A Philosophy of Science Perspective.* Cambridge: Cambridge University Press, 2001.

Chubachi, Shigeru. "Preliminary Result of Ozone Observations at Syowa Station from February 1982 to January 1983." In *Memoirs of National Institute of Polar Research*, special issue no. 34: 13–19. Proceedings of the Sixth Symposium on Polar Meteorology and Glaciology. Tokyo: National Institute of Polar Research, 1984.

Crutzen, Paul, and Frank Arnold. "Nitric Acid Cloud Formation in the Cold Antarctic Stratosphere: a Major Cause for the Springtime 'Ozone Hole'." *Nature* 324, (18 December 1986): 651–655.

Davidson, Donald. "What Metaphors Mean." *Critical Inquiry* 5 (1978): 31–47.

Dommann, Monika. "Vom Bild zum Wissen: eine Bestandsaufnahme wissenschaftshistorischer Bildforschung." *Gesnerus* 61 (2004): 77–89.

Du Carla–Boniface, Marcellin. *Expression des nivellemens; ou, Méthode nouvelle pour marquer rigoureusement sur les cartes terrestres et marines les hauteurs et les configurations du terrein.* Paris: L. Cellot, 1782.

Farman, Joseph C., Brian G. Gardiner and Jonathan D. Shanklin. "Large Losses of Total Ozone in Antarctica Reveal Seasonal ClOx/NOx Interaction." *Nature* 315, (1985): 207–210.

Farman, Joseph C., and Richard A. Hamilton. "Measurements of Atmospheric Ozone at the Argentine Islands and Halley Bay, 1957–1972." *British Antarctic Survey Scientific Report,* no. 90 (1975).

Fleck, Ludwik. "Schauen, sehen, wissen." In *Erfahrung und Tatsache. Gesammelte Aufsätze*, 147–174. Frankfurt am Main: Suhrkamp, 1983.

—. "Über die wissenschaftliche Beobachtung und die Wahrnehmung im allgemeinen." In Ludwik Fleck, *Erfahrung und Tatsache. Gesammelte Aufsätze*, 59–83. Frankfurt am Main: Suhrkamp, 1983.

Fleig Albert J., K. F. Klenk, P. K. Bhartia, and D. Gordon. *User's Guide for the Total-Ozone Mapping Spectrometer (TOMS) Instrument First-Year Ozone-T Data Set*, Washington, D.C.: NASA, 1982.

Goddard Space Flight Center. *Animated Atlas of TOMS Ozone Data, 1978–1988*. VHS (NTSC), 40 min., NSSDC Photo Lab, 1989.

Graber, Frédéric. "Le nivellement. Une mesure de l'action autour de 1800." *Histoire & Mesure* 21 (2006): 29–54.

Grundmann, Reiner. *Transnationale Umweltpolitik zum Schutz der Ozonschicht.* Frankfurt am Main: Campus Verlag, 1995.

Gugerli, David, and Daniel Speich. *Topografien der Nation. Politik, kartografische Ordnung und Landschaft im 19. Jahrhundert*. Zürich: Chronos Verlag, 2002.

Haas, Peter M. *Saving the Mediterranean: The Politics of International Environmental Cooperation*. New York: Columbia University Press, 1990.

—. "Obtaining International Environmental Protection through Epistemic Consensus." *Millennium* 19 (1990): 347–64.

—. "Banning Chlorofluorocarbons: Epistemic Community Efforts to Protect Stratospheric Ozone." *International Organization* 46 (1992): 187–224.

Hankins, Thomas L. "A 'Large and Graceful Sinuosity.' John Herschel's Graphical Method." *Isis* 97 (2006): 605–33

Harley, J. Brian. "Deconstructing the Map." In *The New Nature of Maps: Essays in the History of Cartography*, edited by Paul Laxton, 149–68. Baltimore: Johns Hopkins University Press, 2001.

Hentschel, Klaus. "Bildpraxis in historischer Perspektive: Neue Bücher zur wissenschaftlichen Bilderzeugung, -bearbeitung und -verwendung." *NTM,*19 (2011): 413–24.

Heymann, Matthias. "Klimakonstruktionen. Von der klassischen Klimatologie zur Klimaforschung." *NTM* 17 (2009): 171–97.

Lambright, Henry. *NASA and the Environment: The Case of Ozone Depletion.* Washington, D.C.: NASA History Division, 2005.

Lenoir, Timothy. *Inscribing Science. Scientific Texts and the Materiality of Communication.* Stanford: Stanford University Press, 1998.

Lynch, Michael. "The Externalized Retina: Selection and Mathematization in the Visual Documentation of Objects in the Life Sciences." *Human Studies* 11 (1988): 201–34.

Mateer, Carlton L., Donald F. Heath, and Arlin J. Krueger. "Estimation of Total Ozone From Satellites Measurements of Backscattered Ultraviolet Earth Radiance." *Journal of Atmospheric Science* 28 (1971): 1307–11.

Mitchell, William J. T. *Picture Theory. Essays on Verbal and Visual Representation.* Chicago: University of Chicago Press, 1994.

Parson, Edward A. *Protecting the Ozone Layer.* New York: Oxford University Press, 2003.

Pearce, Fred. "Joe Farman's Old-Tech Triumph." *New Scientist* 199 (2008): 46–47.

Pukelsheim, Friedrich. "Robustness of Statistical Gossip and the Antarctic Ozone Hole." *Institute of Mathematical Statistics Bulletin* 19 (1990): 540–42.

Reisin, Steven C., ed. *IEEE Geoscience and Remote Sensing Newsletter* 121. New York: Geoscience and Remote Sensing Society, 2001.

Rheinberger, Hans-Jörg. *Experimentalsysteme und epistemische Dinge.* Göttingen: Wallstein Verlag, 2001.

Robinson, Arthur H. *Early Thematic Mapping in the History of Cartography.* Chicago: Chicago University Press, 1982.

Rodgers, Clive D. *Inverse Methods for Atmospheric Sounding.* New York: World Scientific Publishing, 2000.

Schneider, Birgit. "Climate Model Simulation Visualization From a Visual Studies Perspective." *WIREs Climate Change* 3 (2012): 185–93.

—. "Linien als Reisepfade der Erkenntnis. Alexander von Humboldts Isothermenkarte des Klimas." In *KartenWissen. Territoriale Räume zwischen Bild und Diagramm*, edited by Stephan Günzel and Lars Nowak, 173–97. Wiesbaden: Ludwig Reichert Verlag, 2011.

Stolarski, Richard, et al. "Nimbus 7 Satellite Measurements of the Springtime Antarctic Ozone Decrease." *Nature* 322 (1986): 808–11.

—. "The Antarctic Ozone Hole." *Scientific American* 258 (1988): 20–26.

Sullivan, Walter. "Low Ozone Level Found Above Antarctica." *New York Times*, 7 November, 1985. http://www.nytimes.com/1985/11/07/us/low-ozone-level-found-above-antarctica.html

Talman, Fitzhugh. "A List of Meteorological Isograms." *Monthly Weather Review* 43 (1915): 195–98.

Thrower, Norman J. W. "Edmond Halley as a Thematic Geo-Cartographer." *Annals of the Association of American Geographers* 59 (1969): 652–76.
Ungar, Sheldon. "Knowledge, Ignorance, and the Popular Culture: Climate Change versus the Ozone Hole." *Public Understanding of Science* 9 (2000): 297–312.
Wise, Norton. "Making Visible." *Isis,* no. 97 (2006): 75–82.

IMAGE COPYRIGHTS & SOURCES

Figure 1: 1983 NASA.
Figure 2: NASA image 86-HC-289.
Figure 3: Goddard Space Flight Center, *Animated Atlas of TOMS Ozone Data, 1978–1988*, VHS (NTSC), 40 min., NSSDC Photo Lab, 1989, NASA.
Figure 4: Arlin J. Krueger, et al., *The 1987 Airborne Antarctic Ozone Experiment: The Nimbus-7 TOMS Data Atlas*, Washington, D.C.: NASA, 1988, 164.
Figure 5: Historic Maps Collection, Princeton University Library / Public Domain.
Figure 6: © Staatsbibliothek zu Berlin, S.P.K.
Figure 7: © 1985 Nature.

Images for Data Analysis: The Role of Visualization in Climate Research Processes

THOMAS NOCKE

This article provides a situation analysis of scientific data visualizations in climate research processes. It investigates how visualizations are applied from a computer graphics perspective. It builds on findings that have been derived from case studies and from a survey undertaken with climate and climate impact scientists. Based on this, the article discusses the use of visualization techniques and tools in everyday climate research workflows. Beyond standardized methodologies, the application of a new generation of visualization techniques and tools will be presented, focusing on interactive climate data exploration.

1. INTRODUCTION

With increasing computing power, climate modelers generate increasingly large simulation data sets. In addition, enhanced observation techniques such as remote sensing contribute to the growth of spatio-temporal climate related data. To provide an example: the size of climate data stored for the IPCC AR5 report is estimated to be around 10 petabytes, which is 5 times the content of all US academic research libraries. Apart from its size, the complexity and heterogeneity of the climate data is a challenging problem[1] resulting from several factors, including data production from multiple models, multiple future scenarios, with multiple model simulations on manifold spatial grids (2D/3D, equidistant/polar, etc.) and temporal time scales (minutes to centuries). In

1 Edward, *A Vast Machine*, 2010.

addition, a variety of climate-related variables contributes to the complexity of the climate data. They range from physical measures such as temperature, air pressure or precipitation to ecosystem and economy research related measures such as plant net primary production or energy demand.

When trying to identify the underlying properties in the data—such as structures and dependencies between the variables—analysis increasingly becomes the bottleneck. Furthermore, there is a rising demand for methods which support the evaluation of the reliability of conclusions based on (climate) simulation data.

An important method tackling the challenges mentioned is the visual illustration of such data. From a computer graphic perspective, such a visual illustration of the relevant characteristics of a data set is called "visualization." In this sense, visualization is a calculation method which transforms abstract, typically non-geometric information into 2- or 3-dimensional geometric representations (e.g. points, lines, surfaces) that are rendered into images afterwards. The aim of this transformation is to facilitate analysis, comprehension and communication of models, concepts and data. The visualizations need to be constructed in such a way that the observer is enabled to recognize, understand and evaluate the data and its meaning.

Visualization created within scientific contexts targets the provision of meaningful representations that support the exchange of working results and provide scientists with appropriate tools to reveal hidden patterns and relations within the data. The advantage of visualization is that it establishes a direct interface between digital data in a computer and the human perceptual and cognitive abilities, presenting abstract relationships compactly and intuitively.

Visualization techniques are available for different data classes,[2] for instance for volume data,[3] representing three-dimensional phenomena such as CT scans, for flow data[4] representing data with a direction such as air flow around an air plane, and multi-variate data[5] representing multiple data variables simultaneously, such as temperature, humidity, pressure and wind speed. Since 1990, significant advances have been made in enhancing visualization as a flexible, easy-to-use data exploration tool.[6] The innovative approach is the possibility of interacting directly with the image and linking several views to the data. Going

2 Some general examples beyond climate research.

3 Brodlie and Wood, "Recent Advances in Volume Visualization," 2001.

4 Laramee, et al., "The State of the Art in Flow Visualization," 2004.

5 Bürger and Hauser, "Visualization of Multi-variate Scientific Data," 2007.

6 See: Shneiderman, "The Eyes Have It," 1996.

beyond this, the new research field of "visual analytics" that has emerged within the last decade, focuses on coupling human perception abilities with automatized computational methods, allowing new insights into huge, complex data sets.[7]

In meteorology and climatology, visualization has a long tradition.[8] The application of visualization in this context focuses on standardized techniques. Typically, these are plots such as time series graphs, bar charts and colored 2D-maps, ensuring their comprehension by a large audience. However, these plots are mainly restricted to simplified views of the data, and show particular parts of the large data sets only, often leading to a loss of relevant details. In particular, this results from the missing interaction options with such static, "overview" images, not providing the user with the chance to interact with the images or enabling her/him to "zoom and filter, then [to refine the visualization with] details-on-demand."[9] Thus, the traditional, static methods are of limited use for detecting possibly interesting, partly unknown features in the data (in the sense of a data exploration), such as finding model inconsistencies or analyzing specific high-dimensional data patterns of interest.

Against this background, this article investigates the potential arising from interactive visual analytic methods and tools for exploring climate data with especially considering interactive methods. The article is structured as follows: section 2 provides a detailed situation analysis. Then, standardized visualization techniques and tools used in climate research are presented (section 3). Afterwards, the paper presents experiences, results and ideas of recent work on establishing new visual analytics technology in climate science (section 4). Finally, it summarizes the main findings and provides conclusions (section 5).

2. SITUATION ANALYSIS

A major problem in the presentation of data in general and climate data in particular is restricting representations to a certain data section only and/or limited points of view. Many facets of the data stay hidden, for example, the original data that statistical aggregations are based on. This hampers climate experts in their attempts to gain a detailed picture of the properties of the data and to assess its quality. The relationship can be illustrated with the metaphor of an iceberg, in which the visual representation is—as a visible part of the

7 Thomas and Cook, *Illuminating the Path*, 2005.

8 Monmonier, *Air Apparent*, 1999.

9 Shneiderman, ibid.

iceberg—already a compression of the original data that lies below the water surface. Below "this surface" lie statistical measures that are based on ensembles of model simulations, which are based on one or more complex climate models, which in turn are based on a variety of physical, mathematical, information technological and other assumptions and concepts.

To gain an overview of the data visualization practices and requirements in the heterogeneous field of climate and climate impact research, we informally interviewed 76 researchers.[10] A major result of our survey was that there exists a lack of utilizing visualization as an interactive analysis tool in the routine work of climate scientists. Up to now researchers tend to transform data and analysis results into the established vocabulary of classic static visual representations for scientific publications, without regarding new developments.

Possible reasons for adhering to the standardized techniques are manifold. First of all, data size and heterogeneity are most challenging in this context, because they burden climate researchers with the task of making an appropriate choice for the methods or tools to be applied (which need to match both size and data characteristics of a certain climate data set). Currently around 10 tools are specialized for meteorological and climate background, however, the number of tools available with high potential for visual climate data analyses can be estimated at more than 50. In practice, however, we found that climate researchers are familiar with one or two visualization system(s)[11] and depend on the tools at hand used in their scientific surroundings. Thus, the users' flexibility in using interactive visualization techniques is strongly restricted, and therefore they tend to resort to basic "plotting" solutions, such as the already mentioned time series graphs, bar charts and colored 2D-maps.

In addition, one can observe that visual thinking is mainly presentation (and less exploration) driven, questioning if/how the (climate) scientist can present a result with a certain visual metaphor. This leads to the situation that alternative visualizations that might not be known in the scientific community and/or are not immediately self-explanatory—even if they allow to analyze the data much better (e.g., high dimension data relations)—are hardly used in everyday data

10 The interviews were executed with researchers from Potsdam Institute for Climate Impact Research, with a multitude of disciplines related to climate research (meteorologists, biologists, economists, a.o.); see Nocke, et al., "Visualization of Climate and Climate Change Data," 2008.; Tominski, et al., "Information Visualization in Climate Research," 2011.

11 Visualization systems (synonym: visualization tools) are computer programs that generate data visualizations.

exploration situations. This relates to the tension of complexity of the climate data and the underlying knowledge on the one hand and the required simplicity of its visual analysis and representation on the other.

Discussions with the scientists revealed that there is even a kind of mistrust in interactivity. Scientists often are skeptical with respect to the "arbitrariness" of visual representations that have been generated by interactive adjustments by the user—like thresholds or visualization parameters. In the natural sciences, the comparability of visual representations is very important. Therefore, script-based[12] visualization systems that generate reproducible representations are favored over interactive solutions where it is often unclear which parameter settings are required to generate a certain view of the data.

However, in recent years, new technologies such as "Google Maps," "Gapminder," and other web-based visualization services, served as a kind of starter for exploring new ways of analyzing the data interactively. Young scientists are more accustomed to utilizing the interactive features that such tools offer. Thus, nowadays there is a rising acceptance of interactive visualization, however, still mainly for the purpose of presentation. Building on these observations, recent developments have been started that go beyond presentation to provide researchers with interactive visual tools for data exploration and analysis.

The goal of current geo-visualization research is to incorporate such tools into typical scientific workflows. This also includes raising the awareness of well-studied interactive visualization concepts such as multiple visualization views that are connected with respect to user selections[13] and immediate alteration of the data shown based on user interactions.[14]

---

12 Script-based visualization means that the user him/herself writes a high-level sequence based visualization program, for instance "1) load data X, 2) select part Y of data, 3) create diagram type Z, 4) annotate visualization with A, 5) generate image," which is then executed by the visualization system.

13 In visualization research these are called "multiple coordinated views". Selecting certain items in one visualization view leads to an intermediate highlighting (so called brushing) of these items in this view and within all other connected views (linking).

14 So called dynamic queries.

3. STANDARD VISUALIZATION TECHNIQUES AND TOOLS IN CLIMATE RESEARCH PROCESSES

Most frequently, climate data is visualized with reference to its temporality, to identify temporal trends directly using time charts (Figure 1a). For the representation of more abstract data views (typically without time or spatial axes), bar charts, stacked bar charts,[15] box plots[16] and Taylor plots[17] are frequently used to show multiple climate variables (multi-variate) or multiple simulation data sets (ensembles) together (Figure 1c). Being of similar importance, meteorological and climate data are represented in their geo-spatial reference, typically using color and isolines (Figure 1e), with animation to display the time axis. If the distributions of two climate related variables are compared, scatter plots are most frequently used (Figure 1b). For specific sub-domains, such as oceanographic and atmospheric studies, vector field representations are established (Figure 1d). 3D representations, for instance displaying atmospheric humidity distributions, are hardly used in static images due to the problem that typically important parts of the data[18] are not visible. However, there are many 2.5/3D visualization examples in presentation videos (e.g., Figure 1f).

A variety of tools and systems have been developed so far to visualize climate data. They range from general purpose solutions[19] to special purpose solutions for climate data only, providing different levels of interactivity. The choice for a certain tool has consequences by determining the techniques being used, and even lets climate researchers prefer certain color maps which are provided by the systems. So, in many cases, it is possible to tell by the look of a certain graph which tool has been used to create it. These tools can be subdivided into the following five categories: (1) classical script-based weather/climate visualization tools (e.g. Ferret, Grads, GMT), (2) office tools (e.g. Microsoft Excel), (3) geographical information systems (ArcGIS), (4) mathematical/statistical systems

15 In stacked bar charts, bars are divided into subparts to show cumulate effects (typically of several variables), for instance to present different parts of the energy mix (e.g. photovoltaic, nuclear, coal energy production together).

16 McGill, Tukey and Larsen, "Variations of Box Plots," 12–16.

17 See: Taylor, "Summarizing multiple aspects of model performance in a single diagram," 7183–92.

18 For the humidity data as example using a certain part of the clouds.

19 Which can be used for data from other domains as well.

Figure 1: Standard visualization techniques used in climate research: a) temporal variations in global sea mean level (IPCC-WG1 2007, 411); b) scatter plot of cumulative capacity and costs of renewable energies (IPCC-WG2 2007, 272); c) comparison of greenhouse gas emissions of energy production types (IPCC-WG2 2007, 283); d) streamline and arrow visualization of the horizontal ocean velocity field (CLIMBER-3 model); e) geographic distribution of long-term linear trends in mean sea level (IPCC-WG1 2007, p.411); f) atmospheric dust as 2.5d semi-transparent surface and surface albedo color coded on earth surface (historic CLIMBER-2 model ice age simulation).

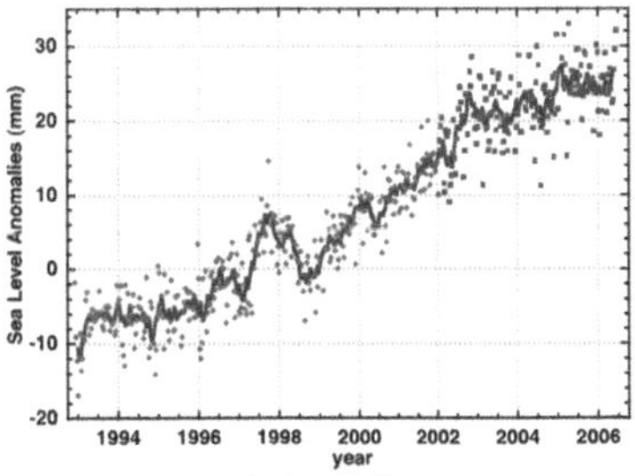

a) time chart

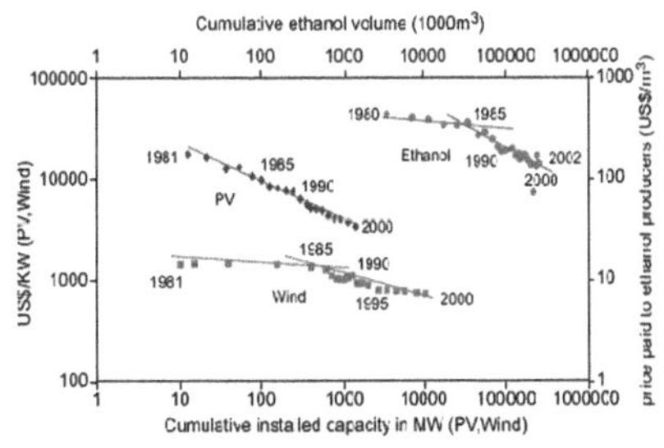

b) scatter plot

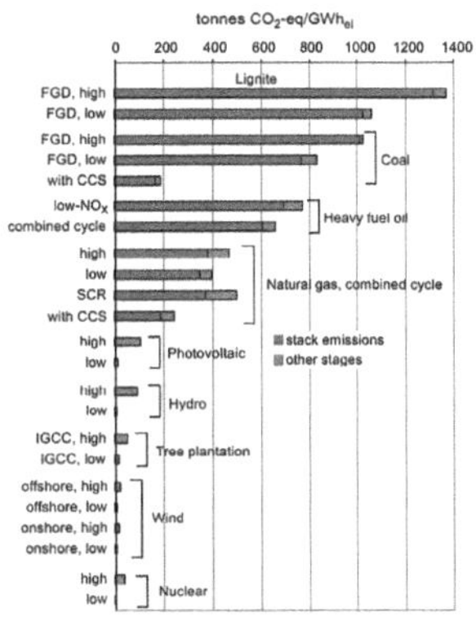

c) bar chart

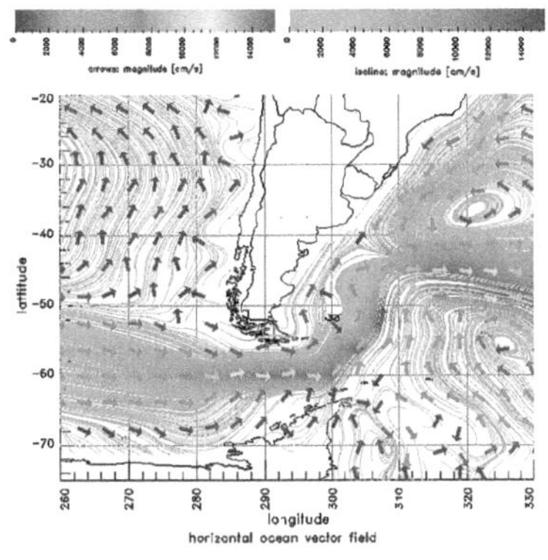

d) 2D map with vector data

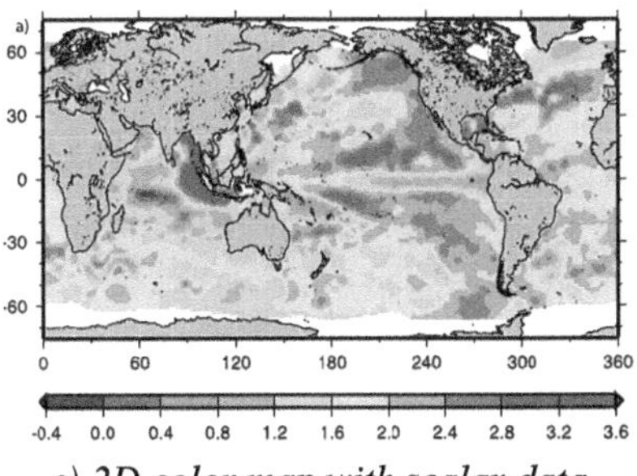

e) 2D color map with scalar data

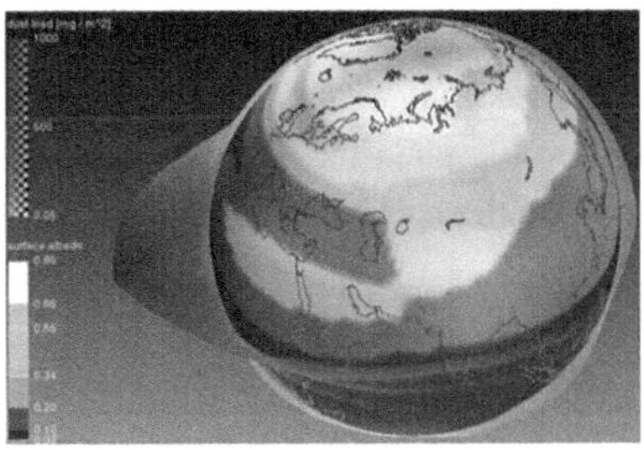

f) 2.5D color map

(e.g. Matlab, R), simple general purpose script-based tools with visualization extensions (Python), and (5) 2D/3D scientific visualization systems (e.g. Paraview, Avizo, OpenDX). In particular, class 1 tools are designed for climate data specific requirements (for the climate data storage formats and typical spatial data grid types). Classes 3, 4 and 5 provide climate specific extensions; however, they are used for other application domains as well.

Summarizing, there is a large spectrum of systems/tools that are currently used in climate research, in particular dealing with the demands of producing static, easily reproducible images.

4. The Visual Analytics Approach to Climate Research

The idea of "visual analytics"[20] is to tightly link automated data analysis methods with interactive visualization methods to strongly improve data analysis tasks. Visualization researcher Daniel Keim defined the aim in the following way: "The goal of visual analytics is the creation of tools and techniques to enable people to synthesize information and derive insight from massive, dynamic, ambiguous, and often conflicting data, [to] detect the expected and discover the unexpected, [to] provide timely, defensible, and understandable assessments and [to] communicate assessment effectively for action."[21] And further: "The challenge is to identify the best automated algorithm for the analysis task at hand, identify its limits which cannot be further automated, and then develop a tightly integrated solution which adequately integrates the best automated analysis algorithms with appropriate visualization and interaction techniques."

The goal, from a visualization scientist's point of view, is to facilitate the use of visual analytics for climate and climate impact research, reducing the obstacles to using such technologies in everyday climate research workflows. Three options to obtain this goal are discussed in the following: the use of alternative visual mappings (section 4.1), the use of sophisticated, interactive tools providing so called linked multi-views (section 4.2) and new strategies to support climate users of such technologies, tailored for their individual data sets and visualization tasks (section 4.3).

20 Thomas and Cook, *Illuminating the Path*, 2005.

21 Keim, et al., "Visual Analytics," 2008.

4.1 Alternative Visualization Techniques

By aiming at visual analysis the large, complex climate data sets, standard mapping of climate data faces some obstacles. For example, Figure 1d depicts easily comprehensible maps, whose interpretation assumes no prior knowledge. However, if several climate impacts are compared, such spatial representations reach their limits, because of the obstacles/difficulties just outlined. For this kind of problem, suitable representation techniques have been developed in the visualization domain, in particular for examining higher-dimensional relationships. In the following, some advanced approaches for visualizing complex data interactively will be discussed.

An important class of techniques in this context is so called polyline representations which represent multiple data values at once by drawing lines between axes.[22] For presentation purposes the circular arrangement of the axes,[23] and for visual analysis tasks the parallel axes arrangement[24] are widespread. They are constructed as follows (see Figure 2): first, the axes of each variable are placed in a certain layout (typically parallel or circular). Second, lines representing a pair of data values for the two variables are drawn from one axis to the other. Depending on the data values, vertical start and end point position are mapped on one of the axes. Then, further axes representing the data variables can be added, and the individual lines between the axes are connected, thereby constructing polylines. Each polyline represents a point in the high dimensional space (of all axes), which then can be compared very effectively, because several data set characteristics such as distribution properties, correlation structures and outliers—even of large data sets—can be identified at a glance.

Figure 2 (bottom) illustrates the potential of the parallel version (parallel coordinates) for a climate data set of German measurement stations. Here, after selecting certain parts of the polylines and highlighting those, the observer gets a direct impression of the high dimensional conditions for extreme temperature situations, for instance medium and slightly above medium data values in "cloud area fraction" and "humidity" in cold months (blue polylines). In contrast, the

22 Each of these polylines represents a multi-dimensional point, e.g. multiple metrological measurements at a certain location and certain time, such the seven values of maximum air temperature, sunshine duration, vapor pressure, humidity, wind speed and precipitation (in Figure 2).

23 So-called radar charts, star plots, cobweb charts, or kiviat diagram.

24 So-called parallel coordinates, see d'Ocagne, *Coordonnées parallèles et axiales,* 1885 and Inselberg, "The Plane with Parallel Coordinates," 1985.

red lines represent warm day conditions, which are related to less "cloud area fraction" and "humidity" conditions.

Figure 2: Visualization technique parallel coordinates: representation of monthly climate measurements in Germany: using transparency for polyline density (top); with additional brushing of high (red) and low (blue) daily maximum temperature values (bottom). (→ color plate 2)

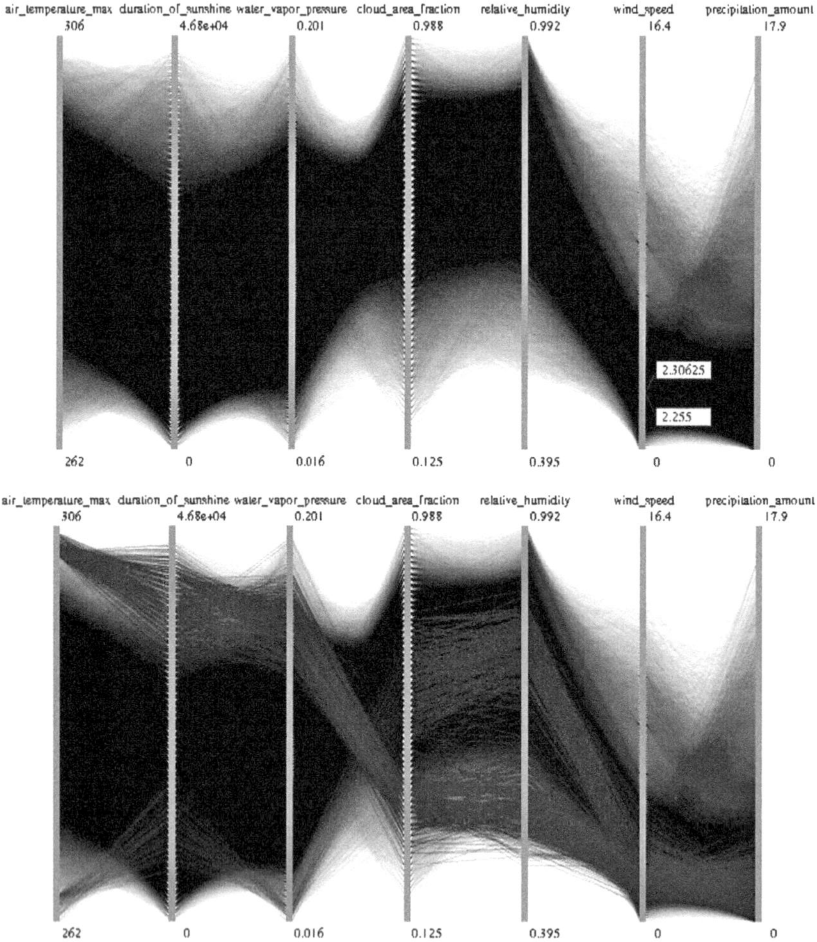

Reducing the clutter of multiple overlaid polylines and thus, making use of this kind of representation, several improvements have been proposed in visualization literature[25]. The most important ones are (1) the use of transparency, highlighting regions of high line density (see Figure 2), and (2) the use of so-called brushes, which highlight polylines by using alternative colors (typically using individual colors to highlight certain subspaces, see Figure 2 bottom).

A second example is sophisticated time visualization techniques[26] as an alternative to simple time charts. One prominent example used for weather / climate time series visualization is "two-tone color mapping"[27], which compactly represents long climate time series (see Figure 3). This technique allows a very compact representation of long time series, representing more details within the same area compared to a single time chart (see Figure 3c). In addition, it highlights patterns (in Figure 3 based on statistical values of the data distribution) which facilitates an easy comparison of time series, avoiding occlusion occurring in standard time charts visualizing multiple time series together ("spaghetti graphs").

In this technique, time is mapped to the x-axis similar to standard time charts. In contrast to time charts, where the data values are represented by single-color primitives (point, line) at certain positions on the y-axis, two-tone coloring draws a two colored bar filling the full height of the chart for each time step. The two colors determine the data interval of the current data value, and the height ratios of the two bar parts are proportional to the distance ratios of the data value to the two interval limits. Figure 3a displays maximum temperature values of the 80s (re-ordered) for an individual measurement station, mapping each day's values within this decade to an area of one-pixel width. The warmer a day is, the darker its bar (parts) get(s), and for colder days vice versa. Figure 3b shows a "fingerprint" of the daily values of 14 meteorological variables of a whole year, e.g. showing that hot conditions ("tmed," "tmax," "tmin" in black) typically match with a long sunshine duration ("son" as well in black).

A third example is the application of recent network visualization developments to climate networks. Such networks are typically constructed by (1) applying measures to relate time series of climate measurement stations or simulation grid points (the nodes of the network), (2) constructing a network linking stations / grid points based on their similarity (the links of the network),

25 See: Novotny and Hauser, "Outlier-Preserving Focus," 2006; McDonnell and Mueller, "Illustrative Parallel Coordinates," 2008.

26 See: Aigner, et al., *Visualization of Time-Oriented Data,* 2011.

27 Saito, et al., "Two-Tone Pseudo Coloring," 2005.

Figure 3: Temporal visualization of daily values of the Potsdam meteorological data series using two-tone color mapping: a) temperature maximum values in the 80s, and b) of 14 variables for the year 1995; c) comparison of two-tone mapping with standardized time chart.

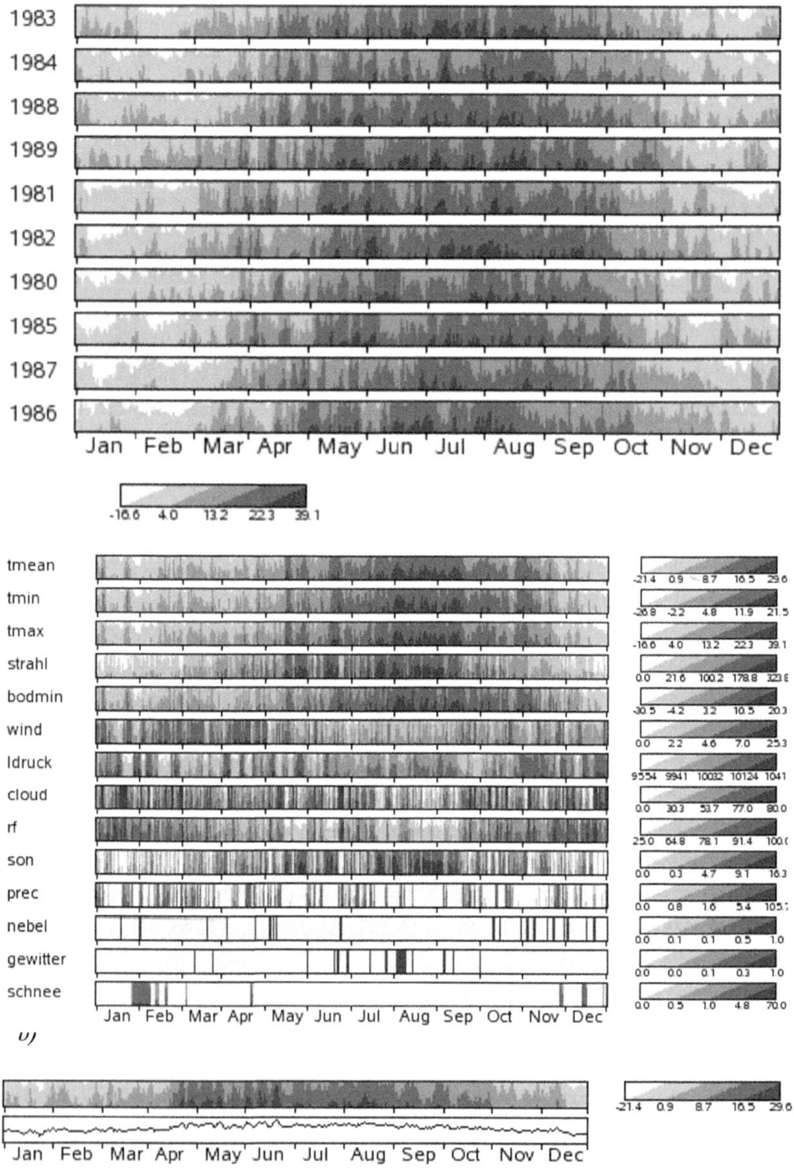

and then by (3) calculating network measures such as betweenness or clustering centrality for both nodes and edges.[28]

Figure 4 illustrates sophisticated visualization solutions for climate networks produced using interactive filtering with the network visualization system CGV.[29] Figure 4a shows a part of a global temperature network with 3D arcs, depicting both long distant teleconnections and regional structures.

The challenges for visual analysis of such networks arise from their geo-reference, which impedes to use standard graph layout algorithms that—to reduce edge overlaps—change the node positions, because the spatial node embedding is essential for the interpretation of the network. To avoid a resulting, so-called "edge clutter", algorithms to bundle edges with similar properties can be applied.[30] To provide an example of such bundles, Figure 4b illustrates the strong cluttering in clustered regions of a precipitation network. Now, using an edge-bundling algorithm (Figure 4c) which lets similar edges attract each other, the general edge structure becomes better comprehensible.

However, many of these advanced visualization techniques are used by visualization experts mostly, since they can hardly be interpreted without prior knowledge. As one solution to this challenge, techniques can be used that construct a link between the meaning of the data and its visual representation. Such techniques visually map complex information relationships intuitively by means of easy to understand metaphors and icons. An example are in the mass media well-established weather maps which integrate information on sunshine, clouds, rain or snow in composite icons on the map very intuitively and compactly.

These easy-to-understand icons can be used for the visualization of climate change and impacts, too, for instance for various factors indicating a potential failure of the maize yield in northeastern Brazil on a map (Figure 5, Nocke et al. 2005[31]). Here, six indicators that include various climatic risks are mapped to different areas of the icon. Thick light parts of the icon represent a low risk value in a certain indicator, while thin dark areas represent high risk values. Alongside

28 See: Donges, et al., "The Backbone of the Climate Network," 2009.

29 Coordinated Graph Visualization (CGV), see as well: Tominski, Abello and Schumann, "CGV," 2009; Tominski, et al., *Information Visualization in Climate Research,* 2011.

30 See: Holten and van Wijk, "Force-Directed Edge Bundling for Graph Visualization," 2009.

31 Nocke et al., "Icon-based Visualization using Mosaic Metaphors," 2005.

Figure 4: Visualization of climate networks: (a) global temperature network; (b) part of German precipitation network with edge clutter and (c) with bundled edges.

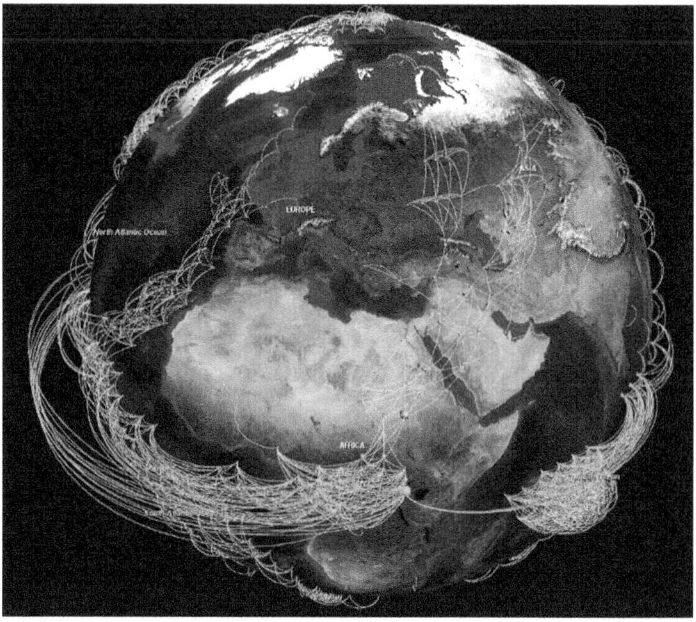

a)

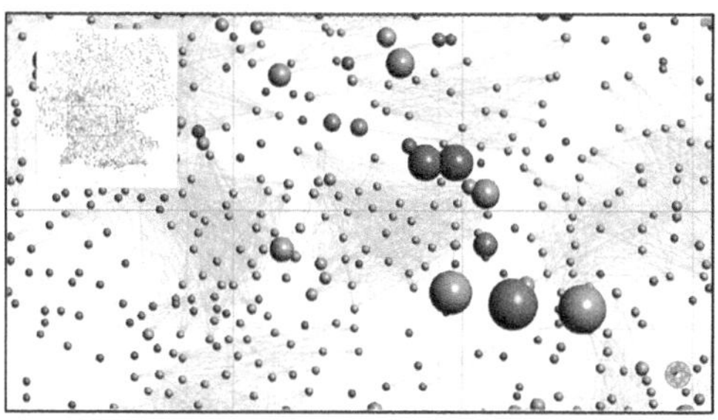

b)

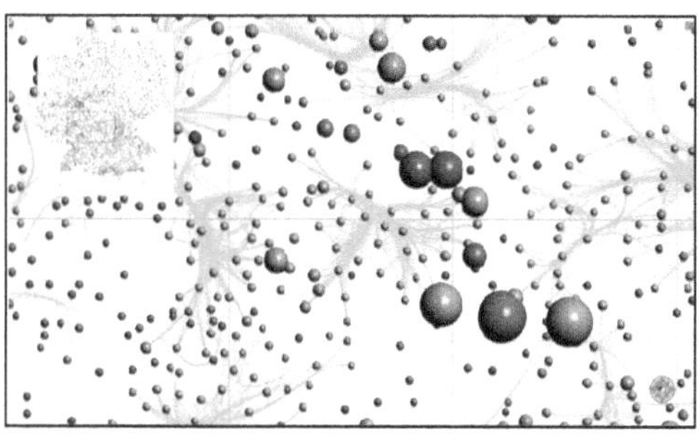

c)

Figure 5: Visualization of multiple climate variables at once: maize harvest risks for northeastern Brazil using multi-variate maize icons.

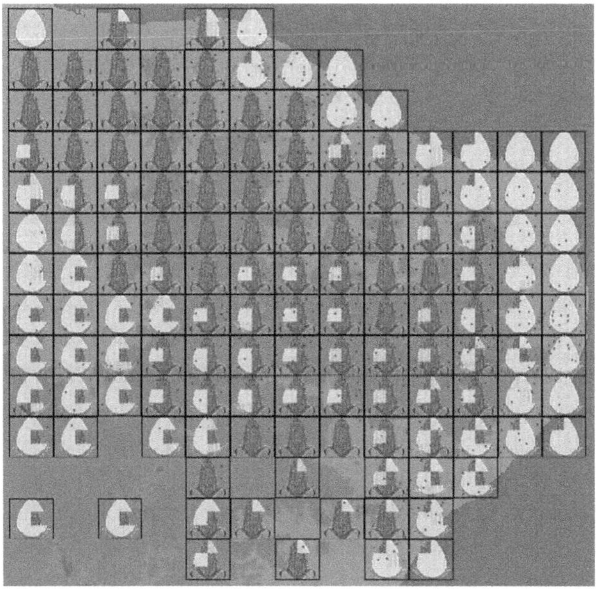

Figure 6: Screen shot of the Visual Analytics tool ComVis, visually analyzing energy model ensemble simulations for different future climate scenarios.

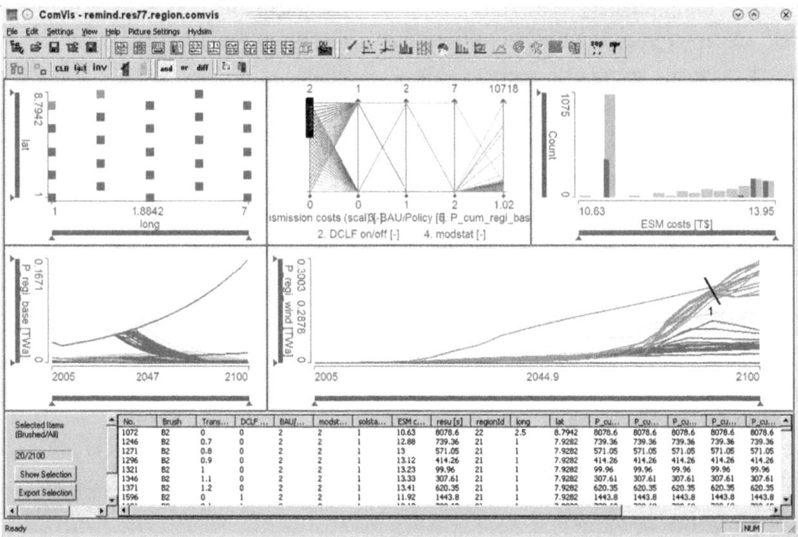

retrieving general distribution patterns this allows to analyze specific risks patterns and the related areas.

Summarizing, climate scientists are typically not familiarized with such visualization tools and require new viewing habits, however, once they were used to these new visual mappings, a faster and even higher degree of insight might become possible.

4.2 Visualization Tools

If a certain visualization technique—as the ones presented in the last section, which could be highly useful for specific tasks—is really applied in climate science practice is determined by the tools at hand and their functionalities. In contrast to the standard tools listed in section 3, visual analytics tools provide two or more of the following extensions: (1) multiple-linked views, (2) a variety of highly interactive visualization techniques, (3) ability to handle large data sets interactively, and/or (4) integration with data processing mechanisms (e.g. calculation of derived data).

As an example for the multiple views, Figure 6 shows an analysis session of an energy transport model ensemble simulation[32] with the ComVis tool.[33] It combines a spatial view (Figure 6 top left), a parallel coordinates view (top center), a histogram view (top left), two time charts (center left and right) and a detailed textual table depicting selected data values (bottom). To provide insight, two parts of the data sets have been interactively selected and highlighted using color coding: high regional wind energy production simulations in the end of the 21st century (green) and high emission cost simulations (red).

Due to its simplicity and flexibility analyzing multi-variate ensemble data sets in multiple linked views, tools such as ComVis have proven their potentials in several climate related studies.

In addition, there are some other examples of Visual Analytics tools finding their way in climate change research processes. To visually analyze large climate spatio-temporal data sets interactively, tools such as SimVis[34], Paraview[35] and Avizo[36] have been used in several studies. They are designed to handle even larger data sets of several gigabytes interactively.

---

32 See: Haller, Ludig and Bauer, "Bridging the Scales," 2012.

33 Matkovic, et al., "ComVis," 2008.

34 Doleisch, Stonawski and Hauser, "SimVis," 2005.

35 http://www.paraview.org.

36 http://www.vsg3d.com/avizo/green.

4.3 Visualization Design for Visual Climate Research

For climate scientists to adopt alternative visualization techniques and tools into their everyday workflows (in addition to the standard climate data visualization tools), there must be strong advantages in doing so. Problems for climate researchers arise from the complexity of most visualization software tools, which leads to relatively high learning effort until a user can produce first expressive visualizations of its data (considering both data import and visual image generation). Furthermore, many tools often do not support visualization quality issues relevant for climate research, such as providing expressive/suitable color maps for climatic variables.[37]

Regarding this situation, the manifold of potentially useful visualization techniques and tools requires a substantial support in selecting and parameterizing the appropriate visualization techniques. To fill this gap, the automated visualization design approaches such as the VisTrailes[38] or SimEnvVis[39] have been developed. To provide an example in more detail, SimEnvVis combines standard and advanced techniques and tools as well as standard parameterizations tailored for climate data. It implements a visualization design wizard as human-computer interface, which provides a step-by-step visualization generation/selection process. In addition, by hiding usage details of the individual visualization tools, it supports climate scientists to focus on the analysis process itself, reducing the risk of visualization users getting frustrated configuring a manifold of visualization options.

Therefore, a two-step, rule-based mechanism is applied, which provides predefined starting points into interactive visualization techniques from different visualization tools. In a first step, the general suitability of a visualization technique is estimated based on data characteristics and other factors (see Figure 7). Then, these techniques are parameterized based on rules. This includes the selection of a visual mapping strategy (e.g. isolines or/and coloring), the selection of a mapping of certain variables to certain visual attributes (e.g. temperature to colors) and the fine-tuning of the parameters (e.g. selection of a color map).

---

37 Default color maps of most systems—and thus often used in scientific imaging—do not fulfill quality standards from cartography and visualization background; e.g. Brewer, "Color Use Guidelines," 1999.

38 http://www.vistrails.org.

39 Nocke, "Visuelles Data Mining," 2007.

Figure 7: Visualization Design Wizard for Climate Data: visualization technique selection. (→ color plate 3)

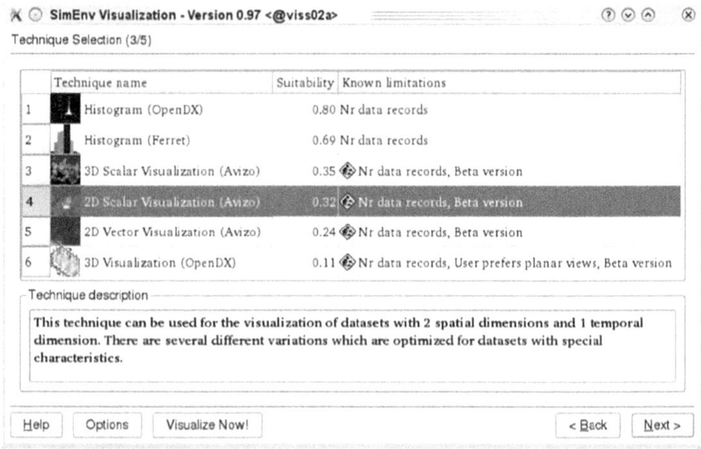

5. CONCLUSION

To summarize: the visualization of climate data is an important means to both get insight in climate and climate related science and to communicate those insights. However, most frequently climate data is processed using conventionalized statistical methods, and visualization is oftentimes used by producing a final static image. This is appropriate for presentation purposes (such as the IPCC report), however, it does not exploit the power of the human visual system in combination with the strengths of computer-based automatic data analysis. Such an in-depth climate data analysis—tightly coupling statistics and visualization—is the subject of ongoing research.

In particular, interactive visual analytics of large, complex time-dependent, geo-referenced climate data is still a challenging problem. On the one hand, visual analytics methods are not yet widespread in climate research. On the other hand, the combined application of interactive methods provided by visualization and geographic information systems and of automated statistical mining methods is still hampered for climate science users.

For bridging the gap between complex visualization systems and climate scientists and decision makers, methods from human-computer interaction and visualization design can be applied to climate data visualization, providing easy-to-use graphical user interfaces even for complex data sets and a multitude of interactive visual representations.[40]

ACKNOWLEDGMENTS

The author wishes to thank Tamina Christ for carrying out and evaluating the survey, and all people supporting climate data visualization (Christian Tominski, Kresimir Matkovic, Helmut Doleisch, Magnus Heitzler, Michael Flechsig, Markus Wrobel, Conrad Thiede, Mario Baalcke). The author acknowledges financial support by the Federal Ministry for Education and Research via the Potsdam Research Cluster for Georisk Analysis, Environmental Change and Sustainability (PROGRESS).

BIBLIOGRAPHY

Aigner, Wolfgang, Silvia Miksch, Heidrun Schumann and Christian Tominski. *Visualization of Time-Oriented Data*, London: Springer, 2011.

Brewer, Cynthia A. "Color Use Guidelines for Data Representation." *Proceedings of the Section on Statistical Graphics*, 55–60. Alexandria VA: American Statistical Association, 1999.

Brodlie, Ken and Jason Wood. "Recent Advances in Volume Visualization." *Computer Graphics Forum* 20, no. 2 (2001): 775–792.

Bürger, Raphael and Helwig Hauser. "Visualization of Multi-variate Scientific Data." In Schmalstieg and Bittner (editors): Proceedings of EuroGraphics 2007 (State of the Art Reports), 2007: 117–134.

Doleisch, Helmut, Georg Stonawski and Helwig Hauser. SimVis: Interactive Visual Analysis of Simulation Results. In "*Proceedings of the 2nd NAFEMS CFD-Seminar "Simulation of Complex Flows—Application and Trends*," 2005.

Donges, Jonathan F., Yong Zou, Norbert Marwan and Jürgen Kurths. "The Backbone of the Climate Network." *Europhysics Letters* 87, no. 4 (2009): 48007.

Edwards, Paul N. *A Vast Machine. Computer Models, Climate Data, and the Politics of Global Warming*, Cambridge, Mass.: MIT Press, 2010.

40 Wrobel, et al., "Graphical User Interface Design," 2008. ; Nocke, et al., "Visualization of Climate and Climate Change Data," 2008.

Galton, Francis. *Meteographics, or, Methods of Mapping the Weather*. London: Macmillan, 1863.

Haller, Markus, Sylvie Ludig and Nico Bauer. "Bridging the Scales: A Conceptual Model for Coordinated Expansion of Renewable Power Generation, Transmission and Storage." *Renewable and Sustainable Energy Reviews* 16, no. 5 (2012): 2687–2695.

Holten, Danny and Jarke van Wijk. "Force-Directed Edge Bundling for Graph Visualization." *Computer Graphics Forum* 28, no. 3 (2009): 983–990.

Inselberg, Alfred. "The Plane with Parallel Coordinates." *Visual Computer* 1, no 4 (1985): 69–91.

IPCC-SR. 2007: *Climate Change 2007: Synthesis Report*. Contribution of Working Groups I, II and III to the Fourth Assessment Report of the Intergovernmental Panel on Climate Change [Core Writing Team, Pachauri, Rajendra K. and Reisinger, Andy. (editors)]. IPCC, Geneva, Switzerland, 104 pp.

IPCC-WG1. 2007: *Climate Change 2007: The Physical Science Basis*. Contribution of Working Group I to the Fourth Assessment Report of the Intergovernmental Panel on Climate Change [Solomon, Susan, Dahe Qin, Martin J. Manning, Zhenlin Chen, Melinda Marquis, Kristen Averyt, Melinda M. B. Tignor and Henry LeRoy Miller Jr. (editors)]. Cambridge, UK and New York, NY, USA: Cambridge University Press.

IPCC-WG2. 2007: *Climate Change 2007: Impacts, Adaptation and Vulnerability*. Contribution of Working Group II to the Fourth Assessment Report of the Intergovernmental Panel on Climate Change [Martin L. Parry, Osvaldo F. Canziani, Jean P. Palutikof, Paul J. van der Linden and Clair E. Hanson (editors)]. Cambridge, UK and New York, NY, USA: Cambridge University Press.

Keim, Daniel, Gennady Andrienko, Jean-Daniel Fekete, Carsten Görg, Jörn Kohlhammer and Guy Melançon. 2008. "Visual Analytics: Definition, Process, and Challenges." *Information Visualization*. 4950:154–175.

Laramee, Robert S., Helwig Hauser, Helmut Doleisch, Benjamin Vrolijk, Fritz H. Post and Daniel Weiskopf. "The State of the Art in Flow Visualization: Dense and Texture-Based Techniques". *Computer Graphics Forum* 23, no. 2 (2004): 203–221.

Matkovic, Kresimir, Wolfgang Freiler, Denis Gracanin and Helwig Hauser. "ComVis: A Coordinated Multiple Views System for Prototyping New Visualization Technology." In *Proceedings of the 12th International Conference Information Visualisation*. July 2008, 215–220.

McDonnell, Kevin T. and Klaus Mueller. "Illustrative Parallel Coordinates." *Computer Graphics Forum* 27, no. 6 (2008): 1031–1038, doi: 10.1111/j.1467-8659.2008.01239.

McGill, Robert, John W. Tukey and Wayne A. Larsen. "Variations of Box Plots." *The American Statistician* 32, no. 1 (February 1978): 12–16, doi:10.2307/2683468.

Monmonier, Mark. *Air Apparent: How Meteorologists Learned to Map, Predict and Dramatize Weather*. Chicago: University of Chicago Press, 1999.

Nocke, Thomas, Stefan Schlechtweg and Heidrun Schumann. "Icon-based Visualization using Mosaic Metaphors." In *Proceedings of the 9th International Conference Information Visualisation*, 103–109. Los Alamitos, Calif.: IEEE Computer Society Press, 2005.

Nocke, Thomas. "Visuelles Data Mining und Visualisierungsdesign für die Klimaforschung" *(Visual Data Mining and Visualization Design for Climate Research)*. PhD diss. (in German), University of Rostock, 2007.

Nocke, Thomas, Michael Flechsig and Uwe Böhm. "Visual Exploration and Evaluation of Climate-Related Simulation Data." In *Proceedings of the 39th conference on Winter simulation: 40 years! The best is yet to come*, 703–711. NJ: IEEE Press Piscataway, 2007.

Nocke, Thomas, Till Sterzel, Michael Böttinger and Markus Wrobel. "Visualization of Climate and Climate Change Data: An Overview." In *Digital Earth Summit on Geoinformatics 2008: Tools for Global Change Research*, edited by Ehlers, Manfred, Kai Behncke, Friedrich-Wilhelm Gerstengarbe, Florian Hillen, Lothar Koppers, Ludwig Stroink and Joachim Wächter, 226-32. Heidelberg: Wichmann, 2008.

Nocke, Thoms, Ursula Heyder, Stefan Petri, Katrin Vohland, Markus Wrobel and Wolfgang Lucht. "Visualization of Biosphere Changes in the Context of Climate Change", In *Information Technology and Climate Change. 2nd International Conference IT for empowerment*, edited by Wohlgemuth, Volker, 29–36. Berlin: Trafo Wissenschaftsverlag, 2009.

Novotny, M. and Helwig Hauser. "Outlier-Preserving Focus+Context Visualization in Parallel Coordinates." *IEEE Transactions on Visualization and Computer Graphics* 12, no. 5 (2006): 893–900.

d'Ocagne, Maurice. *Coordonnées Parallèles et Axiales: Méthode de Transformation Géométrique et Procédé Nouveau de Calcul Graphique Déduits de la Considération des Coordonnées Parallèles*. Paris: Gauthier-Villars, 1885.

Saito, Takafumi, Hiroko Nakamura Miyamura, Mitsuyoshi Yamamoto, Hiroko Saito and Yuka Hoshiya. "Two-Tone Pseudo Coloring—Compact Visualization for One-Dimensional Data." In *Proceedings of the 11th IEEE*

Symposium on Information Visualization 2005, edited by John Stasko, 173-80. Piscataway, NJ.: IEEE, 2005.

Shneiderman, Ben. "The Eyes Have It: A Task by Data Type Taxonomy for Information Visualization." In *Proceedings of the IEEE Symposium on Visual Languages '96*, edited by IEEE Symposium on Visual Languages, 336pp. Los Alamitos, Calif.: IEEE Computer Society Press, 1996.

Taylor, Karl E. "Summarizing Multiple Aspects of Model Performance in a Single Diagram." *Journal of Geophysical Research: Atmospheres* 106, no. D7 (2001): 7183-92.

Thomas, Jim J. and Kristin A. Cook. *Illuminating the Path: The Research and Development Agenda for Visual Analytics.* Los Alamitos, Calif.: IEEE Computer Society Press, 2005.

Tominski, Christian, James Abello and Heidrun Schumann. "CGV—An Interactive Graph Visualization System." *Computers & Graphics* 33, no. 6 (2009): 660–78.

Tominski, Christian, Jonathan F. Donges and Thomas Nocke. "Information Visualization in Climate Research." In *Proceedings of the International Conference on Information Visualization*, edited by Banissi, Ebad, Stefan Bertschi, Remo Burkhard, Urska Cvek, Martin Eppler, Camilla Forsell and Georges Grinstein, 298-305. Los Alamitos, Calif.: IEEE Computer Society Press, 2011.

Wrobel, Markus, Thomas Nocke, Michael Flechsig and Achim Glauer. 2008. "Graphical User Interface Design for Climate Impact Research Data Retrieval." In: *Proceedings of the iEMSs Fourth Biennial Meeting: International Congress on Environmental Modelling and Software,* edited by Sànchez-Marrè, Miquel, Javier Béjar, Joaquim Comas, Andrea E. Rizozoli and Giorgio Guariso, 1669–76. International Environmental Modelling and Software Society. Barcelona: Universidad Politécnica de Cataluña, 2008.

IMAGE COPYRIGHTS & SOURCES

Figure 1a: © IPCC-WG1 2007.
Figure 1b: ©IPCC-WG2 2007.
Figure 1c: ©IPCC-WG2 2007.
Figure 1d: © Thomas Nocke.
Figure 1e: © IPCC-WG1 2007.
Figure 1f: © Thomas Nocke & Stefan Petri.
Figure 2: © Thomas Nocke.

Figure 3: © Thomas Nocke.
Figure 4: © Christian Tominski.
Figure 5: © Mario Baalcke & Thomas Nocke.
Figure 6: © Kresimir Matkovic & Thomas Nocke.
Figure 7: © Thomas Nocke.

Chapter 2
Communicating Results: The Status of Climate Expert Graphs in IPCC Reports

"Tricks," Hockey Sticks, and the Myth of Natural Inscription: How the Visual Rhetoric of Climategate Conflated Climate with Character

LYNDA WALSH

This essay employs the recent "Climategate" controversy as an opportunity to examine public attitudes about the proper role of rhetoric in the creation and presentation of scientific graphics. Applying methods from classical rhetoric, semiotics, and gestalt psychology, Walsh examines the formal and informal inquiries into the construction of the infamous "hockey stick" family of graphics, arriving at three conclusions: first, that the hockey stick graphs implied the argument that humans were causing unprecedented global warming; second, that the controversial "tricks" associated with these graphs were not statistical techniques but were rather rhetorical interventions aimed at reinforcing the gestalt perception of the trend line (and thus facilitating the implied anthropogenic argument); third, that Climategate's revelation of the "tricks" disrupted a powerful popular misconception about scientific graphics: the myth of natural inscription. As a result, graphs that were assumed to display climate came to be read as performing the character of their authors.

INTRODUCTION

The infamous "hockey stick" graph of global temperature is one of the most contentious scientific graphics ever published. The first version of this graph appeared in a 1998 Nature paper by Michael Mann, Raymond Bradley, and Malcolm Hughes called "Global-Scale Temperature Patterns and Climate Forcing Over the Past Six Centuries" (MBH98). A version with a longer time scale appeared in 2001 in the United Nation's Intergovernmental Panel on Climate Change (IPCC's) *Third Assessment Report* (TAR). These graphs presented global temperature over time as a combination of direct historical temperature measurements with indirect measurements of pre-historical temperature via proxies such as tree-ring width and ice-core thickness. All data series showed an anomalous increase in global temperature beginning around the time of the Industrial Revolution and steepening dramatically around the middle of the 20th century. The striking trend line catalyzed megabytes of discussion in the cybersphere and three formal inquiries into the professional conduct of their authors, particularly Michael Mann of the Pennsylvania State University.

But the "hockey stick" graphic family became even more notorious, if that were possible, in late 2009 when a cache of emails discussing it was hacked from the prestigious Climate Research Unit (CRU) at East Anglia University in England—and briefly again when a new cache was released in 2011.[1] The hacked emails showed Mann and other IPCC lead authors debating how to display proxy tree-ring temperature data that did not exhibit the same steep increase post-1960 as did the other data series. Famously, the authors of the emails discussed "tricks" to "hide the decline" in the problematic tree-ring series so it would not "create a distraction/detraction" from the "tidy story" of unprecedented warming over the last millennium, thus providing unwarranted "fodder" for "skeptics."[2]

The purpose of this chapter is not to re-open the question of the IPCC authors' legal or ethical conduct in making the visual choices they did regarding the "hockey stick" and related proxy-series graphs; at least seven international inquiries, as of this writing, have exonerated them of all charges of professional wrongdoing. Nor do I engage the minutiae of the various technical "tricks" underpinning the proxy-series graphs, such as pass filters, smoothing, and various methods for weighting and correcting data series. Rather, I propose to explore a question left unanswered by all of these inquiries, a question that continues to haunt not only the IPCC but any scientific body responsible for creating public

---

1 Since these emails were illegally obtained, the provenance of any accessible database of them on the Internet is suspect. I will be referring in this chapter to a database located at www.foia2011.org. While it was undoubtedly posted by parties hostile to the CRU authors, it is nevertheless the most comprehensive and searchable database of the CRU emails that I know of. I have carefully compared the wording of the emails I use from this database with all other quotations of those emails that I could find online since 2009 and have found no significant discrepancies. I will refer to these emails by their generally used document numbers.

2 See emails 093803154, 094277707; 093801812 in database www.foia2011.org (see footnote 1).

documents about controversial scientific problems: what role does and should rhetoric play in the creation and publication of graphics like these? Many commenters on Climategate and the proxy-series graphs, including climate scientists themselves, have argued that persuasion should play no role whatsoever and that the graphs should strictly represent scientific data. Other scientists and observers have argued that some amount of selectivity and design are inevitable in the creation of scientific graphics for lay audiences, and that the only germane questions to ask concern the level and intention of these rhetorical interventions.

In addressing this question with respect to the hockey stick graphs and the Climategate debate, I will float an assumption based on scholarship in the sociology and rhetoric of science—that rhetoric is not a sort of finish that can be applied (or not) to scientific information but is fundamentally integrated in the process of creating scientific graphics for any audience. With that assumption in place, I will make three primary claims about the visual "tricks" of Climategate: that the graphics in question, like all scientific graphics, were argumentative; thus, that the "tricks" of Climategate were not statistical but rather rhetorical choices aimed at facilitating the viewer's gestalt perception of unprecedented global warming; and finally, that once these "tricks" were revealed, they disrupted a pervasive and powerful popular myth about scientific graphics—that they are inscribed by nature itself. I will conclude by arguing that critics interpreted this disruption as a usurpation of natural authority by the IPCC authors. For these viewers, the proxy-series graphs displayed not climate but their authors' ethos—their character and credibility as scientific authorities.

BACKGROUND: CLIMATEGATE

The initial cache of emails hacked from servers at the CRU covered a period from 1996–2009, but the most attention by far has been paid to a series of emails from fall 1999 that discussed the drafting of proxy-series graphs for the TAR and related publications. These emails circulated among a handful of lead IPCC authors—primarily Phil Jones, Keith Briffa, and Chris Folland of the CRU and Michael Mann from Penn State—and concerned a single line in a single graph. The graph was TAR Figure 2.21 (Figure 1), and the line in question represented Keith Briffa's dendroclimatic (using tree-ring width and density proxy data) reconstruction of northern-hemispheric summer temperatures over about 1,000

Figure 1: The infamous "hockey stick" graph from the Third Assessment Report of the United Nations Intergovernmental Panel on Climate Change (2001).
(→ color plate 4)

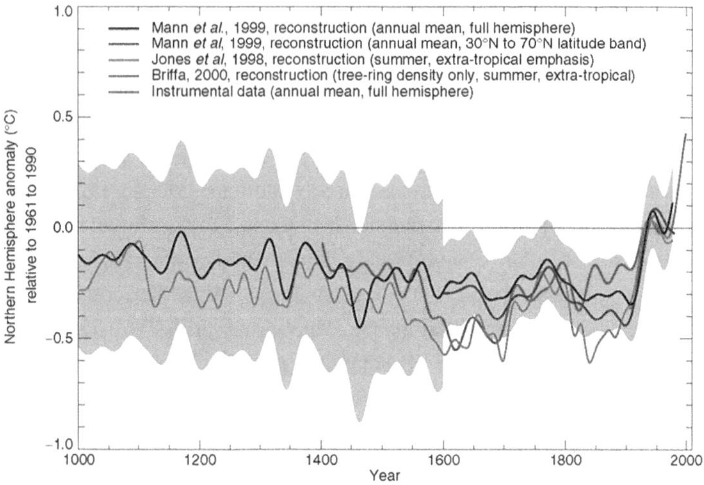

Figure 2.21: Comparison of warm-season (Jones *et al.*, 1998) and annual mean (Mann *et al.*, 1998, 1999) multi-proxy-based and warm season tree-ring-based (Briffa, 2000) millennial Northern Hemisphere temperature reconstructions. The recent instrumental annual mean Northern Hemisphere temperature record to 1999 is shown for comparison. Also shown is an extra-tropical sampling of the Mann *et al.* (1999) temperature pattern reconstructions more directly comparable in its latitudinal sampling to the Jones *et al.* series. The self-consistently estimated two standard error limits (shaded region) for the smoothed Mann *et al.* (1999) series are shown. The horizontal zero line denotes the 1961 to 1990 reference period mean temperature. All series were smoothed with a 40-year Hamming-weights lowpass filter, with boundary constraints imposed by padding the series with its mean values during the first and last 25 years.

years[3]). Dendroclimatic reconstructions use tree-ring width and density measurements as proxies for temperature over epochs for which there were no direct records of temperature, and they are generally presented as line graphs. The line in Briffa 2000 presented a problem called "divergence": whereas the dendroclimatic reconstruction tracked other pre-historic proxies and direct measurements reasonably well up to the mid-twentieth century, starting in about 1960, Briffa's line went sharply down while the other lines all went sharply up. This divergence was apparently visible in a "zero-order" draft of TAR Fig. 2.21 that the lead authors had viewed at their Tanzania meeting in early September 1999.[4]

---

3 cf. Briffa et al., "Reduced Sensitivity of Recent Tree-growth," 1998.

4 A version of this graph, which Stephen McIntyre purports to have obtained through the IPCC review process, appears on his skeptical blog post (December 10, 1999). As there is no way of authenticating that version as the IPCC TAR zero-order draft of Fig. 2.21, and as the IPCC authors are not required to release unpublished data under FOIA, I have not included a version of the graphic in this paper. Nevertheless, the

In the weeks following the meeting, an intense email discussion circulated among the key authors over how to handle the divergence. Referencing an earlier conversation not recorded in the cache, Jones started the conversation on September 22 by saying that the divergent Briffa reconstruction should not be excluded, as Mann had apparently recently recommended. Jones argued the Briffa series was valuable due to its hemispheric scale and went on to recommend two graphs side-by-side—one showing the multi-proxy series that agreed quite well, and one showing Briffa's divergent series. Jones finished by admitting that his two-graph proposal "might make it somewhat awkward for the reader trying to put them into context."[5]

Folland replied, stating that a proxy-series graph was a "clear favourite" for the IPCC TAR Summary for Policymakers (SPM), a lay executive summary that accompanies every IPCC report. However, he had concerns that the divergent Briffa series

"somewhat contradicts the multiproxy curve and dilutes the message rather significantly. We want the truth. Mike [Mann] thinks it lies nearer his result (which seems in accord with what we know about worldwide mountain glaciers and, less clearly, suspect about solar variations). The tree ring results may still suffer from lack of multicentury time scale variance. This is probably the most important issue to resolve in Chapter 2 at present."[6]

At this point, Briffa weighed in. He stated he would go along with whatever "the general concensus" was about the way the graph should look in the TAR SPM; however, he wanted the dendroclimatic data explicitly addressed in Chapter 2. He was against the two-graph solution because he felt it privileged Mann's proxy series prematurely: Briffa believed there was still a large amount of uncertainty in the proxies and that all series were necessary if this uncertainty was to be reduced. He acknowledged,

"There is still a potential problem with non-linear responses in the very recent period of some biological proxies (or perhaps a fertilisation through high CO_2 or nitrate input). I know there is pressure to present a nice tidy story as regards 'apparent unprecedented

---

post-1961 divergence of the Briffa series is referred to repeatedly in the FOIA-obtained email cache, and the problem is explicitly addressed in the following scientific articles: Briffa et al, "Reduced Sensitivity of Recent Tree-growth," 1998; Briffa et al., "Annual Climate Variability in the Holocene," 1999.

5 093803154 in database www.foia2011.org (see footnote 1).

6 093803154 in database www.foia2011.org (see footnote 1).

warming in a thousand years or more in the proxy data' but in reality the situation is not quite so simple. We don't have a lot of proxies that come right up to date and those that do (at least a significant number of tree proxies) some unexpected changes in response that do not match the recent warming. I do not think it wise that this issue be ignored in the chapter.

For the record, I do believe that the proxy data do show unusually warm conditions in recent decades. I am not sure that this unusual warming is so clear in the summer responsive data. I believe that the recent warmth was probably matched about 1000 years ago. I do not believe that global mean annual temperatures have simply cooled progressively over thousands of years as Mike appears to and I contend that that there is strong evidence for major changes in climate over the Holocene (not Milankovich) that require explanation and that could represent part of the current or future background variability of our climate. I think the Venice meeting will be a good place to air these issues."[7]

A few hours later, Mann chimed in, joking that he'd "walked into this hornet's nest" as he came in to work on Eastern Standard Time. After reassuring Briffa that he did not think of his own series as the gold standard and would not use his position as lead author on Chapter 2 to coerce others into thinking so, he recommended that the Briffa series be kept in TAR Figure 2.21. He went on,

"That having been said, it does raise a conundrum: We demonstrate [...] that the major discrepancies between Phil's and our series can be explained in terms of spatial sampling/latitudinal emphasis [...]. But that explanation certainly can't rectify why Keith's series, which has similar seasonality *and* latitudinal emphasis to Phil's series, differs in large part in exactly the opposite direction that Phil's does from ours. This is the problem we all picked up on (everyone in the room at IPCC was in agreement that this was a problem and a potential distraction/detraction from the reasonably concensus viewpoint we'd like to show w/ the Jones et al and Mann et al series[)].
So, if we show Keith's series in this plot, we have to comment that "something else" is responsible for the discrepancies in this case [...]. Otherwise, the skeptics have an [sic] field day casting doubt on our ability to understand the factors that influence these estimates and, thus, can undermine faith in the paleoestimates. I don't think that doubt is scientifically justified, and I'd hate to be the one to have to give it fodder!"[8]

7 093803154 in database www.foia2011.org (see footnote 1).

8 093801812 in database www.foia2011.org (see footnote 1).

Over the next several weeks, the TAR Chapter 2 authors wrestled with how to interpolate the Briffa series in the graph without "diluting the message" of unprecedented recent warming. The work was particularly urgent as the team wished to put some version of the TAR draft figure on the cover of the World Meteorological Organization's (WMO) Statement of the Status of the Global Climate in 1999. Regarding this version of the graph, on November 16 Jones sent to Ray Bradley and Malcolm Hughes the email containing the sentence that would be treated as the "smoking gun" of Climategate: "I've just completed Mike's Nature trick of adding in the real temps to each series for the last 20 years (ie from 1981 onwards) amd [sic] from 1961 for Keith's to hide the decline."[9]

This "trick" effectively erased the Briffa divergence from the WMO cover graph (Figure 2): an interior caption alluded to "uncertainties" in the early parts of the data series and listed references, some of which explicitly addressed the divergence problem. But nowhere in the WMO report itself did the authors state they had replaced the post-1961 tree-ring data in Briffa's series with instrumental temperature data.[10]

The divergence was handled in a different way in the final TAR Fig. 2.21, as is readily apparent (Figure 1): the updated Briffa series (in green) was cut off at 1961 and tucked behind the other series to make the truncation less obvious. The caption indicated that all series were smoothed and were padded with their "mean values" during the first and last 25 years—a typical statistical technique used to keep smoothed series from trending anomalously up or down at their ends—but nowhere was the truncation of the Briffa series explicitly addressed or justified. A paragraph in section 2.3.2.1 of the chapter cautioned the reader that "several important caveats must be borne in mind when using tree-ring data for palaeoclimate reconstructions" and went on to mention sampling bias, regionality, the effects of non-climatic growth trends, and temporal variability. In this last category, the divergence issue was alluded to with the following language—"high latitude tree-ring density variations have changed in their response to temperature in recent decades, associated with possible non-climatic factors"—and with the citation of the Briffa et al. 1998 Nature letter discussing the divergence issue. The paragraph concluded, "For these reasons, investigators have increasingly found tree-ring data most useful when supplemented by other types of proxy information in "multi-proxy" estimates of past temperature change."[11]

9 0942077707 in database www.foia2011.org (see footnote 1).

10 World Meteorological Organization, WMO statement on the status of the Global Climate in 1999.

11 Intergovernmental Panel on Climate Change, *Third Assessment Report,* 2001.

Figure 2: The cover image of the World Meteorological Organization's Statement on the Status of the Global Climate Report in 1999. (→ color plate 5)

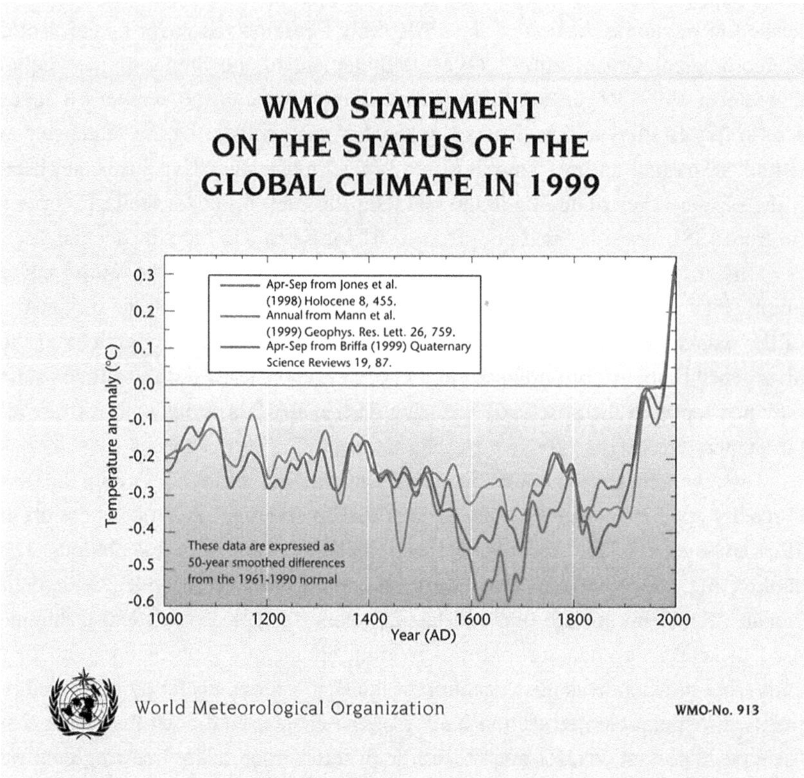

Original caption: "Front cover: Northern Hemisphere temperatures were reconstructed for the past 1000 years (up to 1999) using palaeoclimatic records (tree rings, corals, ice cores, lake sediments, etc.), along with historical and long instrumental records. The data are shown as 50-year smoothed differences from the 1961–1990 normal. Uncertainties are greater in the early part of the millennium (see page 4 for further information). For more details, readers are referred to the PAGES newsletter (vol. 7, No. 1: March 1999, also available at http://www.pages.unibe.ch) and the National Geophysical Data Center (http://www.ngdc.noaa.gov). (Sources of data: P.D. Jones, K.R. Briffa and T.J. Osborn, University of East Anglia, UK; M.E. Mann, University of Virginia, USA; R.S. Bradley, University of Massachusetts, USA; M.K. Hughes, University of Arizona, USA; and the Hadley Centre, The Met. Office)."

When the TAR was published in 2001, there was enough skepticism about the dramatic message of its proxy-series graphs—particularly Fig. 2.20, the "hockey stick" based on MBH 1998, and Fig. 2.21—to goad United States Representative James Inhofe to demand a federal inquiry into Michael Mann's use of public funding. Despite this inquiry and two following it (by the National Research Council and the University of Virginia), no formal charges were filed against Mann, and the controversy died down everywhere except in a handful of skeptical blogs like Climate Audit, run by Canadian mining consultant Steve McIntyre. In 2007 the popular reception of the IPCC's Fourth Assessment Report, as well as their shared Nobel Peace Prize with Al Gore, seemed to shift media coverage slightly in the IPCC's favor.[12] But the CRU hack in November 2009, especially given its timing a few weeks before the UN Climate Change Conference at Copenhagen, cracked back open the informal "case" against the proxy-series graphs published in MBH 1998, the WMO report, and the TAR—and thus reinvigorated arguments denying anthropogenic global warming (AGW).

The informal case was first prosecuted, as has increasingly become the norm with climate science publications, in the blogosphere. The discussion here raged back and forth for about a month between supportive blogs such as Gavin Schmidt's Real Climate and skeptical outlets such as Climate Audit and Watts up with That?. Mass media outlets picked up on the story within a week of the November 17 cache release; the label "Climategate" is generally acknowledged to have first been applied by Telegraph blogger James Delingpole (November 20, 2009). Formal inquiries began in December with the Muir-Russell Independent Inquiry in Britain and ended with the National Science Foundation's investigation of Mann in 2011. All told, seven formal inquiries found no evidence of scientific or ethical wrongdoing by the key players; however, a few scolded the key players for lack of transparency with their data and resistance to fulfilling Freedom of Information Act requests.[13]

Now that the basic timeline of events has been established, I am going to examine in more depth the discussion—both formal and informal—of the "tricks" Jones et al. used to "hide the decline" in the Briffa series. While this discussion ranged widely, nearly all of the critiques had this in common: they espoused a

12 Walsh, "The Technology of Prophecy," 2013.

13 Many nations have Freedom of Information laws requiring institutions receiving public funds to respond to requests from members of the public for access to data acquired with those funds. The United States passed its Freedom of Information Act in 1966; the United Kingdom, in 2000.

model of science–public communication that segregated argumentation and data presentation.

FORMAL INQUIRIES INTO "TRICKS" TO "HIDE THE DECLINE"

The formal inquiries into the CRU hack were conducted from 2010–2011 by (roughly in chronological order) the United Kingdom House of Commons Science and Technology Committee, two University of East Anglia independent review panels, the Pennsylvania State University, the U.S. Environmental Protection Agency (EPA), the U.S. Department of Commerce, and the U.S. National Science Foundation (NSF). Concerning the "trick," all the formal inquiries that explicitly addressed this issue supported climate scientists' claim that the word indicated a legitimate statistical technique: as a representative example, the House of Commons report concluded the word "appears to be a colloquialism for a 'neat' method of handling data."[14] This committee concluded similarly that "hide the decline" referred to the discarding of unreliable data. In fact, almost all the inquiries treated these techniques as strictly technical issues; some of the committees claimed they were not in a position to pass judgment on them for that reason. For example, the Penn State inquiry passed off a formal question about Mann's scientific conduct to the NSF, who themselves referred it to the wider field of scientific inquiry, saying questions about the "tricks" were "appropriate for scientific debate [...]. Such scientific debate is ongoing but does not, in itself, constitute evidence of research misconduct."[15]

The only rhetorical situation in which the "tricks" weren't read as strictly technical involved the WMO 1999 cover. Here, a few committees acknowledged that while the statistics behind the proxy-series graphs appeared legitimate, nevertheless, the graphs also reflected a rhetorical choice that might call for judgment—namely, the failure to disclose the splicing of temperature data onto the post-1961 Briffa series. The EPA deferred this judgment to the WMO: "While it would have been clearer about its sources if the graph had overlaid the instrumental record using a separate color rather than merging the instrumental and proxy records together, this is an issue for WMO to address, considering the

14 House of Commons Science and Technology Committee, The Disclosure of Climate Data from the Climatic Research Unit at the University of East Anglia, 20.

15 National Science Foundation Office of the Inspector General, Closeout Memorandum, 3, 2011.

purpose of the report and the role of the graph on the cover."[16] The Muir-Russell report was less deferential. It slammed the WMO cover as "misleading." But again it drew a distinction between science and rhetoric: "We do not find that it is misleading to curtail reconstructions at some point *per se*, or to splice data, but we believe that both of these procedures should have been made plain—ideally in the figure but certainly clearly described in either the caption or the text."[17] In summary, the formal inquiries all attempted to segregate scientific-technical, rhetorical, and legal questions and generally restricted themselves to settling the last category.

INFORMAL INQUIRIES INTO "TRICKS" TO "HIDE THE DECLINE"

The informal inquiry into the CRU hack was less straightforward to compile than the formal reports. I assembled my archive by performing a Google blogs search for the keywords "climategate trick hide the decline" for a year following the November 2009 hack and taking the first 100 blog posts returned. Of these posts roughly two thirds were either neutral or supportive of the IPCC, and the remaining third were critical. The supportive posts followed the lines of argument I indicated above for the formal inquiries. Here, I turn my attention to the critical posts.

Many of these posts began by quoting one of the various defenses of the term "trick" as a legitimate technical solution. Then, the blog authors zeroed in on the words "trick" and "hide." They negated the technical defenses of "trick" and doubled it with words like "trickery," "chicanery," "deception," "cherry-pick," "trick or cheat," "guile," and "fraud." For "hide," which they doubled with related terms such as "conceal" and "obscure," several commenters developed an enthymeme or abbreviated argument that ran like this: if there was nothing untoward about the "trick," there would have been no need to "hide" anything from readers. While this argument confused the "trick" with the "decline," it clearly indicated suspicion of the IPCC authors' motivations.

But the more interesting arguments that developed in these critiques related to the rightful relationship of science and persuasion. Without exception, the critics seemed to support a model in which science and persuasion were a zero-sum game. This model was reflected in a series of antitheses: "data" opposed to

16 Walsh, "The Technology of Prophecy," 2013, 222.

17 Ibid., 13.

"argument," "dogma," and "politics"; "show" vs. "hide," "manipulate," and "push"; "science" vs. "politics"; and "real" and "actual" vs. "imaginary," "dramatic," and "desired." Several critics went on to read arguments in the proxy-series graphics: "...that anthropogenic emissions are largely responsible for global warming" *(Canada Free Press)*, that the U.S. "has selfishly imperiled the planet by inducing climate change" *(Science and Public Policy Institute)*, "[that] global warming...is occurring and that humans are responsible" *(Real Clear Politics)*, and "if we do not act quickly there is no hope" *(Scared Monkeys)*. Thus, the informal Climategate inquiry found the IPCC authors guilty because they mixed science and persuasion.

If we grant that the seven formal, independent, international inquiries into the ethics of these scientists arrived at the correct determination, then we can assume that the "tricks" the IPCC scientists employed in the proxy-series graphics fell within the bounds of normal, ethical technical practice. But this assumption does not and should not enjoin us to dismiss the extensive public criticism of the "tricks." In fact, I would argue this informal inquiry was in fact *trying a different case* than the formal inquiries were. While the formal inquiries referenced established scientific and legal practices in the handling and representation of data, the informal inquiries focused on the scientists' right (or lack thereof) to make arguments and the related perception that by articulating AGW they were pushing a political cause.

The informal charges were not purely opportunistic. A close reading of the proxy-series graphs reveals that a) they do in fact present arguments, as do all scientific graphics; and b) the IPCC authors made rhetorical choices both in data analysis and graphical design to reinforce those arguments.

ANALYSIS: ARGUMENT, FORECLOSURE, AND GESTALT

A substantial body of work in the rhetoric and sociology of science has established that extra-scientific criteria enter at the ground floor of scientific work: from the funding sources courted, to the selection of phenomena of interest to study, to the engineering of instrumentation, to the interpretation of inscriptions made by these instruments, to the reporting of work to peers or external stakeholders—scientists make judgment calls that reference social values.[18]

---

18 Blakeslee, "Newcomers Getting to Know and Speak to Audiences," 2000; Brasseur, *Visualizing Technical Information,* 2003; Gilbert and Mulkay, *Opening Pandora's Box,* 1984; Latour, "Drawing Things Together," 1990; Lynch, "Discipline and the

Admitting this is not tantamount to saying that all scientific activity is somehow part of a political agenda. The funding bodies, peers, and laboratory apparatus mentioned above all place important checks on individual biases and explain in large part the persistence of science as a respected and vitally useful form of life in the West. Nevertheless, this body of scholarship does encourage us to view with skepticism the position articulated by many Climategate disputants that rhetoric and science can or should be non-overlapping enterprises.

What about the proxy-series graphs—can we see evidence of rhetorical choices and argumentation in them? We can, if we agree that persuasive arguments include not just the "shoulds," "goods," and "bads" of policy but also arguments that point out causes or effects, establish definitions, and select and order data for display. These lower-level arguments are well documented in scientific graphics.[19] A series of researchers have noted that the majority of scientific graphics provide "synoptic" snapshots of natural systems; these synoptic views imply the argument that the scientist has comprehended and to some extent controls the system.[20] Other studies have focused on how scientists select lines of argument in composing graphics, thus disciplining their audiences to take certain interpretive approaches to their data while foreclosing other approaches.[21] Finally, a group of researchers has noted that the salient rhetorical feature of graphs such as the proxy-series graphs is the trend line, which is recognized instantly by gestalt and prompts the viewer to seek a causal explanation for the trend.[22] Such

---

Material Form of Images," 1985; Myers, *Writing Biology,* 1990; Rymer, "Scientific Composing Processes," 1998.

19 There is a standing debate about whether scientific graphics can make arguments alone or whether they must be combined with text to make arguments. A summary of the case for purely visual arguments can be found in (Birdsell, and Groarke, "Toward a theory of visual argument," 1996); a case against them is summarized in (Gross, "Presence as a Consequence of Verbal-Visual Interaction," 2009). The proxy–series graphs all include significant textual elements in their axis labels, legends, and attached captions (the IPCC graphics cannot be downloaded without these elements), so they are not a test case that can settle this debate.

20 Barton and Barton, "Modes of Power in Technical and Professional Visuals," 1993, 138–162; Latour, "Drawing Things Together," 1990, 19–68; Prelli, "Visualizing a Bounded Sea," 2006, 90–120; Rosner, "Theories of Visual Rhetoric," 2001, 391–413.

21 Dyehouse, "A Textbook Case Revisited," 2011; Fahnestock, *Rhetorical Figures in Science,* 1999; Latour, ibid.

22 Barton and Barton, "Modes of Power in Technical and Professional Visuals," 1993, 138–162; Bastide, "The Iconography of Scientific Texts," 1990, 187–229; Brasseur,

simplifying trend lines perform scientists' mandate to search for natural causes and thereby to discipline natural chaos.

We can observe similar argumentative dynamics operating in the proxy-series graphs. First, let us examine the dimensionality of the graphs depicted in Figures 1 and 2: they plot time against temperature (actually deviation by degrees from global averages). The resulting "hockey stick" trend line invites causal explanation: why does temperature suddenly begin to increase in the 18th century? What phenomenon began during that time that could cause this change in the trend line? The answer, which was provided in the text of any document in which the proxy-series graphs appeared, is carbon forcing due to human industry. Thus, through the IPCC authors' choice of dimensions, the proxy-series graphs visually structure an enthymeme—an abbreviated argument—whose minor premise is indicated by the trend line ("Global warming has been increasing since the 18th century and has recently begun accelerating in an unprecedented manner"), whose major premise or warrant the viewer must supply from her store of historical knowledge ("Human industry has been on the rise since the 18th century"), and whose conclusion is the AGW argument ("Therefore, humans have caused global warming"). The enthymeme is not the only lens through which this argument becomes visible: Kress & van Leeuwen have provided an alternate frame for elucidating the arguments implied by trend lines—the narrative, a conventional "vector" of visual arrangement that implies an agent (human industry, in this case) and a patient or receiving object (global warming).[23]

This analysis of dimensionality seems self-evident until we remember that it could have been very different. As Mary Rosner has argued, we have become so acculturated to the dominant visual genres of science that we forget they could have been otherwise; thus, we efface the rhetorical agency of the people who created them.[24] That the authors of Figures 1 and 2 selected time versus temperature (deviation) is therefore neither a foregone conclusion nor an accident; it is a deliberate rhetorical choice. The datasets that the IPCC authors combined in these graphs all referenced multiple dimensions (geographical space, time, thickness, density, age, season, degrees of temperature, etc.). Among all of those dimensions latent in the dataset, time and temperature were the only two that

Visualizing Technical Information, 2003; Gross, "Presence as a Consequence of Verbal-Visual Interaction," 2009.

23 Birdsell and Groarke, "Toward a Theory of Visual Argument," 1996; Kimball, "London Through Rose-Colored Graphics," 2006, 42.

24 Rosner, "Theories of Visual Rhetoric", 2001.

would frame the "hockey stick" historical trend and goad the viewer to consider the causal effects of human industry.

Next, let us consider the IPCC authors' decisions with respect to the scope, or what gestalt psychologists would call the "enclosure," of the proxy-series graphs. The x-axis comprehends about 1,000 years; the y-axis indicates deviation from global average temperature and only extends up or down by two degrees Celsius. Together the axes construct a space that implies the definitional argument that *this is climate*—that this is the space in which all the interesting climatic events have occurred. Again, these decisions were not emergent from the data, but were deliberate rhetorical choices. These choices were made in part to counter public arguments that the Medieval Warm Period (MWP, roughly 1000–1400 A.D.) was warmer than today; therefore, current global warming is a) cyclic, b) not dangerous, and c) not caused by the Industrial Revolution. The CRU email caches make it abundantly clear the IPCC authors have long felt pressure to refute these arguments.[25] But in order to do so, they had to extend the time scale of their global temperature graphs back to 1000 A.D. and include prehistoric proxies for temperature. As we have seen, while these rhetorical choices may have solved one persuasive problem, they opened up another in the form of the Briffa series divergence.

While we don't have as much documentation on the scoping decisions involving the y-axis of the proxy-series graphs, we can observe that these decisions created two rhetorical effects: focusing on deviation rather than absolute temperature created a more clearly ascendant trend line, i.e. graphing the mean global temperature over the same timescale would have yielded a thick black band with little obvious trend. At the same time, the y-axis has been exploded to cover only 1–2 degrees. Celsius in order to disambiguate the warming trend, i.e. an upward anomaly of 0.4 °C would have been functionally invisible on a y-axis scaled to handle the amplitude of mean global temperature. Again, we tend to dismiss these scalar parameters as emergent from the genre of the line graph, but they are in fact set by people, either during the graphic design process or earlier in the setting of defaults for graphing programs. And every such decision has its rationale and consequences: in this case as with most cases of linear trend data, both the goal and the result are the simplification of massive amounts of data into a trend line that implies a causal argument.

Finally, we should consider the trend lines themselves more carefully. How do they come to dominate graphs that contain all sorts of competing visual

25 For a summary of these concerns, see email chain 0209 among Mann, Hughes, and Edward Cook.

information: axes, labels, scales, values, titles, etc.? Perceptual research on line graphs has established that viewers first look for "the curve" or trend line, which they hope will serve as an abstract of the complex information displayed in the rest of the graph.[26] This research has also established that the more simple and clear the trend line, the easier it is for the viewer to resolve it via *Prägnanz* (a gestalt term referring to the nearly instantaneous process by which the human perceptual system assembles discrete visual elements into a form).

Rhetorically speaking, trend lines indicate an argument relating two independent dimensions—time and crime, salaries and debt, time and temperature, etc. A verbal expression of this argument would take more time and space to lay out due to the linear nature of writing; in the interim the viewer could become distracted or misinterpret particular sentences and thus never arrive at recognition of the argument. By contrast, *Prägnanz* of the trend line takes not much longer than a glance.

If we return to the trend lines in the proxy-series graphs with this perceptual background, it becomes clear why the divergent Briffa series was viewed as "diluting the message." Its strongly divergent trend line interfered with the *Prägnanz* principles of continuity and proximity and thus stymied the process by which the viewer inferred the minor premise "Global warming has been increasing since the 18th century and has recently begun accelerating in an unprecedented manner." All of the changes made to the proxy-series graphs following the zero-order draft increased continuity and proximity: splicing temperatures onto the divergent Briffa series to force it to trend with the convergent multiproxy series in the WMO graph (Figure 2, green line); likewise, truncating the Briffa series at 1960 and tucking its trend line behind the rest of the convergent "spaghetti" in TAR Fig. 2.21 (Figure 1, green line) to render the truncation functionally invisible.

The extended debate visible in the CRU email chain over these choices shows us the true stakes of scientific graphics: either the viewer seizes the argument nearly instantaneously via *Prägnanz*, or he enters into a period of trying to make sense of the message of the graph using reason and logic. Into this period can intervene distractions, personal values, and deficits in technical knowledge, all of which interfere with the uptake of the argument the author has designed into the graphic. This is the most striking difference between visual and verbal argument modes—they function via *Prägnanz* and by reasoning, respectively. If a scientific graph is to be worth the space it takes, it must facilitate *Prägnanz*, it

26 Brasseur, *Visualizing Technical Information*, 2003, 27–36.

must tell a "tidy story."[27] Otherwise, the authors run the risk of introducing a fundamental ambiguity into their argument.

There is substantial evidence in the Climategate emails and the other IPCC advisory documents that authors were deeply concerned about non-specialist audiences' ability to grasp the complexities of the climate modeling process and the different forms of uncertainty that attend it.[28] From this perspective, we can appreciate how important the proxy-series graphs were and why the authors insisted on working over the trend lines until they were maximally continuous and proximal: their convergence was the linchpin in the public case for unprecedented global warming.

At the beginning of this section, I reviewed the literature on scientific visual argument, which has established that all scientific graphics present arguments and do so by a sophisticated method of disciplining the viewer toward gestalt. Key aspects of this method include the choice of dimensions (which foreclose on alternative interpretations of the dataset), scope (which sets the limits of the system in question), and scale (which emphasizes trends in the data). By looking at these aspects in the IPCC proxy-series graphs, we saw the authors facilitating *Prägnanz* of the trend line and its implied argument ("Since the 18th century, global warming has been increasing and, recently, has been accelerating in an unprecedented manner"); by doing so, they were also facilitating the viewer's inference of AGW.

So in conclusion, many of the lay critics of the IPCC authors were right: the latter were, in fact, "pushing" a message. But every scientific paper pushes a message, so what was different about this one, and why was it so incendiary? We could note that this is a message with sweeping political, economic, and social consequences and chalk the critics' resistance up to that. However, there remains a dynamic in the reception of the CRU emails and the proxy-series graphs that requires deeper explanation: the predictable way in which critics moved from inferring the graphs' causal argument to attacking the ethos of the scientists who designed it. Graphs that purported to show climate suddenly began to be read as mapping problematic character.

27 Kimball, "London Through Rose-Colored Graphics," 2006.

28 Walsh, "The Technology of Prophecy," 2013.

CONCLUSION: THE CONFLATION OF CLIMATE AND CHARACTER VIA THE MYTH OF NATURAL INSCRIPTION

In the informal inquiry, if a critic found an argument in the proxy-series graphics, s/he in nearly every case used it to attack the ethos of the climate scientists. The best summary of this line of thought appeared on the *Econlog* blog, in which a commenter charged readers, "If you previously thought that climate science was not very politicized, you should change your mind—and discount mainstream research accordingly." That strong claim might be surprising, but even stronger ethical attacks were in evidence in the reception data, with critics calling the scientists "conspirators" guilty of "fraud," "lies," and "con[s]." One commenter even went so far as to draw an extended comparison between the IPCC scientists and "child predators" who abuse their authority to acquire victims *(Dan Tocchini)*. These kind of rhetorical attacks are well documented in the history of public climate arguments since the 1990s and occasionally turn into something more grave: *Popular Science* recently interviewed climate scientists and found many had received threats ranging from anthrax scares to threats of violence to them or their children.[29]

How do we go from the perception of an unpopular scientific argument ("Humans are causing global warming") to ethos attacks and anthrax threats? The key lies in the popular belief that science and politics should not mix—more specifically, in a pervasive myth about scientific graphics that I call the myth of natural inscription.

Bruno Latour and Françoise Bastide first detailed the social-epistemological dynamics of scientific graphics. Based on his ethnographic studies of scientists at work, Latour noted that the essence of a scientific graphic, such as a map, was the "immutable mobile" it carried from actor to actor in the network of scientific social relations.[30] The map's representation of a landform is *immutable* because it could be scaled up or down without losing the utility and integrity of the landform; it is *mobile* because it could be carried back to the geographical location where it was made and compared with it by the navigators who first employed it at a distance.

The scientists who make the map, or a micrograph of a bacterium, may retain an acute awareness that the immutable mobile is not itself a natural object. However, it is more difficult for their peers to recognize the layers of inscription that mediate between the graphic and nature. Bastide charted these difficulties in her

29 Clyne, "The Battle over Climate Change," 2012.

30 Latour and Woolgar, *Laboratory Life,* 1986.

study of the exposé of a fraudulent radiograph. She found that biochemists generally treated radiographs as if they were inscribed directly by nature on the film via the instrumentation. The proper place for the human researcher was at the end of this "chain of visualization," as the passive receiver of the inscription. When a biochemist's peers received evidence that he had inserted himself earlier in the chain, he was charged with fraud, even though the peers were themselves all engaged in inserting mediating layers of decisions, instrumentation, and judgments in the chain of visualization of their own work.[31]

Now, certainly all insertions in the chain of visualization are not equal: there are allowable and forbidden ones. Adjusting the contrast of a spectrograph to make the interesting bands more visible may be allowable under some circumstances; substituting a radiograph of a known element for a graph claimed to depict a new one is quite clearly out of bounds. My points here are merely that a) there is not a hard and fast line between these licit and illicit interventions in the chain of visualization[32]; and b) if scientists themselves can maintain at some level the myth that their graphics come immediately from natural objects, we can certainly expect lay audiences to maintain this myth, particularly when their only contact with an invisible phenomenon such as a bacterium—or climate—is the scientific graphic itself.

If we return to the creation and reception of the proxy-series graphs, we can see how climate becomes read as character due to the myth of natural inscription. If we look at TAR Fig. 2.21 (Figure 1), we don't see the IPCC authors inserted anywhere in the chain of visualization. At first all we see are inscriptions made by computer programs. If we follow the directions of the series labels to seek the source of these inscriptions, we see only mathematical formulas operating on numbers. If we try to see past the numbers to the objects they represent or "draw together," in Latour's terms, ultimately we arrive in our imaginations at nature itself—trees, glaciers, clouds, rain, the sun.[33] The IPCC authors have arranged all of the traces in such a way that they point away from themselves and toward nature; the implication is that nature's finger has traced the spaghetti onto TAR Fig. 2.21.

Naturally, this is a fiction, but it is one we all swear to because as long as we do, these graphs bring us truths from beyond the fray of human politics that will settle the very real problems we face—rising sea-level, droughts, famines, and increasingly destructive storms. But when we receive evidence, as from the CRU

---

31 Bastide, "The Iconography of Scientific Texts," 1990.

32 Frow, "Drawing a Line," 2012.

33 Latour, "Drawing Things Together," 1990.

hack, that the fingers of humans have traced these graphs along with or instead of nature's finger, the myth of natural inscription is disrupted. Following this rupture, we could retreat and regroup around traditional models of expertise from other spheres of life, where wise men and women tell us what to do based on their personal acumen and experience. But the problems of nature are beyond the scope of personal expertise, and we know it. So, when we have demanded of scientists that they make the finger of nature write on the wall for us, and instead we see their fingers writing, even for a second, what they have written becomes illegitimate, and they become imposters.

This is exactly what happened in Climategate when the IPCC's "tricks" were revealed. For instance, the *Volokh* blog made it clear that a *cordon sanitaire* between natural inscription and scientific judgment had been violated: the author denigrated the "tricks" as "a method applied to adjust the data (which by itself is not inherently wrong) to give a desired, pre-determined output." S/he went on, "If you don't think the CRU 'trick' is illegitimate, ask yourself: would they have applied...[it] to the dataset, had it already agreed with their desired and pre-determined results?" The blogger at Uncommon Descent made his/her argument for illicit intervention in the chain of visualization via a vivid extension of the hockey stick metaphor:

> "It seems that the practice by climatologists who need to evidence AGW is to keep in mind that if they're not getting what they want to show from one set of data through time, just use it until it becomes troublesome and switch to another set of data, midstream, for the rest of the time, and don't tell anyone that it's a hodgepodge. Just cover the damage up with some paint, and make sure to say 'buying as is, no refunds or returns.' This deception wouldn't work for selling a hockey stick, much less for selling Anthropic [sic] Global Warming that has been painted to look like a hockey stick."

Indeed, the myth of natural inscription seemed to inspire some of the most creative and humorous responses in Climategate. My personal favorite was the cheeky label the blogger at *I Hate the Media* gave the proxy-series graphs: "Mann-made global warming."

Again, I would argue that these comments should not be interpreted as ignorance but as evidence that a) the myth of natural inscription is still very powerful and b) when scientists make public choices that disrupt this myth, they are punished. This is the best explanation for the unique tone of the ethical attacks against climate scientists in Climategate: the terms "con," "fraud," "conspirator," "cheat," and even and especially "child predator" all connote a very specific kind of ethical charge—damaging usurpation of legitimate authority. In this case, the legitimate authority was nature itself.

All told, the critics' informal charge against the IPCC authors was not just that they were "pushing the cause"—it was that they had exploited the myth of natural inscription to hide their rhetorical agency. By effacing or downplaying explanations of their choices to reinforce the trend line, the IPCC authors at first amplified the authority of their gestalt argument for AGW. However, when their rhetorical interventions were revealed via the CRU hack, critics rightly reproached them for allowing and even encouraging viewers to interpret human judgments about data as the untrammeled inscription of nature. Even Daily Show comedian Jon Stewart, an outspoken progressive and supporter of climate change mitigation, reacted with astonishment to news that the CRU only kept "value-added" tree-ring data in their archives: "Value-added data? What is that, numbers fortified with art? Truth plus, now with lemon?"[34]

We can see that even for sympathetic, educated viewers like Stewart, the myth of natural inscription persists. Many of us—scientists and lay voters alike—cling to the misconceptions that nature's finger can unproblematically write the future on the wall for us and that any human intervention in the chain of visualization is suspect. This remains a grave problem for the public understanding of climate science. So although it was undeniable that Briffa's series "diluted the message" of the proxy-series graphs, by "hiding" their rhetorical interventions to strengthen the trend line and the AGW argument, the IPCC authors missed a rare opportunity—the chance to open a dialogue about the tough rhetorical choices involved in presenting all good science and thus to work toward the demise of the damaging myth of natural inscription. Instead, they got the opportunity—not so very rare these days—to take time away from crucial climate research to defend that research and their character against legal inquiry.

BIBLIOGRAPHY

Barton, Ben F. and Marthalee Barton. "Modes of Power in Technical and Professional Visuals." *Journal of Business and Technical Communication* 7, no. 1 (1993): 138–62, doi: 10.1177/1050651993007001007.

Bastide, Françoise. "The Iconography of Scientific Texts: Principles of Analysis." In *Representation in Scientific Practice*, edited by M. Lynch and S. Woolgar, 187–229. Cambridge, MA: MIT Press, 1990.

34 "Scientists Hide Global Warming Data," *The Daily Show with Jon Stewart*, 2009.

Birdsell, David S. and Leo Groarke. "Toward a Theory of Visual Argument." *Argumentation & Advocacy* 33, no. 1 (1996): 1.

Blakeslee, A.M. "Newcomers Getting to Know and Speak to Audiences." In *Interacting With Audiences: Social Influences on the Production of Scientific Writing*, 83-105. Taylor & Francis, 2000.

Brasseur, Lee E. *Visualizing Technical Information: A Cultural Critique*. Amityville, NY: Baywood Publishing, 2003.

Briffa, K. R., F. H. Schweingruber, P. D. Jones, T. J. Osborn, and S. G. Shiyatov,. "Reduced Sensitivity of Recent Tree-growth to Temperature at High Northern Latitudes." *Nature* 391, no. 6668 (1998): 678–82.

Briffa, Keith R. "Annual Climate Variability in the Holocene: Interpreting the Message of Ancient Trees." *Quaternary Science Reviews* 19, (1999): 87–105.

Clynes, Tom. "The Battle over Climate Change." *Popular Science*, June 21, 2012. www.popsci.com/science/article/2012-06/battle-over-climate change?single-page-view=true.

Delingpole, James. Climategate: The Final Nail in the Coffin of Anthropogenic Global Warming, November 20, 2009. blogs.telegraph.co.uk/news/james-delingpole/100017393/climategate-the-final-nail-in-the-coffin-of-anthropogenic-global-warming/.

Dyehouse, Jeremiah. "A Textbook Case Revisited": Visual Rhetoric and Series Patterning in the American Museum of Natural History's Horse Evolution Displays." In *Technical Communication Quarterly*, 2011.

Fahnestock, Jeanne. *Rhetorical Figures in Science*. New York: Oxford University Press, 1999.

Frow, Emma K. "Drawing a Line: Setting Guidelines for Digital Image Processing in Scientific Journal Articles." *Social Studies of Science* 42, no. 3 (2012): 369–92.

Gilbert, G. Nigel and Michael Mulkay. *Opening Pandora's Box*. Cambridge, UK: Cambridge University Press, 1984.

Gross, Alan A. "Presence as a Consequence of Verbal-Visual Interaction: A Theoretical Approach." In *Rhetoric Review*, Taylor & Francis Ltd, 2009.

Gross, Alan G. "Toward a Theory of Verbal-Visual Interaction: The Case of Lavoisier." *Rhetoric Society Quarterly* 39, no. 2 (2009): 147–69.

House of Commons Science and Technology Committee. The Disclosure of Climate Data from the Climatic Research Unit at the University of East Anglia. London, UK: House of Commons, 2010.

Intergovernmental Panel on Climate Change. "Third Assessment Report—Working Group 1: The Scientific Basis." In Christopher K. Folland, et al., editors, Cambridge, United Kingdom and New York, NY, USA: Cambridge University Press, 2001. www.ipcc.ch/ipccreports/tar/wg1/index.php?idp=48.

Kimball, Miles. "London Through Rose-Colored Graphics: Visual Rhetoric and Information Graphic Design in Charles Booth's Maps of London Poverty." *Journal of Technical Writing and Communication* 36, no. 4 (2006): 353–81, doi: 10.2190/k561-40p2-5422-ptg2.

Kress, Gunther and Theo Van Leeuwen. *Reading Images: The Grammar of Visual Design*. New York: Routledge, 1996, 2006.

Latour, Bruno. "Drawing Things Together." In *Representation in Scientific Practice*, edited by M. Lynch and S. Woolgar, 19–68. Cambridge, MA: MIT Press, 1990.

Latour, Bruno and Steve Woolgar. *Laboratory Life: The Construction of Scientific Facts*. Princeton, CT: Princeton UP, 1986.

Lynch, Michael. "Discipline and the Material Form of Images: An Analysis of Scientific Visibility." *Social Studies of Science* 15, (1985): 37–66.

McIntyre, Stephen. IPCC and the "Trick". *Climate Audit*, December 10, 2009, climateaudit.org/2009/12/10/ipcc-and-the-trick/.

Russell, Sir Muir, Geoffrey Boulton, Peter Clarke, David Eyton, and James Norton. The Independent Climate Change E-mails Review. University of East Anglia, 2010.

Myers, G. *Writing Biology: Texts in the Social Construction of Scientific Knowledge*. Madison: University of Wisconsin Press, 1990.

National Science Foundation Office of the Inspector General. Closeout Memorandum Case Number: A09120086, 2011.

Prelli, Lawrence J. "Visualizing a Bounded Sea: A Case Study in Rhetorical Taxis." In *Rhetorics of Display*, edited by Lawrence J. Prelli, 90–120. Columbia, SC: University of South Carolina Press, 2006.

Rosner, Mary. "Theories of Visual Rhetoric: Looking at the Human Genome." *Journal of Technical Writing & Communication* 31, (2001): 391–413.

Rymer, Jone. "Scientific Composing Processes: How Eminent Scientists Write Journal Articles." In *Advances in Writing Research*, edited by David Jolliffe, 211-250. Norwood, NJ: Ablex, 1998.

"Scientists Hide Global Warming Data." *The Daily Show. Narrated by Jon Stewart*. New York City, NY: Comedy Central, December 1, 2009.

United States Environmental Protection Agency. EPA's Response to the Petitions to Reconsider the Endangerment and Cause or Contribute Findings for Greenhouse Gases under Section 202(a) of the Clean Air Act. Washington, D.C.: United States Environmental Protection Agency Office of Atmospheric Systems Climate Change Division, 2010.

Walsh, Lynda. "The Technology of Prophecy: The IPCC and Climate Change 2007." In *Scientists as Prophets: A Rhetorical Genealogy*. New York: Oxford University Press, 2013.
World Meteorological Organization. WMO Statement on the Status of the Global Climate in 1999. Geneva, CH: WMO, 2000.

IMAGE COPYRIGHTS & SOURCES

Figure 1: © IPCC. 2001. http://www.ipcc.ch/ipccreports/tar/wg1/fig2-21.htm.
Figure 2: © World Meteorological Organization. 2000.

The Color of Risk: Expert Judgment and Diagrammatic Reasoning in the IPCC's 'Burning Embers'*

MARTIN MAHONY, MIKE HULME

This chapter tracks the emergence of a novel visual convention in the representation of the risks associated with climate change—the "reasons for concern" or "burning embers" diagram. By drawing on interpretive document and interview analysis, along with insights from science & technology studies (STS) concerning the role of visualization processes in scientific discourse, we explore how this image has gained a particular prominence in climate change debates. This discursive power has resulted both in material mobility and epistemic transformation as the diagram itself has become a tool for a variety of actors to reason with. We argue that the diagram has become something of an "icon of late modernity," bringing to light a number of challenges associated with attempts to know and visualize abstract concepts such as risk and danger, including the social organization of knowledge production and the role of expert judgment in contexts where science is asked to retreat from normativity.

1. SEEING CLIMATE CHANGE

How can climate change be visualized? The anthropogenic modification of the atmosphere's radiative properties through the emission of greenhouse gases and

* A version of this chapter was originally published as: Mahony, M. & Hulme, M. (2012) "The Colour of Risk: An Exploration of the IPCC's "Burning Embers" Diagram." *Spontaneous Generations: A Journal for the History and Philosophy of Science.* http://spontaneousgenerations.library.utoronto.ca/index.php/Spontaneous Generations/article/view/16075.

aerosols is an almost impossibly intangible, abstract, and remote phenomenon, distant in both space and time in many people's perceptions.[1] While the human-caused depletion of the planet's protective ozone layer became manifest in the figuratively visible "ozone hole" over the Antarctic, the complex causation and uncertain present and future impacts of climate change have generated "a mess of competing visual narratives characterized by suggestive shapes drawn by the plotted lines of story-laden graphs."[2] The notion that graphical representations of climate change are "story-laden" is not to undermine their relationship to physical realities. Rather, it points both to the social processes of their construction and to their appropriation of culturally-embedded representational conventions in the ongoing struggle to render climate change "meaningful."[3] While graphic data representations have proven to be useful heuristics for coming to terms with the complex dynamics of the atmosphere, photographs have often been employed to visualize the possible impacts of climate change. Common tropes of stranded polar bears, flash flooding and parched soil can be found accompanying media coverage of climate change.[4] Such discursive coupling is suggestive of direct causal relationships between climate change and the pictured impacts, even as the scientific debate over the attribution and prediction of extreme weather events appears irresolvable with any certainty.[5]

It is into this representational milieu that the Intergovernmental Panel on Climate Change (IPCC) introduced the so-called "burning embers" diagram in 2001 (Figure 1). The diagram seeks to summarize a number of "reasons for concern" linked to the prospect of rising global temperatures. The left hand side of the figure, which appeared in the Working Group II *Summary for Policymakers* (2001), shows projections of global mean temperature (GMT) change up to 2100 based on various emissions scenarios and the results of numerous climate simulations. It is suggested that GMT could rise by up to 6°C by 2100. The right hand side of the figure schematically represents the level of danger associated with these rises in mean temperature above 1990 levels for five categories of concern. The change in color from white to yellow to red is taken to denote risks of increasing magnitude, severity or geographic spread, and

1 O'Neill and Nicholson-Cole, "Fear Won't Do It," 2009.

2 Hamblyn and Callanan, *Data Soliloquies*, 2009, 43.

3 Schneider, "Image Politics," 2011; Doyle, *Mediating Climate Change*, 2012.

4 Doyle, "Picturing the Clima(c)tic," 2007; Manzo, "Imaging Vulnerability," 2010.

5 IPCC, *Special Report*, 2011.

it is this color palette which gave rise to the moniker "burning embers" among the diagram's creators.[6]

Figure 1: Projected temperature changes under different emissions scenarios (left) and "reasons for concern" or "burning embers" (right). (→ color plate 6)

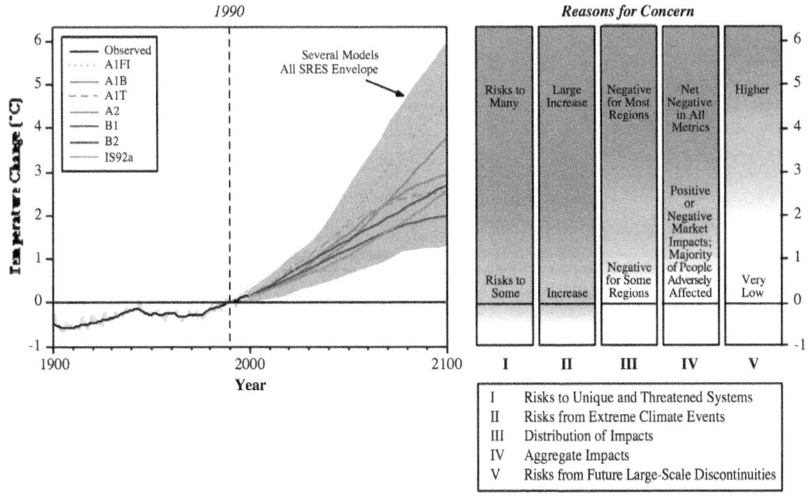

The "burning embers" diagram has become an icon of the climate change debate, "underpinned by a large number of scientific analyses and legitimated through publication and republication."[7] As is shown below, the diagram has found traction among both scientific actors and non-governmental organizations, and has also surfaced in governmental settings such as submissions to the *Stern Review on the Economics of Climate Change*, produced for the UK Government. The diagram has functioned as a powerful "boundary object,"[8] facilitating

6 The following discussion is based on interviews with 11 scientists who were involved with the development of the diagram between 1999 and 2009. The disciplinary backgrounds and interests of the scientists varied between climate modeling, climate impacts analysis, economics and ecology. The interviews were conducted between May and December 2011. Interview data was supplemented by textual analysis of the publications in which the diagram appeared or was discussed.

7 Liverman, "Conventions of Climate Change," 2009, 285.

8 Star and Griesemer, "Institutional Ecology, 'Translations' and Boundary Objects," 1989.

exchange and interaction between the worlds of science and politics—a function which has defined the broader institutional history of the IPCC.[9]

The IPCC occupies a unique position at the science-policy boundary, and constitutes a hybrid space where scientific and political actors together regulate the flow of information across that boundary.[10] IPCC reports are charged with being "policy-relevant and yet policy-neutral, never policy-prescriptive,"[11] and thus IPCC authors face a constant challenge in regulating the boundary between epistemic statements of what *is* and normative statements of what *ought to be*.[12] The chapter which gave rise to the "burning embers" sought to address the question of what might constitute "dangerous anthropogenic interference with the climate system"[13]—a question which the authors acknowledged to be loaded with value judgments. This diagrammatic form thus offers an opportunity to study how the interpenetration of the epistemic and the normative—long recognized by science & technology studies (STS) scholars[14]—is handled under the rubric and norms of relevance and neutrality.

2. THE COLOR OF RISK

In its original form in Chapter 19 of Working Group II's contribution to the IPCC's Third Assessment Report (TAR), the diagram appeared in greyscale (as did all diagrams in the main body of the report) and laid-out horizontally. This original version of the diagram was based on months of work by individuals on the writing team, who assessed the literature on climate change impacts across the five "reasons for concern." This categorization of impacts was a result of earlier deliberations within the team that led to a desire to synthesize the textual information in the form of an accessible visualization. The diagrammatic format demanded that projected impacts be tied to specific temperature points and, as literature on climate change impacts was sparse at the time, the authors had to exercise personal judgment in determining where the color shifts should take place.

9 Miller, "Climate Science and the Making of a Global Political Order," 2004.

10 see Hulme, *Why we Disagree about Climate Change*, 2009.

11 IPCC, Organization, 2013; see also Moss, "The IPCC," 1995.

12 Walsh, "Visual Strategies to Integrate Ethos," 2009.

13 UN, *Framework Convention on Climate Change*, 1992.

14 e.g. Jasanoff, *States of Knowledge*, 2004; Oels, "Rendering Climate Change Governable," 2005; Hulme, ibid.

This kind of risk visualization was not without precedent in the climate change literature. A decade earlier, Rijsberman and Swart (1990) and Vellinga and Swart (1991) presented the "traffic light" system of risk visualization and management (Figure 2). This employed three discrete blocks of color—green, amber and red to represent increasing rates of GMT and sea level rise and their risk corollaries (although again, printing practices reduced the colors to shades of grey). The aim of this visual device was to propose targets for temperature stabilization. The transition from green to amber occurs with a 1°C rise above pre-industrial levels, while the red light is associated with a 2°C rise. The authors argue that the:

"goal of our effort must be, therefore, to go for the green light, and in any case, to fully avoid the red light. To avoid the red light means that we want to limit the GMT rise to well below 2°C with respect to the pre-industrial level and that we want to limit the sea level rise to well below 50cms."[15]

This was the first time that maximum temperature change was used as a means of normatively framing a target-based approach to global climate policy,[16] and was a key moment in the establishment of the 2°C target as an anchoring device for scientific and political discussions of climate change.[17]

Figure 2: The "traffic light" approach to risk management, Vellinga and Swart (1991, 131).

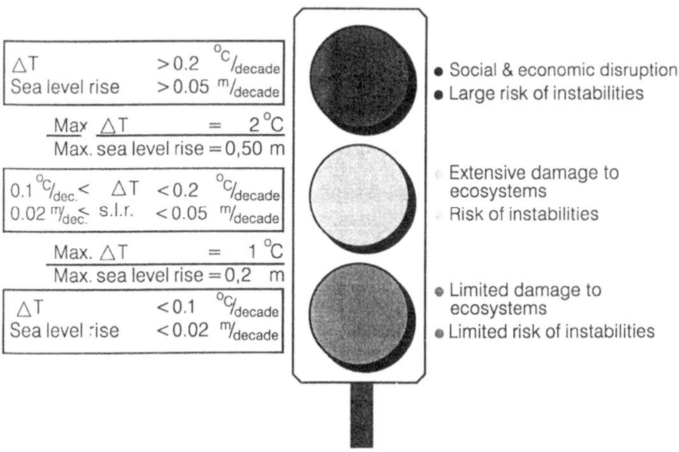

15 Vellinga and Swart, "The Greenhouse Marathon," 1991, 131.

16 cf. Nordhaus, "Economic Growth and Climate: the Carbon-Dioxide Problem," 1977.

17 Randalls, "History of the 2°C Climate Target," 2010.

Diagrams such as these combine expert judgments of observational evidence, future predictions, and normative judgments of risk. Unlike quantitative scientific visualizations, which commonly aim at an ideal of unmediated representation or an "analogue" of physical reality,[18] here the viewer's interpretation is guided explicitly by the design choices and their attendant normative elements. The desirability of the bottom scenario is prefigured in the green traffic light; likewise, the undesirability in the red. The normative content of the temperature and sea level rise scenarios is not left to the viewer's interpretation. The familiarity of the traffic light—and the images it conjures of momentum continued, tempered or abated—directs interpretation towards this seductively linear notion of pathways and targets. This notion has since been criticized for its reductiveness and its tendency to distract from the politically and ethically complex task of reducing emissions.[19] According to this construction, it is science that is almost literally directing the traffic and showing the way.

The "traffic lights" had a direct influence on the development of the "burning embers" diagram. During the early stages of the writing team's deliberations, a diagram was proposed which employed a similar transition from green to red along each "line of evidence" column, as they were known before "reasons for concern." However, this palette was dismissed, as it was thought that the green element indicated an absence of risk or even safety for some levels of climate change: something which the authors took as being contrary both to their own understandings of the risks and to the message they sought to convey (interview, December 1, 2011). The neutrality of white was thus employed as the baseline on which to build the negativity of red. The ordering of the columns was also decided based on a combination of epistemic and aesthetic considerations. Neither fundamentally scientific nor axiological reasons are given for the ordering. Rather, this composition was deemed the most visually appealing, producing an upward-trending diagonal in the emergence of yellow, from column two to column five (interview, November 4, 2011). It is a graphical design choice, but one which provides visual and rhetorical echoes of the rising forms of many high-profile climate visualizations, from Michael Mann's famously controversial "hockey stick" temperature chart on the pages of IPCC reports to Al Gore's dramatically exponential CO_2 concentrations in *An Inconvenient Truth.*[20]

18 Daston and Galison, *Objectivity*, 2007; Barthes, *Image, Music, Text*, 1977.

19 Randalls, "History of the 2°C Climate Target," 2010.

20 Hamblyn, *Data Soliloquies*, 2009; Schneider, "Image Politics," 2011.

As in the "traffic lights," a sense of danger is pre-figured in the "burning embers" color pallete. The color red has been graphically associated with high temperatures since the 19th century,[21] but its connotative associations with danger, fear, violence and passion have a much longer and more engrained lineage in Western cultures. The embedded nature of this scientific diagram within these political discourses and cultural conventions is the source of its meaning. It is also the locus from which it has achieved mobility as an actant within the networks that tie together science, politics, culture and ethics, further blurring their already permeable boundaries while undergoing a number of epistemic transformations.

3. Epistemic Transformations

The "burning embers" diagram achieved a great deal of visibility following its publication in 2001 and has arguably become one of a few iconic scientific visualizations giving illustration to the climate change debate (Liverman 2009).[22] In a particularly interesting use of the "burning embers" diagram, Mastrandrea and Schneider (2004) employ the image as a foundation for a probabilistic assessment of the chances of avoiding "dangerous" climate change under certain policy initiatives. A cumulative density function of the threshold of dangerous anthropogenic interference (DAI) is constructed by placing a data point at the level at which each column turns red (Figure 3). The authors justify this strategy by stating that each column represents the judgment of "dozens of IPCC lead authors' examination of climate impacts literature," and therefore that the red zones represent "a consensus estimate of DAI" (Mastrandrea and Schneider 2004, 572). The "dangerous" temperature thresholds are then used to explore the sensitivity of projections of DAI to three model parameters[23] enabling the authors to claim that the probability of DAI can be reduced from around 45% to near zero by increasing "policy controls."

---

21 Schneider, ibid.

22 see e.g.: West Coast Climate Equity, "Present Targets for CO_2," 2010; Greenpeace Australia Pacific, "Treading on Burning Embers," 2009; Climate Change Food Security, "Policy Making," 2013; UNEP, *Climate Change Science Compendium 2009*, 2009; Yohe, "'Reasons for Concern'," 2010.

23 The model parameters investigated are: the estimated climate sensitivity to a doubling of CO_2 concentrations, projected economic damages, and the discount rate, i.e. the way present costs and benefits are weighed-up against future costs and benefits.

Mastrandrea and Schneider's analysis transforms the burning embers' blurred, uncertain judgments of future climate impacts into a quantitative profile of risk and danger as the global temperature moves up the scale from its late twentieth century baseline. The temperature thresholds for radical changes in social and natural systems, drawn initially from climate impacts studies and then amalgamated and obscured in color, re-emerge as new points; average thresholds, calculated not from the collected-together numbers of the impacts literature, but from the shifting colors of their graphical approximation. Point becomes blur, blur becomes point.

Figure 3: Mastrandrea and Schneider's adaptation of the "burning embers" diagram. The thresholds of dangerous climate change are marked by the black points and connecting line, positioned where each column begins turning red. From Schneider and Mastrandrea (2005, 15729). (→ color plate 7)

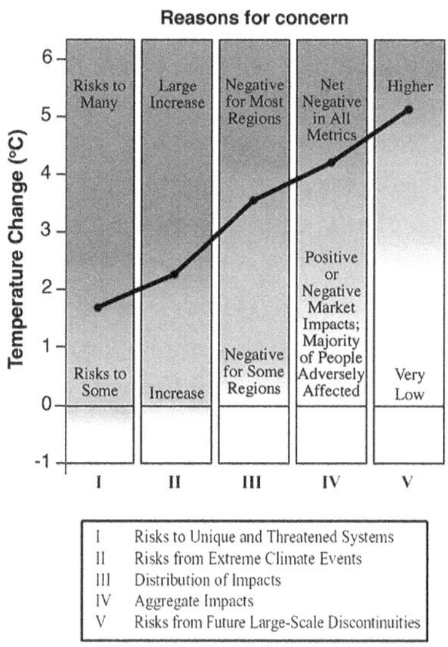

This epistemic transformation illustrates both the challenges of visualizing risk, and the power of consensus in addressing complex environmental issues. The visualization of risk involves not only an attempt to capture and represent physical processes and phenomena. It also represents their interaction with social systems, certain interpretations of the meaning of that interaction, and the social and political capacity to respond to an emerging danger, should it be deemed to be of sufficient magnitude and urgency. The calculation of risk is thus often a task bestowed upon those with the necessary technical expertise to comprehend the complex, multi-faceted nature of anthropogenically "manufactured risks."[24] The concept of "risk" itself "has come to stand as one of the focal points of feelings of fear, anxiety and uncertainty" pertaining to the future."[25] Its calculation must therefore involve grappling with the epistemological, ontological and ethical uncertainties which are constitutive of any effort to project what is known into the future, and then to draw on such projections to reflect on how society should be directed in the present.[26] Such knowledge will always be incomplete and partial—it will vary between experts, social constituencies, and cultures.[27] The social organization of knowledge therefore becomes a key source of epistemic authority, with assessment, synthesis and consensus being central strategies for the application of scientific expertise to questions of societal risk.

The IPCC is mandated to produce consensus. Despite the pursuit of consensus being arguably "a source of both strength and vulnerability for the IPCC," the notion of "consensual science" carries great discursive power in climate change debates.[28] For many analysts,[29] the attainment of consensus has been politically instrumental in convincing the world of the need to act on climate change and thus for the advancement of international climate policy.[30] Consensus can also be said to be generative of community identities, especially those articulated around epistemic norms and values.[31] Although gaining authority through its representation of a form of consensus, the "burning embers" diagram has not always attracted a broader consensus beyond the epistemic community that innovated and developed the diagram.

The IPCC chapter that gave rise to the original "burning embers" was re-mandated for the Fourth Assessment Report (AR4), albeit with many new authors. Although the "reasons for concern" analytical framework persisted and was updated textually in the IPCC's Synthesis Report,[32] the "burning embers"

24 Giddens, "Risk and Responsibility," 1999.

25 Lupton, *Risk*, 1999, 12.

26 Beck, *Risk Society*, 1992; Felt and Wynne, *Taking European Knowledge Society Seriously*, 2007.

27 Lupton, *Risk*, 1999.

28 Hulme and Mahony, "Climate change," 2010, 711.

29 e.g. Edwards and Schneider, "The 1995 IPCC Report," 1997.

30 see also Guston, "On Consensus and Voting in Science," 2006; Grassl, "A Discernible Human Influence on the Global Climate," 2009.

31 Horst and Irwin, "Nations at Ease with Radical Knowledge," 2010; Haas, "Introduction," 1992.

32 IPCC, *Climate Change 2007: Synthesis Report*, 2007.

diagram was absent from the final report. In interviews, authors of the chapter reported a reluctance to wholly import the analytical framings from the TAR, as the AR4 team was required to assess a rapidly evolving and expanding literature. However, towards the end of the writing process, it was decided amongst some authors that an update to the "burning embers" diagram would be appropriate. An updated version of the burning embers diagram was thus presented for inclusion at the Working Group II plenary session. However, the lack of a version of the diagram in the underlying chapter later opened space for procedural objections from government delegations, with the late Steve Schneider, coordinating lead author of the chapter, reporting that "four fossil fuel dependent countries accepted the text but refused the figure," seemingly on the grounds that it was "too much of a judgment."[33] A combination of these governmental protestations, the tight timescales of IPCC drafting processes, and certain objections to this particular analytical framing within the Working Group II hierarchy conspired to see the updated embers excluded from the AR4. The updated diagram was eventually published by a group largely consisting of chapter authors in the *Proceedings of the National Academy of Sciences* (Figure 4).[34]

Figure 4: Updated 'Reasons for Concern.'

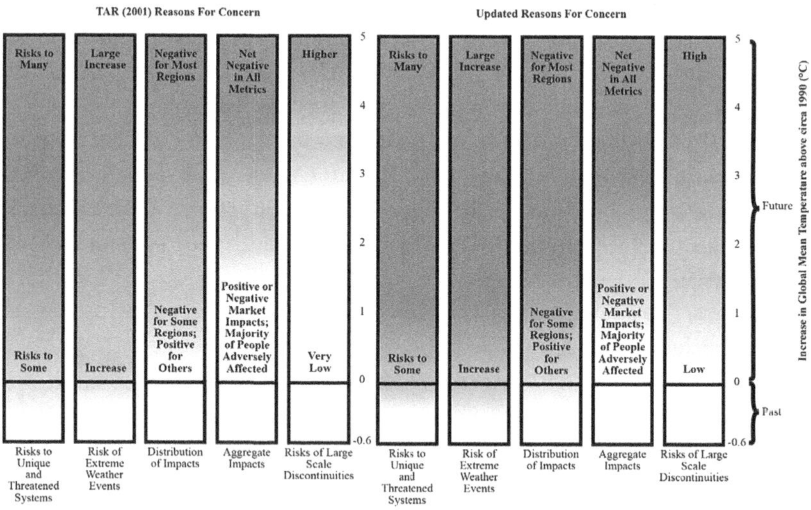

33 quoted in Revkin, "Why 2007 IPCC Report Lacked 'Embers'," 2011.

34 Smith et al., "Assessing Dangerous Climate," 2009; see Figure 4.

Schneider's claim that the diagram represented too much of a "judgment" for some parties emphasizes the challenge of negotiating the boundary between description and prescription. The preservation of this boundary is inscribed both in the IPCC's mandate and in the norms of much contemporary scientific practice[35]. The shift from the left- to the right-hand side of Figure 4, with the visually striking descent of the red, does not portray a change in the ontological status of the risks between 2001 and 2009, but rather maps the changing content of scientific understandings and judgments. The diagram seeks to represent the consensual amalgamation of these judgments, and the authors openly relate the potential for subjectivity in this mode of knowledge production and synthesis (Smith et al. 2009). However, the cognitive and social-epistemological processes that are generative of such judgments are largely indiscernible to the outside observer.

4. An Icon of Late Modernity

The "burning embers" diagram is a collage of space and time[36] with GMT standing in for an indeterminate temporality, while the global is collapsed into the limited dimensionality of graphematic space. This level of abstraction has been a source of criticism. For example, Liverman (2009) argues that the diagram elides the complex geographies of climate change impacts in its effort to present a globalized conceptual space. The dominant "global gaze" of climate science is not an epistemological inevitability, but is rather the result of the complex intertwining of science and politics.[37] For instance, since its inception the "reasons for concern" framework has sought to address the principle enshrined in Article 2 of the United Nations Framework Convention on Climate Change (UNFCCC)—the avoidance of "dangerous anthropogenic interference (DAI) with the climate system."[38]

While attempting to avoid overt policy prescriptiveness, the "reasons for concern" framework has sought to provide illustrative guidance on what might be considered a dangerous level of global mean temperature rise. However, the framework seeks to address only the dangers associated with anthropogenic

35 Shapin, *The Scientific Life*, 2008; Walsh, "Visual Strategies," 2009.

36 Schneider, "Image Politics," 2011.

37 e.g. Miller, "Climate Science and the Making of a Global Political Order," 2004; Oels, "Rendering climate change governable," 2005.

38 UN, *Framework Convention on Climate Change*, 1992.

climate change, rather than those associated with natural climate variability. For example, it is suggested in all versions of the diagram that at some point below 1990 temperature levels, the risks associated with extreme weather events were "virtually" zero or "neutral." Of course, extreme weather happened long before 1990, but the "reasons for concern" framework seeks to address only that which may be attributable to human actions. This inverse purification of the "human" from the "natural"[39] is a function of the diagram's direct engagement with the policy question of "dangerous anthropogenic interference." It thus functions as a heuristic for the dangers associated with an imagined, human-made climate of linear trends and direct causalities, rather than a complex, hybrid climate where cycles, trends and social trajectories interact chaotically in perhaps unknowable ways.[40]

As an "epistemic thing,"[41] the "burning embers" diagram does not simply fulfill a representational role, but rather functions as an object *within* a system of enquiry. In Figures 3 and 5, for example, the diagram has formed the basis of new knowledge claims about the prospect of dangerous climate change. However, instead of Rheinberger's tightly bounded and regulated laboratory spaces, the "burning embers" function in a much wider arena. There the conditions "of the possibility of things becoming epistemic things"[42] are as much political and discursive as they are determined by the materiality of scientific enquiry.[43] The "burning embers" diagram is thus a hybrid form: representational and heuristic, forensic and epideictic[44]; the outcome of an institutionalized yet indeterminate encounter between object and subject. This hybridity, while posing challenges to certain scientific norms, is emblematic of the complex interweaving of competing epistemologies with the challenges of intractable uncertainty that characterizes late modern "risk societies."[45]

---

39 cf. Latour, *We Have Never Been Modern*, 1993.

40 see Hulme et al., "Is Weather Event Attribution Necessary for Adaptation Funding?" 2011, for an example of the political implications of this purification.

41 Rheinberger, *Toward a History of Epistemic Things*, 1997.

42 Rheinberger, "Experimental Systems, Graphematic Spaces," 1998, 297.

43 Jasanoff, *States of Knowledge*, 4; Foucault, *Archaeology of Knowledge*, 2004.

44 see Walsh, "Visual Strategies to Integrate," 2009, for a discussion of climatological imagery through the lens of rhetoric theory.

45 Beck, *Risk Society: Towards a New Modernity*, 1992.

Figure 5: The "burning embers" as they appear in Richardson et al.'s (2009, 16) synthesis of an international scientific conference which took place in the run-up to the Copenhagen climate negotiations of December 2009. The positioning of the "2°C guardrail" at around 1.4°C represents the discrepancy between pre-industrial and 1990 temperature baselines, the latter being preferred in IPCC assessments.

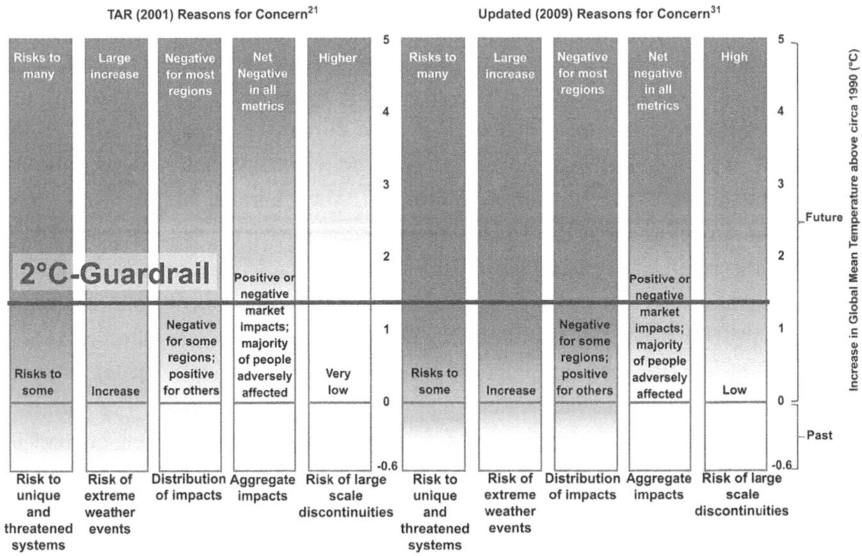

For theorists such as Ulrich Beck and Anthony Giddens, industrialized societies are experiencing conditions of late modernity. This represents a continuation or radicalization of the institutional, economic and cultural changes wrought by modernization to a point where socioeconomic processes generate hazards of a scale which require modernity to reflect on itself; to challenge its assumptions of progress and interminable growth.[46] Late modern societies are thus preoccupied with the future. This preoccupation most often takes the form of the calculation of hazard probabilities and of the social acceptability of risks in order that they may be managed or "controlled." However, late modernity is also characterized by risks of a sort that belie easy calculation, spatio-temporal delineation or straightforward democratic appraisal. The blurring, evocative colors of the "burning embers" exemplify this paradoxical societal relationship with risk as an object of scientific enquiry and political concern; the semiotic functions of the

46 Giddens, *The Consequences of Modernity*, 1991; Beck, ibid.

colors suggest that significant dangers confront humanity, while their blurred transitions point to the irrevocable uncertainties that accompany such threats.

It is in the distinctive societal milieu of late modernity that the "burning embers" diagram may be said to function much like an Expressionist painting. During the early twentieth century, the Expressionist movement pursued an artistic style that prioritized subjective experience, meaning and emotion in a direct rebuttal to realist and naturalist representational paradigms.[47] Likewise, the "burning embers" seeks not to figuratively represent a phenomenon (the changing climate), but rather its intangible effects.[48] These effects, be they heightened levels of danger or risk, are quickly translated into affect through the use of literary and visual conventions such as the emotionally charged color pallet. The Expressionist movement arose in Germany partly in response to conditions of social crisis and upheaval.[49] The "burning embers" too, feed certain anxieties about the future; we can sense ourselves walking powerlessly into the red heat, a fate made all the more inevitable as the red zone creeps towards the colorless safety of the baseline. In the case of this diagram, scientific visualization is not the disinterested gaze of technical apparatus. Rather, it is a suite of social-epistemic practices situated firmly within a set of cultural discourses in the uncertain, reflexive time-space of late modernity. The semiotic, epistemic and social elements of such constructions cannot be understood in isolation, or even analytically delineated. Here they are mutually constitutive; combining and re-combining in a particular graphematic space to produce a mobile and evolving visual convention.

5. CONCLUSION

This volume is broadly concerned with the question of how climate change—as an inherently *in*visible phenomenon—can be visualized. The example of the "burning embers" diagram itself raises the question of whether the exercise of subjective expert reasoning is compatible with the demands of diagrammatic reasoning. It has been argued that the notion of risk is highly complex in epistemic and normative terms, especially when considered in the context of

47 Willett, *Expressionism*, 1970.

48 Coincidentally, one of the most famous examples of expressionist architecture—Erich Mendelsohn's Einstein Tower in Potsdam—lies just a few yards from the meeting room where the "burning embers" diagram was first conceived by IPCC authors.

49 Whitford, *Expressionism*, 1970.

climate change.[50] A perfectly "objective" assessment (in the sense of an ostensible absence of knowing and judging subjects) of the risks posed by a changing climate would be impossible, and the authors of the "burning embers" diagram are right to acknowledge the inevitable subjectivity of such judgments. As an *heuristic* tool, the diagram functions well in its suggestion of when (or, more precisely, at what temperature) danger might be encountered under a changing climate, as evidenced by the variety of uses to which the diagram has been put. As *representation*, the diagram is weakened by the opacity of what exactly is being represented. In the translation from assessment of scientific literature to diagrammatic form, a wide body of scientific knowledge is condensed into a suggestive array of color with the somewhat inevitable loss of what Latour (1999) terms "reference"—the traces, marks and symbols which tie together mind and world.

Despite the widespread rhetorical policing of the boundary between description and prescription, this case makes clear that the communication of climate change through visualization relies not only on translation, but also on what Walsh (2009) terms a "performance of continuity" across the is/ought divide. Highlighting the normative underpinnings of this continuity is perhaps incompatible with the demands of diagrammatic reasoning and the limitations of graphematic space. Our knowledge of the complexity of the climate system is growing and different normative stances on climate change are proliferating, for example in judgments about what might constitute a "reason for concern." In this context, scientists working at the science-policy interface may need to find new, creative ways of communicating their findings. Recognizing and communicating epistemic uncertainty and normative diversity will be central to the success of such efforts.

ACKNOWLEDGMENTS

The authors wish to thank all those who participated in the study for their time and insights. Thanks also to Helen Pallett for her comments and to the reviewers and editors of the original iteration of this chapter in *Spontaneous Generations*. Martin Mahony gratefully acknowledges the support of UK Economic & Social Research Council PhD studentship number ES/H01110/1.

50 Hulme, "Climate Change: What Do We Know About the IPCC?" 2010.

BIBLIOGRAPHY

Barthes, Roland. *Image, Music, Text.* London: Fontana, 1977.

Beck, Ulrich. *Risk Society: Towards a New Modernity.* London: Sage Publications, 1992.

Daston, Lorraine and Peter Galison. *Objectivity.* Cambridge: Zone Books, 2007.

Doyle, Julie. "Picturing the Clima(c)tic: Greenpeace and the Representational Politics of Climate Change Communication." *Science as Culture* 16, no. 2 (2007): 129–50.

—. *Mediating Climate Change.* Farnham: Ashgate, 2011.

Edwards, Paul N., and Stephen H. Schneider. "The 1995 IPCC Report: Broad Consensus or 'Scientific Cleansing'?" *International Studies* 1, no. 1 (1997): 3–9.

Felt, Ulrike and Bryan Wynne. *Taking European Knowledge Society Seriously: Report of the Expert Group on Science and Governance to the Science, Economy and Society Directorate.* Brussels: European Commission, 2007.

Foucault, Michel. *Archaeology of Knowledge.* London: Routledge, 2007.

Gage, John. *Color and Culture: Practice and Meaning from Antiquity to Abstraction.* Berkeley: University of California Press, 1999.

Giddens, Anthony. *The Consequences of Modernity.* Stanford: Stanford University Press, 1991.

—. "Risk and Responsibility." *Modern Law Review* 62, no. 1 (1999): 1–10.

Grassl, Hartmut. "A Discernible Human Influence on the Global Climate: How the IPCC Affected Climate Politics." *GAIA—Ecological Perspectives for Science and Society* 18, no. 3 (2009): 255–56.

Greenpeace Australia Pacific. 2009. "Treading on Burning Embers." Accessed 21 February 2013. http://www.greenpeace.org.au/blog/?p=875.

Guston, David H. "On Consensus and Voting in Science: From Asilomar to the National Toxicology Program." In *The New Political Sociology of Science: Institutions Networks, and Power*, edited by Scott Frickel and Kelly Moore, 373–404. Madison: University of Wisconsin Press, 2006.

Haas, Peter M. "Introduction: Epistemic Communities and International Policy Coordination." *International Organization* 46, no. 1 (1992): 1–35.

Hamblyn, Richard. "The Whistleblower and the Canary: Rhetorical Constructions of Climate Change." *Journal of Historical Geography* 35, no. 2 (2009): 223–36.

Hamblyn, Richard and Martin John Callanan. *Data Soliloquies.* London: UCL Environment Institute, 2009.

Horst, Maja and Alan Irwin. "Nations at Ease with Radical Knowledge: On Consensus, Consensusing and false Consensusness." *Social Studies of Science* 40, no. 1 (2010): 105–26.

Hulme, Mike, Saffron J. O'Neill and Suraje Dessai. "Is Weather Event Attribution Necessary for Adaptation Funding?" *Science* 334, no. 6057 (2011): 764–65.

Hulme, Mike, and Martin Mahony. "Climate Change: What do We Know About the IPCC?" *Progress in Physical Geography* 34, no. 5 (2010): 705–18.

Hulme, Mike. *Why we Disagree About Climate Change: Understanding Controversy, Inaction and Opportunity.* Cambridge: Cambridge University Press, 2009.

IPCC. *Working Group II: Impacts, Adaptation and Vulnerability. Summary for Policymakers.* Cambridge: Cambridge University Press, 2001.

IPCC. *Climate Change 2007: Synthesis Report*, edited by R. Pachauri and A. Reisinger. Geneva, 2007.

IPCC. *Special Report on Managing the Risks of Extreme Events and Disasters to Advance Climate Change Adaptation. Summary for Policymakers.* Geneva, 2011.

Jasanoff, Sheila. *States of Knowledge: The Co–production of Science and Social Order.* London: Routledge, 2004.

Latour, Bruno. *We Have Never Been Modern.* Cambridge, MA: Harvard University Press, 1993.

Latour, Bruno. *Pandora's Hope: Essays on the Reality of Science Studies.* Cambridge: Harvard University Press, 1999.

Liverman, Diana M. "Conventions of Climate Change: Constructions of Danger and the Dispossession of the Atmosphere." *Journal of Historical Geography* 35, no. 2 (2009): 279–96.

Lupton, Deborah. *Risk.* London: Routledge, 1999.

Manzo, Kate. "Imaging Vulnerability: the Iconography of Climate Change." *Area* 42, no. 1 (2010): 96–107.

Mastrandrea, Michael D. and Stephen H. Schneider. "Probabilistic Integrated Assessment of 'Dangerous' Climate Change." *Science* 304, no. 5670 (2004): 571–75.

Miller, Clark A. "Climate Science and the Making of a Global Political Order." In *States of Knowledge: The Co–production of Science and the Social Order*, edited by Sheila Jasanoff, 46–66. London: Routledge, 2004.

Moss, Richard H. "The IPCC: Policy Relevant (not driven) Scientific Assessment." A comment on Sonja Boehmer-Christiansen's 'Global Climate Protection Policy: the Limits of Scientific Advice'. *Global Environmental Change* 5, no. 3 (1995): 171–174.

Nordhaus, William D. "Economic Growth and Climate: the Carbon–Dioxide Problem." *American Economic Review* 67, no. 1 (1977): 341–46.

Nowotny, Helga, Peter Scott, and Michael Gibbons. *Re-Thinking Science: Knowledge and the Public in an Age of Uncertainty*. London: Wiley-Blackwell, 2001.

O'Neill, Saffron and Sophie Nicholson-Cole. "'Fear Won't Do it': Promoting Positive Engagement with Climate Change through Visual and Iconic Representations." *Science Communication* 30, no. 3 (2009): 355–79.

Oels, Angela. "Rendering Climate Change Governable: From Biopower to Advanced Liberal Government?" *Journal of Environmental Policy & Planning* 7, no. 3 (2005): 185–207.

"Organization." IPCC. Accessed February 20, 2013. http://www.ipcc.ch/organization/organization.shtml#.UpXYSNJNKSo Accessed November 27, 2013.

Randalls, Samuel. "History of the 2°C Climate Target." *Wiley Interdisciplinary Reviews: Climate Change* 1, no. 4 (2010): 598–605.

Revkin, Andrew. 2009. "Why 2007 IPCC Report Lacked 'Embers'." *New York Times Dot Earth Blog*. Accessed 5 December 2011. http://dotearth.blogs.-nytimes.com/2009/02/26/why-2007-ipcc-report-lacked-embers/.

Rheinberger, Hans-Jörg. *Toward a History of Epistemic Things: Synthesizing Proteins in the Test Tube*. Stanford: Stanford University Press, 1997.

—. "Experimental Systems, Graphematic Spaces." In *Inscribing Science: Scientific Texts and the Materiality of Communication*, edited by Timothy Lenoir, 285–303. Stanford: Stanford University Press, 1998.

Richardson, Katherine, Will Steffen, Hans Joachim Schellnhuber, Joseph Alcamo, Terry Barker, Daniel M. Kammen, Rik Leemans et al. *Climate Change: Global Risks, Challenges and Decisions. Synthesis Report*. University of Copenhagen, 2009.

Rijsberman, Frank R. and Rob Swart. *Targets and Indicators of Climate Change: Report of Working Group II of the Advisory Group on Greenhouse Gases*. Stockholm: Stockholm Environment Institute, 1990.

Schneider, Birgit. "Image Politics: Picturing Uncertainty: The Role of Images in Climatology and Climate Policy." In *Climate Change and Policy: The Calculability of Climate Change and the Challenge of Uncertainty*, edited by Gabriele Gramelsberger and Johann Feichter, 191–209. London: Springer, 2011.

Schneider, Stephen H., S. Semenov, A. Patwardhan, I. Burton, C.H.D Magadza, M. Oppenheimer, A.B. Pittock, et al. "Assessing Key Vulnerabilities and the Risk from Climate Change." In *Climate Change 2007: Impacts, Adaptation and Vulnerability. Contribution of Working Group II to the Fourth Assessment Report of the Intergovernmental Panel on Climate Change*, edited by P.J. van der Linden, C.E. Hanson, M.L. Parry, O.F. Canziani, and J.P. Palutikof, 779–810. Cambridge: Cambridge University Press, 2007.

Schneider, Stephen H. and Michael D Mastrandrea. "Probabilistic Assessment of 'Dangerous' Climate Change and Emissions Pathways." *Proceedings of the National Academy of Sciences of the United States of America* 102, no. 44 (2005): 15728–35.

"Science: The Many Adverse Impacts." Climate Change Food Security. Accessed 27 July 2012. http://www.climatechange-foodsecurity.org/many_impacts.html.

Shapin, Steven. *The Scientific Life: A Moral History of a Late Modern Vocation*. Chicago, IL: University of Chicago Press, 2008.

Smith, Joel B., Stephen H. Schneider, Michael Oppenheimer, Gary W. Yohe, William Hare, Michael D. Mastrandrea, Anand Patwardhan, et al. "Assessing Dangerous Climate Change Through an Update of the Intergovernmental Panel on Climate Change (IPCC) 'Reasons for Concern.'" *Proceedings of the National Academy of Sciences of the United States of America* 106, no. 11 (2009): 4133–37.

Star, Susan L. and James R. Griesemer. "Institutional Ecology, 'Translations' and Boundary Objects: Amateurs and Professionals in Berkeley's Museum of Vertebrate Geology, 1907–1939." *Social Studies of Science* 19, no. 3 (1989): 387–420.

UN. *Framework Convention on Climate Change*. New York, 1992.

UNEP. *Climate Change Science Compendium 2009*. Nairobi, 2009.

Vellinga, Pier and Rob Swart. "The Greenhouse Marathon: Proposal for a Global Strategy." In *Climate Change: Science, Impacts and Policy*, edited by J. Jager and H.L. Ferguson, 129–34. Cambridge: Cambridge University Press, 1991.

Walsh, Lynda. "Visual Strategies to Integrate Ethos across the 'is/ought' Divide in the IPCC's Climate Change (2007) Summary for Policy Makers." *Poroi* 6, no. 2 (2009): 33–61.

West Coast Climate Equity. 2010. "Present Targets for CO_2 Emission Cuts Will Not Prevent a 4C Global Temperature Rise." Accessed 21 February 2013. http://westcoastclimateequity.org/2010/07/07/present-targets-for-co2-emission-cuts-will-not-prevent-a-4c-global-temperature-rise/.

Whitford, Frank. *Expressionism.* London: Hamlyn, 1970.

Willet, John. *Expressionism.* London: World University Library, 1970.

Yohe, Gary. "'Reasons for Concern' (About Climate Change) in the United States." *Climatic Change* 99, no. 1–2 (2010): 295–302.

IMAGE COPYRIGHTS

Chapter 3
Images of Climate Change in the Press and on the Web

Between Risk, Beauty and the Sublime: The Visualization of Climate Change in Media Coverage during COP 15 in Copenhagen 2009

ELKE GRITTMANN

Climate change is one of the top issues in media coverage. Based on approaches of iconography, aesthetics and visual framing, this paper examines how climate change is framed visually and how media discourse on climate change is shaped by powerful visual iconography. Media are highly event-driven: thus, the visual coverage on climate change and the negotiation during and around the United Nations Climate Change Conference in Copenhagen in 2009 (COP 15) offers a good case study how the topic is communicated visually. This article focuses on visual coverage in German print media. It will be shown that the German case is of international interest. Given that media affect public perception of climate change issues, the role of visuals in this process will be explored. By referring to Entman's framing approach, the analysis examines which pictures become part of the argumentation in media discourse on climate change. The results of the analysis indicate that the visuals support the urgency of action claimed by the media. They transform the topic into a visual argumentation that focuses on the problem, its causes and possible solutions. However, this article argues that visuals are not only a matter of knowledge. By adopting the aesthetics of beauty and the sublime, the visualization also tries to affect the audience emotionally and ethically.

In August 1986, *Der Spiegel*, the most important news magazine in Germany, presented the issue climate change on its cover and thus on its news agenda. The title "Die Klima-Katastrophe" ("The Climate Catastrophe") was illustrated by a photomontage that gave a first impression of environmental disasters to come. At that time, it showed the famous Cologne Cathedral in the center of the cover, partly drowned in the river Rhine, while the river had flooded the whole territory (Figure 1).

Figure 1: Climate change as national climate catastrophe: The Spiegel Cover „Die Klimakatastrophe“ (The Climate Catastrophe) from 1986 refers to a well known symbol for the German nation and culture in West Germany before the reunification, the Cathedral in Cologne.

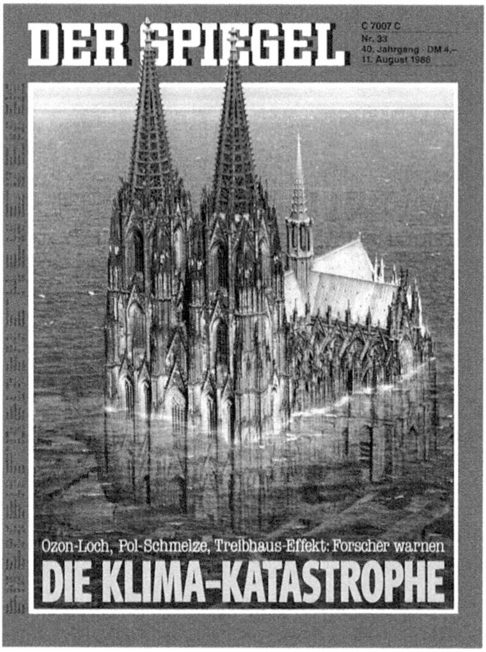

Weingart, Engels and Pansegrau described the picture as a nightmare vision.[1] However, the meaning is more sophisticated. Until the fall of the Berlin Wall in 1989 and the reunification of East and West Germany in 1990, Cologne Cathedral was one of the most famous national landmarks in West Germany;[2] it has been a national symbol since the 19th century. At the same time, it reflects the country's Christian history. Furthermore, "Father Rhine" has been at the very heart of many legends and myths for centuries.

Thus, the picture is a powerful visualization of potential climate change impacts on German society, drawing on an iconography that is well-known in

1 Weingart, Engels and Pansegrau, *Von der Hypothese zur Katastrophe,* 2008, 23.

2 Nipperdey, "Der Kölner Dom als Nationaldenkmal," 1981, 595.

Germany. The symbolic meaning is obvious: climate change can cause the fall of the German nation, and subsequently, the Christian Western civilization and its culture. However, the *Spiegel* did not have to invent the motif to visualize the projected impacts of climate change. Periods of flood and drought along the Rhine are part of the history of the river. When the *Spiegel* covered another flood just three years earlier in an article "Alle 1000 Jahre" ("Every 1000 Years") on May 2, 1983, they selected a similar motif. The Spiegel cover from 1986 is a good example of media attempting to influence public discourse by providing meaningful socio-cultural visuals.

Since then, using this kind of image has become a familiar way of visualizing climate change. Furthermore, climate change has become one of the top issues for the German public. The media have played no small role in that. As climate scientists Hans von Storch and Werner Krauss pointed out, the Spiegel cover "[...] became an iconic image for the German attitude towards climate change [...]" and was used as an argument in the debate on climate catastrophe.[3] According to Libby Lester and Simon Cottle it "[...] is hard to imagine, that climate change could have become the perceived 'global crisis' that it has in recent years without the help of media images symbolizing the harmful impacts of climate change on people, communities, and environments around the world"[4]. Visuals have played a crucial role in raising public awareness of the problem.

Within the last decade, visuals have received more and more attention in communication research. Recent studies on the visualization of climate change in the British[5] and Canadian press[6], national and/or international TV coverage[7] found a strong emphasis on impacts of climate change. In her analysis of Swedish media, Höijer[8] found that the images draw on emotions. Research on

3 von Storch and Krauss, "Culture Contributes to Perceptions of Climate Change," 2005.

4 Lester and Cottle, "Visualizing Climate Change," 2009, 921.

5 Smith and Joffe, "Climate Change in the British Press," 2009.

6 DiFrancesco and Young, "Seeing Climate Change," 2010.

7 Doyle, *Mediating Climate Change,* 2011; Lester and Cottle, "Visualizing Climate Change," 2009; Thiele, "Von Eisbären, dem Patient Erde, Weltrettungs- und Weltuntergangsszenarien," 2011.

8 Höijer, "Emotional Anchoring and Objectification in the Media Reporting of Climate Change," 2010.

advertising and picture archives[9], as well as campaigns by NGOs[10] and the IPCC[11] has revealed that these pictures also promote broader ideas. They focus in particular on the vulnerability of nature and people[12] or on the risk and threat to humanity, nature and planet Earth[13]. Similar to the *Spiegel* picture of the Rhine flooding Cologne Cathedral, the visual iconography of climate change picks up well-known cultural icons and iconography rendering the invisible visible. Furthermore, they raise questions of neocolonialism and global justice.[14]

This article contributes to the academic discussions on the growing importance of visuals in media and in public climate change discourse. It examines the visual representation of climate change within media discourse during and around the United Nations Climate Change Conference in Copenhagen in 2009 (COP 15). With reference to iconography, aesthetics and visual framing approaches, the main question is how climate change is constructed visually in media coverage.

The research perspective is two-fold: On one hand, the focus is on how visuals uncover and provide meaning at the level of knowledge and aesthetics. On the other, the question is how visuals are used in media discourse on climate change and what function they have in this discourse. Thus, the first question is how visuals frame the discourse of climate change iconographically and aesthetically. Second, it will be asked how media discourse on climate change frames visuals.

To provide answers to these questions, the visual coverage in German print media has been analyzed. This case study has been chosen for several reasons: first, the media are highly event-driven,[15] thus media coverage offers a good case study on how the topic is communicated visually. Second, the summit received significant media attention before and during the event[16] because it expected a

9 Hansen and Machin, "Visually Branding the Environment," 2008; Linder, "Cashing-in on Risk Claims," 2006.

10 Doyle, "Picturing the Clima(c)tic," 2007; Doyle, *Mediating Climate Change,* 2011; Manzo, "Imaging Vulnerability," 2010; Manzo, "Beyond the Polar-bear," 2010.

11 Doyle, *Mediating Climate Change,* 2011.

12 Manzo, "Imaging Vulnerability," 2010.

13 Lester and Cottle, "Visualizing Climate Change," 2009; Smith and Joffe, "Climate Change in the British Press," 2009.

14 Eide, "Visualizing a Global Crisis," 2012.

15 Eide and Kunelius, "Domesticating Global Moments," 2010.

16 See: Arlt and Wolling, "Die Presseberichterstattung über die Weltklimakonferenz in Kopenhagen," 2012; Eide, Kunelius and Kumpu, *Global Climate,* 2010.

new mandatory international agreement to succeed the Kyoto protocol that was to expire at the end of 2012. Finally, the German case is of international interest, since there is a broader consensus of science, politics and the public on the anthropogenic influences of climate change.

Furthermore, the following arguments contribute to the importance of the German case: there is no other large economy that has integrated green technology into its structure to the extent that Germany has. Due to the awareness of climate change, green ideas and the Green party have become stable factors in political ideas and institutions. Even Angela Merkel of the ruling conservative party relied on an image as "climate chancellor," when she entered office. That image successfully made its way into the public sphere. And the German industry, the automobile industry in particular, is one of the most influential lobbyists against too much regulation among the members of the European Union. Thus, the visual coverage on climate change and the COP 15 event was of particular interest concerning if and how the media covered the issue at this important moment.

VISUALIZING CLIMATE CHANGE: HOW RISK BECOMES MEANINGFUL REALITY

From a scientific point of view, climate change is a subject of risk communication. According to sociologist Ulrich Beck, "risk is not the same as catastrophe, but the anticipation of the future catastrophe in the presence. As a result, risk leads a dubious, insidious, would-be, fictitious, allusive existence: it is existent *and* non-existent, present *and* absent, doubtful and real."[17] The problem definition, global warming, and the projection of the future impacts of climate change rely on scientific data and models.[18] Due to the many factors that influence climate and climate change, these scientific results are always subject to uncertainties. In spite of this, the report by the Intergovernmental Panel on Climate Change in 2007 included a chapter on "observed climate change" where, in addition to projected impacts (IPCC 2007),[19] the uncertainties are also

17 Beck, "Critical Theory of World Risk Society," 2009, 3.

18 See: Gramelsberger and Feichter, "Modelling the Climate System," 2011.

19 According to Beck, geopolitical lines of conflict can be observed between different risk perception cultures, e.g. in Europe and the United States, Beck, "Critical Theory of World Risk Society," 2009, 10. Also representations of climate change in the media as a highly social and cultural constructed product differs significantly between

outlined. The tension between observation and projection, knowledge and uncertainty is one of the key challenges for climate change research, politics and public communication.

Climate change has become a public issue, analyzed by scientists, discussed by politicians on different local, national and international levels, claimed by NGOs, supported by celebrities and contested or used as marketing tool by industrial branches and economy. It has become a subject of social and political negotiation, one that the media have put on the public agenda.

Since the media are a main source of public knowledge about climate change,[20] how they cover the issue is important. Politicians in particular need public support to legitimize their decisions, measures of mitigation or adaption. Lester and Cottle argue that "Television visuals can play an important role in the development of global environmental awareness and contribute to a sense of ecological citizenship and associated rights and responsibilities […]."[21] This seems also to be true for visual coverage in print news media. Pictures are perceived as reliable sources in media coverage. Credibility is a key value of journalism in society. This credibility depends on the professional ideology of objective, truthful coverage. In the process of news coverage pictures play a crucial role. They are used to authenticate[22] and to promote environmental communication in media discourses. The visible still serves as "evidence of truth"[23] in media coverage. Katy Parry points out that news pictures are characterized by an aesthetic of realism. So it is not only science "[…] that places vision and sight as the primary sense for achieving objectivity and rationality […]", but journalism too.[24]

Journalism does not only refer to scientific results and scientific visual constructions and genres such as the 'hockey stick' graph[25] to visualize climate change as a problem. The key medium in news coverage and reportage is photography, the documentary image. As Stuart Hall points out "News photos

---

different nations, cf. Brossard, Shanahan, and McComas, "Are Issue-Cycles Culturally Constructed?" 2004.

20 See: Taddicken and Neverla, "Klimawandel aus Sicht der Mediennutzer," 2011.

21 Lester and Cottle, "Visualizing Climate Change," 2009, 922.

22 Doyle, *Mediating Climate Change,* 2011.

23 Doyle, *Mediating Climate Change,* 2011, 4.

24 Parry, "A Visual Framing Analysis of British Press Photography During the 2006 Israel-Lebanon Conflict," 2010, 1188.

25 Schneider, "Die Kurve als Evidenzerzeuger des Klimatischen Wandels am Beispiel des 'Hockey-Stick-Graphen'," 2009; see article by Lynda Walsh in this book.

have a specific way of passing themselves off as aspects of 'nature' [...],"[26] Thus, journalism's claim to truth and authenticity is still a core value in news photography and journalistic practice.[27] Within the photographic 'truth claim,'[28] an essential and ontological status is ascribed to pictures. This ascribed relation to reality makes photography a powerful medium in public discourse.

However, this is the crux of the matter: as a subject of risk communication,[29] the impact of climate change cannot be determined with absolute certainty. This situation has proven to be a challenge for newsroom editors.[30] While covering events and breaking news, journalism prefers the "here and now,"[31] yet while covering climate change, editors face uncertainties: Is there a second source to emphasize climate change or contradict? How can we reduce the complexity of climate change and the lack of reason-consequence schemes? Does a single event prove the existence or non-existence of climate change?

Whereas the media rely on the results of scientific research, they also have to amplify the invisible abstract concept of climate change with meaningful pictures.

As with texts, visual coverage is not only a true depiction or mirror of the world, but a social and cultural construction,[32] Pictures convey deeper meanings, positions or *Weltanschauungen* ('world views'[33]). Thus, they make the abstract topic of climate change meaningful.

According to visual semiotics and iconography approaches, pictures not only have a denotative, but also a connotative and ideological level of meaning.[34] In the news, pictures are used to make "real" events and developments visible while connoting meaning. This meaning is not only conveyed by the motif as a subject of knowledge. The use of extraordinary viewpoints and perspectives,

---

26 Hall, "The Determinations of News Photographs," 1973, 188.

27 See: Broersma, "The Unbearable Limitations of Journalism," 2010.

28 Gunning, "What's the Point of an Index," 2004.

29 Beck, "Critical Theory of World Risk Society," 2009; Bødker and Neverla, "Introduction," 2012.

30 Boykoff and Boykoff, "Climate Change and Journalistic Norms," 2007; Maurer, "Wie Journalisten mit Ungewissheit Umgehen," 2011.

31 Doyle, "Picturing the Clima(c)tic," 2007, 129.

32 Barthes, "The Photographic Message," 1977.

33 Panofsky, "Iconography and Iconology," 1970.

34 Leeuwen, "Semiotics and Iconography," 2001; Grittmann and Ammann, "Die Methode der Quantitativen Bildtypenanalyse," 2009.

light and colours and other visual techniques, the aesthetic quality can also affect the viewers' perception.

As Chaseten Remillard argues,[35] two prominent discourses with long-standing traditions shape the representations of environmental risk: on one hand, 'nature-as-resource", rooted in the ideology of instrumentalism, the Christian and colonial ethics of domination, and on the other, nature as a space separated from mankind, "nature-as-sublime". The sublime is the idea of the pristine, the incommensurable nature "beyond human comprehension."[36] According to Emily Brady, researcher in philosophy and geography, the concept of the sublime has gained new importance in debates on environmental aesthetics.[37] Referring to Kant, the sublime can be separated into the "'mathematically sublime,' in which the senses and imagination are overwhelmed when confronted with the seemingly infinite magnitude of nature" and the 'dynamically sublime,' in which the awesome power of nature [...] evokes anxious pleasure."[38] She points out that, facing catastrophic changes from climate change, the models of prediction provide a picture of "a great, sublime event."[39] Brady doubts that the concept can be applied to art and thus to pictures.[40] However, following Remillard's and Brady's own argument against sublimity in pictures, it will be argued here, that pictures of the pristine wilderness of nature and vast, overwhelming landscapes indirectly represent the idea of the sublime.

However, the independent meaning of the visual imagery based on its iconography and aesthetics is only one side of the communication process in media coverage. Originating from different traditions and contexts, the pictures are framed verbally at a further level of meaning by the media: the framing of the discourse.

FRAMING VISUALS: PICTURES WITHIN MEDIA DISCOURSE

Framing is one of the most popular concepts in current political communication and science communication research. In recent years, the framing concept has

35 Remillard, "Picturing Environmental Risk," 2011.

36 Ibid., 132.

37 Brady, "The Environmental Sublime," 2012, 171.

38 Ibid., 172.

39 Ibid., 174.

40 Ibid., 172.

been discussed and adopted in visual communication research.[41] The approach of iconography has been integrated and transformed by the visual framing approach. Rodriguez and Dimitrova have defined three levels of visual framing: the denotative, connotative and aesthetic dimensions of pictures.[42] However, what seems to be more important in this context is that visuals are also framed by integrating them into a specific discourse. Thus, it is not only the visual that frames the topic by making some particular aspects more salient than others, by referring to a specific iconography and specific aesthetics, the visual is also framed by a further level of meaning.[43] According to Entman this level of framing can be defined as follows:

"To frame is to *select some aspects of a perceived reality and make them more salient in a communicating text, in such a way as to promote a particular problem definition, causal interpretation, moral evaluation, and/or treatment recommendation,* for the item described. [...] Frames, then, *define problems—determine* what a causal agent is doing with what costs and benefits, usually measured in terms of common cultural values; *diagnose* causes—identify the forces creating the problem; *make moral judgments—evaluate* causal agents and their effects; and *suggest remedies—offer* and justify treatments for the problems and predict their likely effects."[44]

Frames are employed in political and public debate on contested issues. They refer to the public argumentation articulated and cited by different public speakers (e.g. politicians, NGOs, scientists) in media coverage or by the journalists themselves.

Similar to the rhetorical concept of stases, the four frame elements—problem definition, causal interpretation, moral evaluation, and/or treatment recommenddation—constitute "a taxonomy of arguments"[45] within the public debate. Pictures serve as single frame elements in this context by focusing on problem

41 Coleman, "Framing the Pictures in our Heads," 2010; Griffin, "Picturing America's "War on Terrorism" in Afghanistan and Iraq," 2004; Fahmy and Kim, "Picturing the Iraq War," 2008; Parry, "A Visual Framing Analysis of British Press Photography During the 2006 Israel-Lebanon Conflict," 2010; Rodriguez and Dimitrova, "Levels of Visual Framing," 2011; Schwalbe, "Remembering our Shared Past," 2006.

42 Rodriguez and Dimitrova, "Levels of Visual Framing," 2011.

43 Parry, "A Visual Framing Analysis of British Press Photography During the 2006 Israel-Lebanon Conflict," 2010.

44 Entman, "Framing," 1993, 52.

45 Fahnestock and Secor, "The Stases in Scientific and Literary Argument," 1988, 428.

definition *or* causes *or* solutions/treatment recommendation. The interpretation of pictures of the polar bear on an ice shelf as "impact" does not derive from the iconography of the picture, but rather from the applied frame in the media discourse. While these three frame aspects are made visible by motifs, moral evaluation is expressed by visual aesthetics such as the sublime or beautiful views or dark colours, sharp contrasts or emotional expressions (e.g. people mourning or smiling).

However, climate change as a topic is only one level of news coverage. As climate change has become a key subject of political action, the media focus on events, on politics and climate-change policies. They also cover the process of debate. So we can differentiate between (visual) coverage of the topic (climate change as issue of public debate), that can be framed, and coverage of politics (process of policy-making, discussion and public debate).

METHOD AND MATERIAL

The case study presented is based on a qualitative analysis of visual coverage on climate change in the German press from November 15 to December 23, 2009, the period leading up to and including the United Nations Climate Change Conference in Copenhagen in 2009.[46] The research sample consisted of: three nationwide German quality newspapers *die tageszeitung (taz), Süddeutsche Zeitung (SZ)* and *Die Welt,* and the biggest tabloid in Germany, *BILD*. Two weekly news magazines were also included, *Spiegel* and *Stern*, a weekly newspaper: *Die Zeit* and the most popular monthly reportage magazine, *GEO*. Most of them play an important role in intermedia agenda-setting and public awareness. Each of the daily newspapers represents a political guideline (green, labour, conservative) or different genres of journalism (elite, popular). Together, they amount to a broad source for exploring climate change coverage. Units of analysis were all visuals published in articles relating to climate change and/or the Copenhagen climate change summit in headline, lead or legend.

The methodology framework employed was a qualitative image type analysis[47], based on the iconographical approach.[48] This method is useful for

46 The dailies and news magazines mainly reported before and during the summit, the main coverage in magazines was at the end of November and in December. The topic disappeared a few days after the summit.

47 Grittmann and Ammann, "Die Methode der Quantitativen Bildtypenanalyse," 2009.

classifying motifs regarding image types, analyzing their meaning (iconographic analysis) and interpreting inherent cultural values and ideas (iconological interpretation). All photos, diagrams and film stills were classified as specific image types according to their motifs. This process permits conclusions about patterns of journalistic selection. Then a framing analysis, based on Entman's framing approach, was conducted. The frame applied to the picture by the text was analyzed. Thus, it was operated on the picture and the contextual level.

RESULTS: ICONOGRAPHY AND FRAMING VISUALS OF CLIMATE CHANGE

During the period analyzed, climate change was covered broadly in the German press. All media outlets reported on it. Even the reportage magazine *Geo* covered climate change visually, although the magazine does not refer directly to current events. While the summit provided the trigger for covering climate change, most forms of media paid more attention to the issue than to the event itself. They used the event to provide background information on climate change and stories from all parts of the world, hence becoming active agents.

The analysis differentiated between pictures of politics and policy. Pictures of politics focus on the process and the actors of political participation, articulation and decision-making. They include photos of actors and events of the summit. Pictures of policy focus on the topic of political discourse, they cover climate change as issues. The media provide the audience with specific narratives and arguments in political discourses. Pictures of policy serve as frame elements in this argumentative narration.

1. Actors in the Public Discourse on Climate Change

The coverage of the summit draws on a well-known standardized iconography of coverage of international political meetings,[49] with moments of discussion, speeches, and individual portraits of main political actors being the main motifs of the summit, while pictures of demonstrations and symbolic protests by citizens and NGOs represent people's activity. Thus, actors involved in defining and solving the problems are pictured in portraits or certain situations. Among

48 The iconographical approach has been elaborated by art historian Erwin Panofsky (1970a,1970b; cf. Hatt and Klonk 2006).

49 Grittmann, *Das politische Bild*, 2007.

them are scientists, politicians and political leaders, lobbyists and representatives of NGOs, and celebrities. Polluters do not occur as partners in these discussions, just as they were not participants in the discourse.

The visual coverage focuses on the powerful and the elite. In addition, it tends to be domesticated, that means the media focus especially on German political actors.[50] Besides the German Chancellor Angela Merkel and the then Environment Minister, Norbert Röttgen, the political leaders of the USA and China are prominent figures as well. The representatives of the most powerful nations appear as key actors of the meeting. Furthermore, German scientists were interviewed and are depicted as experts, while industrial actors are left out. Thus, the industrialized nations are presented as the leading actors, responsible for finding a solution.

2. Climate Change as (an) Issue

In the visual coverage of climate change, four basic frame perspectives can be identified according to the frame concept: problem definition, causes, treatment recommendation (solution) and moral evaluation. As mentioned above, these frame elements are applied to different motifs. Moral evaluation is evoked by visual aesthetics, for example, the use of light, atmosphere, or sharp contrasts.

In the following, the iconography of climate change will be analyzed according to the three frame elements that refer to the motifs. The fourth frame element—moral evaluation—is not exclusive, but visible simultaneously as a constitutive part of visual aesthetics.

Problem Definition

The dominant frame applied to the visuals in the media discourse on climate change is problem definition. Thus, the media emphasize climate change as a matter of enormous urgency. Diagrams, tables and graphs provide scientific evidence for the existence of climate change. However, they rarely tend to define the problem, global warming and its impacts and consequences, in the discourse. According to Emily Brady, these models of "[...] predicting changes on a huge scale provide a kind of indirect, representational picture of a great, sublime event on a huge temporal and spatial scale."[51]

50 See: Hahn and Herrmann, "The Camouflage of Green Businesses," 2010. ; Roosvall, "Sweden," 2010.

51 Brady, "The Environmental Sublime," 2012, 174.

Catastrophes and Threats

The most obvious consequences of climate change, ones which were proved early on, are the warming of the oceans and—as a consequence—the Arctic melt.[52] Pictures of polar bears on retreating ice sheets have been used extensively by environmental groups to symbolize climate change.[53] They have also become a visual cliché in media coverage on the issue as a symbol for the endangered natural world. However, they were published only a few times in the coverage in 2009, and though melting ice is a still popular motif, new results from the IPCC seem to influence the iconography, too. In 2007, the IPCC stated that " [...] for the first time wide-ranging impacts of change in current climate had been documented [...]"[54]: retreating glaciers became a new popular motif, for example. According to Manzo, these pictures can be interpreted as symbols of vulnerability.[55]

Furthermore, photographs of apparently natural disasters like flooding, hurricanes and drought are included to visualize potential impacts of climate change (Figure 2). By revealing an unbound nature, these pictures evoke the idea of the sublime, of an incommensurable nature.[56] The pictures selected show moments of disasters threatening all continents including Germany. Thus, climate change is constructed as a global problem.

Affected People

Personalized photo stories, like the one in *Stern* magazine, focus on the consequences of climate change. It shows a farmer, cowboys, peasants, a fisherman and other people from selected regions of the world in the center of the picture (Figure 3). While pictures of affected people have been interpreted as "human impact" visuals[57] with a dramatic call for compassion and empathy, many reportage photos focus more on vulnerability. They often refrain from emotional scenes and gestures and depict scenes of daily life. The cognitive function of news photography, then, seems to be more important than the emotional.

52 See: Doyle, "Picturing the Clima(c)tic," 2007.
53 Doyle, "Picturing the Clima(c)tic," 2007.
54 IPCC Working Group II Report, 2007, 14.
55 Manzo, "Imaging Vulnerability," 2010.
56 See: Brady, "The Environmental Sublime," 2012.
57 Lester and Cottle, "Visualizing Climate Change," 2009, 929.

Figure 2: A dust storm, a so called Haboob, is moving toward Khartoum in Sudan. The city seems to be hit in a few seconds. Showing a disaster to come, the photo is framed in a double sense. The single extreme weather event is also discussed as harbinger of disasters caused by climate change.

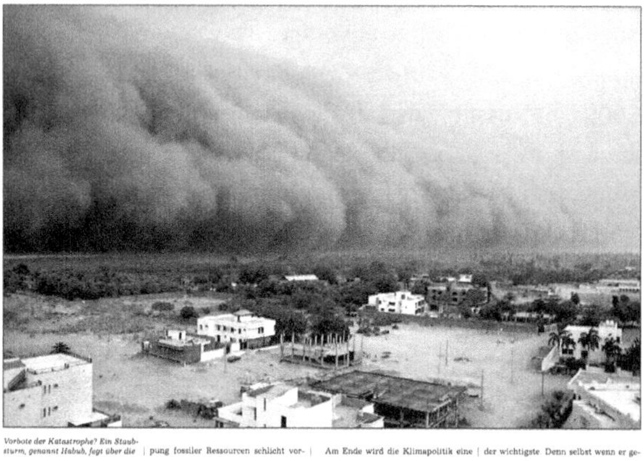

Figure 3: Pride instead of pity: The photo (detail) of a cowboy with his son in Mali portraits one of the people who became affected by climate change in their everyday life. The photo is part of a larger project by Monika Fischer and Mathias Braschler. It has been published among others from the series in the magazine Stern.

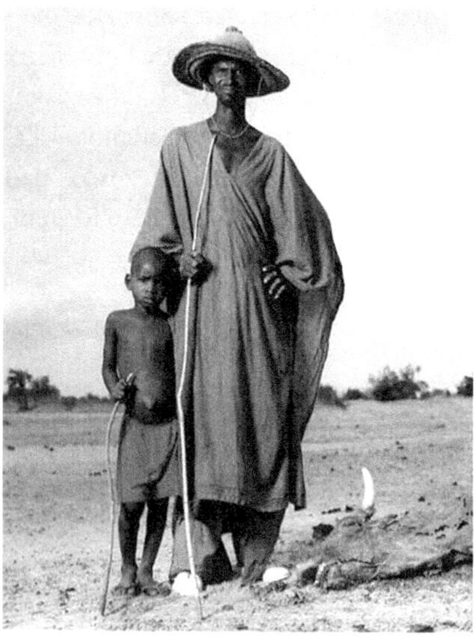

Figure 4: Nature under threat is a common frame in climate change discourse. The Geo magazine applied it on the spectacular photographs by Bernhard Edmaier. The former geologist documents the endless vastness of untouched landscapes in his aerial photography. This photo shows mountains in East Greenland. (→ color plate 8)

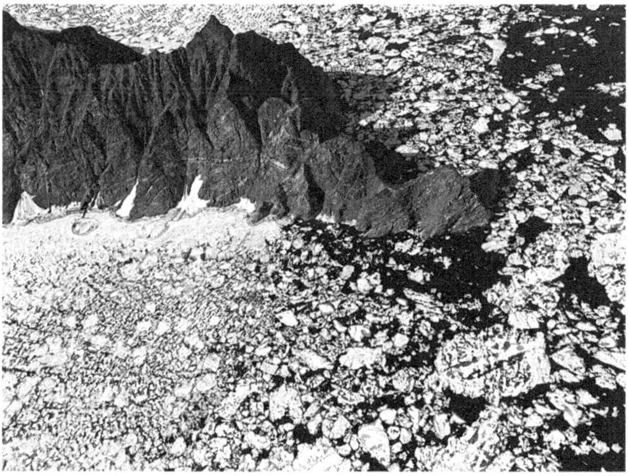

Endangered Nature

Another perspective focuses on the beauty and sublimity of nature and the untouched grace of a pristine landscape (Figure 4). This form of adoration or even worship can be traced back to the era of Enlightenment.[58] As Palmer points out, "[...] nature under threat is the environmental variant of the sublime image of disaster, with all its attendant problems."[59] Images of the landscape are abundant, spectacular ways to characterize nature's power. They are vivid due to the structure and unlimited size of frame and the long shot. They emphasize nature's beauty by careful choice of light and colour. And who would not want to protect this overwhelming pristine wilderness and beauty?[60]

Affected people or endangered nature connote vulnerability, according to Manzo[61]. Vulnerability has become a key term in climate change discourse. As Manzo already mentioned "[...] even without textual referents, connotations of danger and vulnerability can be deduced from visual signifiers of exposure to

58 See: Remillard, "Picturing Environmental Risk," 2011.

59 Palmer, "Photography, Technology, and Ecological Criticism," 2013, 80.

60 See article by Edward Morris and Susannah Sayler in this book.

61 Manzo, "Imaging Vulnerability," 2010.

hazard [...]."[62] According to the IPCC,[63] Africa is the continent most vulnerable to climate variability. And once more, the media seem to follow these statements visually.

Also, the iconic image of the "blue planet" is a popular motif in the media, used to symbolize global crises and to evoke the idea of one world under threat.[64]

POLLUTERS AND CAUSES

The so-called greenhouse gases, especially carbon dioxide, but also methane and nitrous oxide exist independently of any anthropogenic influence. However, their increasing share in the air is caused by the burning of fossil fuels, by transportation and industrial processes, and deforestation and agriculture.[65] Visual coverage follows this argumentation. Motifs of energy production and power generation, traffic, destroyed forests (Figure 5) and cattle breeding are framed as causes of climate change.

Figure 5: The destroyed rainforest in Borneo provides an example for one of the main anthropogenic causes that contributes to climate change. The dark colors and the shot—the sky is almost invisible—support the critical perspective.

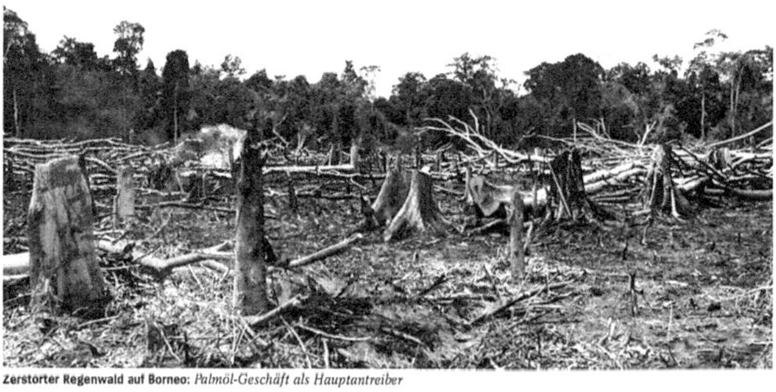

62 Manzo, "Imaging Vulnerability," 2010, 97.

63 IPCC 2001 / 2007, 13.

64 See: Jasanoff, "Image and Imagination," 2000; Manzo, "Imaging Vulnerability," 2010; Lester and Cottle, "Visualizing Climate Change," 2009; Smith and Joffe, "Climate Change in the British Press," 2009.

65 See: IPCC Working Group II Report, 2007.

Power Generation

One standard motif is power plants emitting steam or water vapor, pictured in an overview, with the light environment adding a threatening momentum to the scene. However, depicting a causal agent such as in the *Spiegel* picture is an exception. Usually, power plants are shown as part of the whole; they symbolize the energy industry in western industrialized nations, sometimes pointing the finger at a single polluter.

Transportation

Long shot photos of traffic jams are typical motifs that identify the polluter. The pictures are mostly localized in the USA or in Asian megacities.

Energy Consumption

The long shot on bright lights of big highly industrialized cities is the third conventionalized motif. It represents the high consumption of energy.

Industrial Animal Breeding and Food Production

Cattle breeding is said to be responsible for climate change, to some extent. Cattle emit methane, their numbers represent the rising demand for meat in industrialized countries. Here, the news media often show pictures of cattle herds or industrial meat production (often in China). These pictures are predominantly taken from an overview to emphasize their industrial character.

By choosing these motifs and techniques, causes for environmental change are attributed to industrialized nations and newly industrialized countries, particularly to the USA and China, and their high demand for energy and nourishment. The visualization of causes remains superficial and anonymous—merely by choosing a long shot. Very rarely are companies or individuals required to take responsibility.

SOLUTIONS

In science and politics, adaptation and mitigation have been explored and already applied as key strategies to avoid further impacts of climate change. The causes are ascribed to industrialized countries and so are the solutions.

Figure 6: Photos of renewable energy technologies are used to make solutions visible. The picture of the Wind wheels (Denmark) in a calm sea and under a friendly blue sky evokes a vision of harmony with nature and mankind.

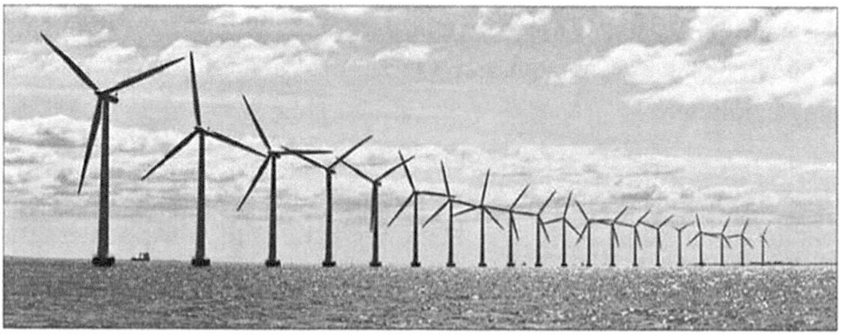

Der Windmühlenpark Middelgrunden in Dänemark ist eine der größten Offshore-Anlagen der Welt. Das Land gewinnt rund 22 Prozent seiner Energie aus Windkraft. Foto: ddp

New Technologies

The visualization of alternative energy production is the main emphasis. Wind wheels in front of the sun or photovoltaics devices are now standard motifs or even stereotypes. They fit in the landscape like natural elements. They are pictured in atmospheric settings with the sun rising or blue skies, etc. (Figure 6). Also, alternative technology is embedded in a pristine and overwhelming landscape. Beauty and the overwhelming, vastness of nature are also evoked in these pictures. In contrast to the polluters, solutions for climate change are linked to single companies, particularly in Germany which allegedly offer help. These pictures literally radiate a high-tech image like the one of a solar technology enterprise in Berlin.[66]

Solutions for Everyday Life

Another visual focus is on domestic energy savers, such as houses with ecological construction or insulation. These motifs visualize forms of adaptation, while forms of mitigation are only presented in photo reportage, such as "Plan B" in *Geo Magazine*.[67]

---

66 *Spiegel*, July 12, 2009, 55.

67 *Geo Magazine*, 2009, 84–99.

SUMMARY

The pictures analyzed in this article refer to a well-known iconography of visual environmental communication. Thus, they contribute to the textual discourse on the subject. Although they seem to be mere depictions of reality or analytical tools of scientific evidence, they also convey deeper meanings.

Pictures not only contribute their own meaning in the media discourse on climate change, they are simultaneously framed by media discourse and used as rhetorical arguments. Visual coverage of climate change does provide a wide variety of meaningful pictures of sublime and beautiful nature, scientific evidence, of modern technology and third world problems. Visual communication of climate change in media discourse is directed by the attribution of problem definition: causes and solutions for climate change defined by scientists, politics and NGOs. The focus on problem definition is reduced to scientific visual evidence on one hand and the iconography of unbound sublime nature on the other: disasters like flooding, hurricanes and drought. However, individual destinies are also shown to make mankind's vulnerability visible. In contrast, references to a longstanding visual tradition of pristine nature are obvious in pictures of sublime landscapes. The beauty of nature is also constructed by photographic techniques. The implicit slogan is: this beauty is endangered but worth saving. Causes are evidenced as pictures of power plants, traffic, deforestation and meat production: the standardized motifs. The industrialized natural world and nature as resource are evaluated critically by visualization, in particular in the way the causes are depicted aesthetically. The reasons for climate change are attributed to industrialized nations or emerging nations such as China. Polluters remain global images largely anonymous.

In contrast to the motifs of endangered nature, affected people and threats, the solution focuses on western sustainable high technology. Thus, the iconography of climate change visual coverage includes a wide range of motifs from causes or consequences to technical solutions and mitigation, while only one story in *GEO* covers adaptation. The main actors in the discourse of climate change are scientists, politicians, lobbyists, celebrities and representatives of NGOs. The coverage is clearly focussed on elite and domesticated settings, while polluters are almost ignored in the visualization of political discussions.

Summarizing the visual strategies, there is the obvious goal of attracting public attention for the problem with its causes and solutions by providing German media coverage on an informational and emotional level. The claim for action is implicitly articulated in visual terms. There is a broad consensus in German media on both the anthropogenic causes of climate change and on the possibility of treating the risk and preventing it by developing and applying new technologies. The whole earth can be preserved. Nature and alternative technology are not contradictions, rather, they coexist in harmony: The sublime and the beautiful are here and now.

BIBLIOGRAPHY

Arlt, Dorothee, and Jens Wolling. "Die Presseberichterstattung über die Weltklimakonferenz in Kopenhagen. Normative Anforderungen und empirische Befunde." *Studies in Communication/Media* 1, no. 2 (2012): 283–297.

Barthes, Roland. "The Photographic Message." Translated by Stephen Heath. In *Image, Music, Text,* edited by Stephen Heath, 15–31. New York: Hill and Wang, 1977.

Beck, Ulrich. "Critical Theory of World Risk Society: A Cosmopolitan Vision." *Constellations* 16, no. 1 (2009): 3–22.

Brady, Emily. "The Environmental Sublime." In *The Sublime. From Antiquity to the Present,* edited by Timothy M. Costelloe, 171–82, Cambridge et al.: Cambridge University Press, 2012.

Bødker, and Neverla. "Introduction. Environmental Journalism." *Journalism Studies* 13, no. 2 (2012): 152–56.

Boykoff, and Boykoff. "Climate Change and Journalistic Norms. A Case Study of US Mass–media Coverage." *Geoforum* 38 (2007): 1190–1204.

Brönnimann, Stefan. "Picturing Climate Change." *Climate Research* 22 (2002): 87–95.

Brossard, Dominique, James Shanahan, and Katherine McComas. "Are Issue-Cycles Culturally Constructed? A Comparison of French and American Coverage of Global Climate Change." *Mass Communication & Society* 7(3) (2004): 359–77.

Broersma, Marcel. "The Unbearable Limitations of Journalism. On Press Critique and Journalism's Claim to Truth." *International Communication Gazette*, 72(1) (2010): 21–33.

Coleman, Renita. "Framing the Pictures in our Heads. Exploring the Framing and Agenda-Setting Effects of Visual Images." In *Doing News Framing Analysis*, edited by Paul D'Angelo, and Jim A. Kuypers, 233–61. New York: Francis Taylor, 2010.

DiFrancesco, Darrin Anne, and Nathan Young. "Seeing Climate Change: The Visual Construction of Global Warming in Canadian National Print Media." *Cultural Geographies* 18, no. 4 (2010): 517–36.

Doyle, Julie. "Picturing the Clima(c)tic: Greenpeace and the Representational Politics of Climate Change Communication." *Science as Culture* 16, no. 2 (2007): 129–50.

—. *Mediating Climate Change*. Farnham, Surrey: Ashgate, 2011.

Eide, Elisabeth. "Visualizing a Global Crisis. Constructing Climate, Future and Present." *Conflict & Communication Online* 11, no. 2 (2012): 1–16.

Eide, Elisabeth, and Risto Kunelius. "Domesticating Global Moments. A Transnational Study on the Coverage of the Bali and Copenhagen Climate Summits." In *Global Climate—Local Journalisms. A Transnational Study of How Media Make Sense of Climate Summits*, edited by Elisabeth Eide, Riso Kunelius, and Ville Kumpu, 11–50. Bochum/Freiburg: projectverlag, 2010.

Eide, Elisabeth, Risto Kunelius, and Ville Kumpu, editors. *Global Climate—Local Journalisms*. Bochum/ Freiburg: projectverlag, 2010.

Entman, Robert M. "Framing. Toward Clarification of a Fractured Paradigm." *Journal of Communication*, 43, no. 4 (1993): 51–58.

Fahmy, Shahira, and Daekyung Kim. "Picturing the Iraq War: Constructing the Image of War in the British and US Press." *International Communication Gazette* 70, no. 6 (2008): 443–462.

Fahnestock, Jeanne, and Marie Secor. "The Stases in Scientific and Literary Argument." *Written Communication* 5 (1988): 427–443.

Gramelsberger, Gabriele, and Johann Feichter. "Modelling the Climate System. An Overview." In *Climate Change and Policy. The Calculability of Climate Change and the Challenge of Uncertainty*, edited by Gabriele Gramelsberger and Johann Feichter, 9–90. Berlin: Springer, 2011.

Griffin, Michael. "Picturing America's 'War on Terrorism' in Afghanistan and Iraq. Photographic Motifs as News Frames." *Journalism* 5, no. 4 (2004): 381–402.

Grittmann, Elke. *Das politische Bild. Fotojournalismus und Pressefotografie in Theorie und Empirie*. Köln: Herbert von Halem Verlag, 2007.

Grittmann, Elke, and Ilona Ammann. "Die Methode der Quantitativen Bildtypenanalyse. Zur Routinisierung der Bildberichterstattung am Beispiel von 9/11 in der Journalistischen Erinnerungskultur." In *Visuelle Stereotype*, edited by Thomas Petersen and Christoph Schwender, 141–58. Köln: Herbert von Halem Verlag, 2009.

Gunning, Tom. "What's the Point of an Index? or, Faking Photographs." *Nordicom Review* 25, no. 1–2 (2004): 39–50.

Hahn, Oliver, and Daniel Herrmann. "The Camouflage of Green Businesses." In *Global Climate—Local Journalisms. A Transnational Study of How Media*

Make Sense of Climate Summits, edited by Elisabeth Eide, Riso Kunelius, and Ville Kumpu, 195–209, Bochum/Freiburg: projectverlag, 2010.

Hall, Stuart "The Determinations of News Photographs." In *The Manufacture of News. Social Problems, Deviance and the Mass Media,* edited by Stanley Cohen, and Jock Young, 176–90. London: Constable, 1973.

Hansen, Anders, and David Machin. "Visually Branding the Environment: Climate Change as a Marketing Opportunity." *Discourse Studies* 10, no. 6 (2008): 777–94.

Hatt, Michael, and Charlotte Klonk. *Art History: A Critical Introduction to its Methods.* Manchester / New York: Manchester University Press, 2006.

Höijer, Birgitta. "Emotional Anchoring and Objectification in the Media Reporting of Climate Change." *Public Understanding of Science* 19, no. 6 (2010): 717–31.

IPCC 2010. *Guidance Note for Lead Authors of the IPCC Fifth Assessment Report on Consistent Treatment of Uncertainties.*

IPCC Working Group II Report. *Climate Change 2007. Impact, Adaptation and Vulnerability*, edited by M.L. Parry et al.. Cambridge/New York: Cambridge University Press, 2007. Accessed 20 November 2013. http://www.-ipcc.ch/publications_and_data/ar4/wg2/en/contents.html.

IPCC Third Assessment Report. *Climate Change 2001—Synthesis Report*, edited by Robert Watson et al. Cambridge: Cambridge University Press, 2001. Accessed 20 November 2013. http://www.grida.no/publications/other-/ipcc_tar/.

Jasanoff, Sheila. "Image and Imagination. The Formation of Global Environmental Consciousness." In *Changing the Atmosphere. Expert Knowledge and Environmental Governance*, edited by Clark Miller and Paul Edwards, 309–37. New York: MIT Press, 2000.

King, Cynthia, and Paul Martin Lester. "Photographic Coverage during the Persian Gulf and Iraqi Wars in Three U.S. Newspapers." *Journalism & Mass Communication Quarterly* 82(3) (2005): 623–37.

Leeuwen, Theo van. "Semiotics and Iconography." In *Handbook of Visual Analysis*, edited by Theo van Leeuwen, and Carey Jewitt, 92–118. London/ Thousand Oaks/New Delhi: Sage, 2001.

Leiserowitz, Anthony. "Climatic Change Risk Perception and Policy Preferences. The Role of Affect, Imagery and Values." *Climatic Change* 77, no. 1–2 (2006): 45–72.

Lester, Libby, and Simon Cottle. "Visualizing Climate Change. Television News and Ecological Citizenship." *International Journal of Communication* 3 (2009): 920–36.

Linder, Stephen H. "Cashing-in on Risk Claims. On the For–profit Inversion of Signifiers for 'Global Warming'." *Social Semiotics* 16 (2006): 103–32.

Manzo, Kate. "Imaging Vulnerability. The Iconography of Climate Change." *Area* 42, no. 1 (2010): 96–107.

—. "Beyond the Polar-bear? Re-Envisioning Climate Change." *Meteorological Applications* 17, no. 2 (2010): 196–208.

Maurer, Marcus. "Wie Journalisten mit Ungewissheit Umgehen. Eine Untersuchung am Beispiel der Berichterstattung über die Folgen des Klimawandels." *Medien & Kommunikationswissenschaft* 59, no. 1 (2011): 60–74.

Neverla, Irene. "The IPCC-reports 1990–2007 in the Media. A Case-Study on the Dialectics Between Journalism and Natural Sciences." Paper presented in the panel "A global dialogue on climate change?" at the ICA–Conference "Global Communication and Social Change" Montreal, May 22–26, 2008.

Nipperdey, Thomas. "Der Kölner Dom als Nationaldenkmal." *Historische Zeitschrift* 233, no. 3 (1981): 595–613.

Palmer, Daniel. "Photography, Technology, and Ecological Criticism: Beyond the Sublime Image of Disaster." *Environmental Conflict and the Media*, edited by Libby Lester and Brett Hutchins, 75–92, New York et.al.: Peter Lang, 2013.

Panofsky, Erwin. "Introduction: The History of Art as a Humanistic Discipline." In *Meaning in the Visual Arts*, 23–50. Harmondsworth: Penguin Books, 1970.

—. "Iconography and Iconology: An Introduction to the Study of Renaissance Art." In *Meaning in the Visual Arts*, 51–81. Harmondsworth: Penguin Books, 1970.

Parry, Katy. "A Visual Framing Analysis of British Press Photography During the 2006 Israel–Lebanon Conflict." *Media, War & Conflict* 3, no. 1 (2010): 67–85.

Remillard. Chaseten. "Picturing Environmental Risk: The Canadian Oil Sands and the National Geographic." The *International Communication Gazette* 73, no. 1–2 (2011): 127–43.

Roosvall, Anna. "Sweden: Between Domestication and Glocalisation." In *Global Climate—Local Journalisms. A Transnational Study of How Media Make Sense of Climate Summits*, edited by Elisabeth Eide, Riso Kunelius, and Ville Kumpu, 309–23, Bochum/Freiburg: projectverlag, 2010.

Rodriguez, Lulu, and Daniela V. Dimitrova. "Levels of Visual Framing." *Journal of Visual Literacy*, 30, no. 1 (2011): 48–65.

Schäfer, Mike, Ana Ivanova, and Andreas Schmidt. "Globaler Klimawandel, Globale Öffentlichkeit? Medienaufmerksamkeit für den Klimawandel in 23

Ländern." *SCM* 1 (2011): 131–48. Accessed 20 November 2013. http://www.scm.nomos.de/fileadmin/scm/doc/SCM_11_01_05.pdf

Smith, N. W., and Helene Joffe. "Climate Change in the British Press. The Role of the Visual." *Journal of Risk Research* 12 (2009): 647–663.

Schneider, Birgit. "Die Kurve als Evidenzerzeuger des Klimatischen Wandels am Beispiel des 'Hockey-Stick-Graphen'." In *Sehnsucht nach Evidenz.* edited by Karin Harrasser et al., 41–55. Bielefeld: transcript, 2009.

Schwalbe, Carol B. "Remembering our Shared Past. Visually Framing the Iraq War on U.S. News Websites." *Journal of Computer-Mediated Communication* 12, no. 1 (2006), article 14. Accessed 20 November 2013. http://jcmc.indiana.edu/vol12/issue1/schwalbe.html.

Taddicken, Monika, and Irene Neverla. "Klimawandel aus Sicht der Mediennutzer. Multifaktorielles Wirkungsmodell der Medienerfahrung zur komplexen Wissensdomäne Klimawandel." *Medien & Kommunikationswissenschaft* 59, no. 4 (2011): 505–25.

Thiele, Matthias. "Von Eisbären, dem Patient Erde, Weltrettungs- und Weltuntergangsszenarien. Eine Interdiskurs- und Dispositivanalyse des Medienereignisses 'Klimagipfel Kopenhagen'." In *Blickregime und Dispositive Audiovisueller Medien*, edited by Nadja Elia-Borer, Samuel Sieber, and Georg Cristoph Thole, 267–94. Bielefeld: transcript, 2011.

Von Storch, Hans, and Werner Krauss. "Culture Contributes to Perceptions of Climate Change." *Niemans Report* 59, no. 4 (2005): 99–102.

Weingart, Peter, Anita Engels, and Petra Pansegrau. "Risks of Communication: Discourses on Climate Change in Science, Politics, and the Mass Media." *Public Understanding of Science* 9 (2000): 261–83.

—. *Von der Hypothese zur Katastrophe. Der Anthropogene Klimawandel im Diskurs Zwischen Wissenschaft, Politik und Massenmedien.* 2nd edition, Opladen: Budrich, 2008.

IMAGE COPYRIGHTS & SOURCES

Figure 1: Die Klima-Katastrophe, Spiegel, August 11, 1986, Cover, photomontage.

Figure 2: Das größte Rad der Welt, Süddeutsche Zeitung, May 12, 2009, 8. Photo: AFP; Text: Michael Bauchmüller.

Figure 3: Das Gesicht des Klimawandels, Stern 51/2009, 124. Photos: Mathias Braschler and Monika Fischer.

Figure 4: Grönlands Wilder Osten, Geo 1/2010, 30–44. Photos: Bernhard Edmaier; Text: Angelika Jung-Hüttl and Bernhard Albrecht, © Bernhard Edmaier.

Figure 5: Das Zwei-Grad-Leben, Spiegel 49/2009, 60. Photo: Reuters, Text: Beste, Ralf, et. al., © Reuters.

Figure 6: Königlicher Biosprit, Süddeutsche Zeitung, November 12, 2009, 9. Photo: ddp; Text: Gerhard Fischer.

Twist and Shout: Images and Graphs in Skeptical Climate Media

BIRGIT SCHNEIDER, THOMAS NOCKE, GEORG FEULNER

In this paper, we present an overview of strategies that use graphics and photographs in skeptical climate media. After showing how photographs are used as emotional teasers or as a way of maligning certain scientists, we give examples of the most widespread figures used in a method called "cherry picking". This describes the strategy of focusing on a small detail among findings while at the same time blocking out the larger context in order to contradict the consensus that anthropogenic emissions are causing global warming (e.g., "But some glaciers are growing!" instead of "Most glaciers around the world are melting."). The use of such misleading graphs, mainly in the form of time series curves, is more widespread than that of deliberately faked graphs. Given the complex nature of climate science, misleading graphs can easily confuse people not involved in climate research, while the publishing conditions of the web are actually helping the skeptics. Examples for the climate 'skeptical' use of images and graphs are discussed in this paper from an interdisciplinary view based on climate science, visual studies and computer graphics. At the same time the example illustrates the precarious distinction of science (epistemology / facts) and politics (values) in modern risk societies, where political decisions are sought to be legitimated in technical terms only.

INTRODUCTION: CLIMATE "SKEPTICS"AGAINST GLOBAL "WARMISTS"

In this article, we focus on images used in climate skeptical media, raising the question of whether there is anything specific about the use of images by climate skeptical media, and if yes, what? We attempt to answer a number of questions: how are illustrations chosen? What function do they play in climate skeptical argumentation? In what ways are photographs deployed? Are scientific graphs predominantly reproduced one-to-one from scientific sources that rely on peer-reviewed science, or are they modified by the climate skeptical actors? In what ways are they typically modified? How many cases exist where graphs were intentionally tampered with? And finally: are certain image types used in preference to others to highlight the arguments of climate change skeptics?

We should clearly state our understanding of the broad term "climate skepticism"[1] at the outset, to avoid any misunderstanding. We will analyze the images of such climate skeptical media published by actors who *strategically* cast doubt on the findings of peer-reviewed, mainstream climate science. This has been the case since climate modeler James E. Hansen, then director of the Goddard Institute for Space Studies, announced in 1988 that anthropogenic global warming was observable; the Intergovernmental Panel on Climate Change (IPCC) was created in the same year. Since then, organized campaigns of denial have been undertaken to systematically obfuscate the causes and effects of global warming and by this, oppose climate science, the most effective campaigns being started in the United States.[2] The strategy of these campaigns is based on the aim of casting doubt on global warming among the general public by deliberately highlighting opposing minority views and uncertainties to give the impression that "the science is not settled." It is the term "cherry picking" in the context of climate skepticism that stands for isolating pieces of evidence while rejecting all other data that do not fit the desired picture.[3] Often, it employs not only the strategy of misrepresenting and distorting scientific research, but also attacking climate researchers personally; thus deliberately discrediting the scientific work of climate researchers.

Most of the protagonists and institutions we look at are not involved in climate research, they do not actively practice climate science and do not take part in scientific reviewing processes in order to put their arguments up for scientific discussion. The organizations they work for are partly funded by the energy

1 For an overview on different typologies that have been developed to differentiate the field of climate skeptics see Painter, *Poles Apart*, 2011, 18–19.

2 See Oreskes and Conway, *Merchants of Doubt*, 2011.

3 See Washington and Cook, *Climate Change Denial*, 2011; Cook, *The Scientific Guide to Global Warming Skepticism*, 2010; Painter, *Poles Apart*, 2011, 7–34; Norgaard, *Living in Denial*, 2011; McCright, "Dealing with Climate Change Contrarians," 2008, 200–212; Leuschner, *Die Glaubwürdigkeit der Wissenschaft*, 2011, 77–110; Weingart et al., *Von der Hypothese zur Katastrophe*, 2008, 141–153; Greitmeyer, "Beware of Climate Skeptic Films," 2013.

industry or liberal parties, or they exist in close contact to them: it is the conservative think tank "The Heartland Institute", for example, which plays a key role in bundling up activities. Because their main activities are aimed at the public, media and politics, they can be called climate skeptic activists. Their opponents are not only environmentalists like Al Gore, but above all mainstream climate research related to the IPCC and their protagonists who, in their opinion, overshoot their scientific mandate by warning society about their findings. Because climate researchers take responsibility for the political meaning of their findings, climate skeptics sometimes call them "radical global warming alarmist[s],"[4] "global warming campaigners" or simply "global warmists."

The distinction between climate change skeptics and climate science may be superficially based on an idealized distinction between "sound science" and "junk science" only. Though at the same time it is a showcase about how climate change is a political problem that cannot be solved on an epistemological basis only—a notion we cannot discuss adequately in this context.[5] By considering the varying motifs of the protagonists in attacking mainstream climate science, the social construction of climate change skeptical knowledge becomes more distinct. Most common are political agendas, ideological beliefs and commercial interests, like the belief in free unregulated markets which, since the finding of anthropogenic global warming caused by burning fossils has consolidated, have been the object of serious political consideration.[6] As in all ideological disputes, it is interesting to see how both sides point the finger at the other, claiming the other to be radically biased and steeped in ideology.

Climate skeptical strategies are distinguished from the "healthy" skepticism every science should be committed to and which is implemented in science by the strict processes of pluralistic peer-reviewing. Skepticism as a science-driving factor is also an integral component of climate science.[7] This is the reason why,

4 Rawls, Alec (blog), January 28, 2005, http://errortheory.blogspot.de/2005/01/greenhouse-alarmists-fight-new-sunspot.html.

5 On the social construction of climate knowledge with a special consideration of climate change skeptics see Demeritt, "Science Studies, Climate Change and the Prospects for a Constructivist Critique, 2006.

6 See Oreskes and Conway, *Merchants of Doubt*, 2011, 213–14 and 240–74.

7 For example, there are many vital, ongoing disputes about the quality of data in climate science and the uncertainties of climate models (resulting from the high complexity of interactions and feedback loops). Methods of preventing bias in the theories are discussed, because the philosophy of science has long known about the problem

in the strictly implemented IPCC process of reviewing, controversial results are discussed and uncertainties are highlighted. These discussions within the scientific community do not, however, affect the main message of observable, human induced global warming, which is attacked by the climate skeptics. But while normal skepticism in research processes means "considering the full body of evidence before coming to a conclusion"[8] what is often observed in climate skeptical argumentation is the "cherry picking" of single pieces of evidence. Cook values this deliberate "agntology"[9] on the epistemological basis: "This isn't skepticism. It is ignoring facts and the science".[10]

In the hands of climate skeptical activists, skepticism is detached from the scientific process of scrutinizing and reviewing and placed on the level of a deliberately deceptive rhetoric instead. One of the major strategies of organized climate skepticism is "doubt-mongering,"[11] as Erik M. Conway and Noami Oreskes have pointed out in their study. "Doubt is crucial to science [...]—but it also makes science vulnerable to misinterpretation, because it is easy to take uncertainties out of context and create the impression that *everything* is unresolved."[12] Since the 1980s, when climate change research entered the sphere of politics because it seriously affects issues of global risk management, "various groups and individuals began to challenge scientific evidence that threatened their commercial interests or ideological beliefs,"[13] beginning with the free market ideology.[14]

As philosopher of science Anna Leuschner states, generally "in public debates scientific credibility is made subject of a discussion, if public interests are concerned and scientific uncertainties exist."[15] Both apply to climate science, which is expected to provide statements about future risks and may be the start-

---

that observers may be biased simply because they ascribe to a certain theory. Cf. Leuschner, *Die Glaubwürdigkeit der Wissenschaft*, 2011, 83–88.

8 Cook, *The Scientific Guide to Global Warming Skepticism*, 2010, 1.

9 On the making of ignorance as a political instrument see Proctor / Schiebinger, "Agnotology," 2008.

10 Cook, *The Scientific Guide to Global Warming Skepticism*, 2010, 1.

11 Oreskes and Conway, *Merchants of Doubt*, 2011, 34.

12 Ibid., 34.

13 Ibid., 34–35.

14 For how very different values like cultural, political, religious ones influence individual views on global warming see Hulme, *Why we Disagree*, 2009.

15 Leuschner, *Die Glaubwürdigkeit der Wissenschaft*, 2011, 67 (translated by the authors).

ing point for political action.[16] This coincides with the observation that statistics as a "technology of trust" become most important when there are political pressures and credibility is in short supply.[17] To make future risk issues disputable, policy makers depend on numbers.[18]

Climate skeptics aim at "outer scientific credibility,"[19] credibility that laypeople attribute to experts, therefore they rely strongly on mass media. Because journalists aim for balanced journalism,[20] which entails providing information about different perspectives and searching for pros and cons, they have often given equal space to the climate skeptics, although the protagonists are very small in number and do not represent the general scientific consensus that anthropogenic climate change is happening. This is not only true for right-wing newspapers: journalists have continuously presented climate skeptical issues as "scientific controversies" and with that, became complicit in helping to produce a distorted image of climate science. But science does not work like a "two-party system"[21]: while the general scientific issue of anthropogenic global warming was already decided in the 1990s, the mass media still represents it as an open discussion, particularly in the U.S. It is the outer scientific credibility of mainstream climate science that is situated at the core trajectory of climate skeptical lobbyists. Skeptics have achieved their aim when the outer scientific credibility of climate science is destroyed. This happened, for example, in 2009 when emails of IPCC climate scientists were leaked and "climategate" became public knowledge[22] or with the public attack, as early as 1994, of climate scientist Benjamin Santer, the convening lead author of chapter 8 of the Second IPCC Assessment, who was accused of doing "bad science" and tampering with his summary,[23] or in 2013 when on occasion of the new IPCC report climate skep-

16 Cf. Beck, *Risk Society*, 1992.

17 Porter, *Trust in Numbers*, 1995, chapter 5.

18 Such as the important 2° goal. See: Jaeger and Jaeger, "Warum zwei Grad?," 2010, 7–15.

19 "Außerwissenschaftliche Glaubwürdigkeit", Leuschner, *Die Glaubwürdigkeit der Wissenschaft*, 2011, 36.

20 For the "false balance" debate see Painter, *Poles Apart*, 2011, 11; Boykoff and Boykoff, "Balance as Bias," 2004, 125–36.

21 Oreskes and Conway, *Merchants of Doubt*, 2011, 214.

22 See article by Walsh in this volume.

23 Oreskes and Conway, *Merchants of Doubt*, 197–215.

tics got disproportionate air time in the media in comparison to climate scientists.[24]

Are cherry picking or misleading strategies also to be found amongst images? Climate-change skeptics use all types of media to convey their message and to present selected evidence to uphold their story; the most frequent arguments like "warming stopped" or "the sun is the cause of warming trends" are usually accompanied by graphs. One very popular outlet for communicating climate skepticism are climate-skeptical blogs on the Internet. Alongside a vast number of popular scientific books over the last decade, the Internet has been the main organ for distributing doubt.

Two different types of image can be distinguished in climate skeptical media. Images defining the context or general atmosphere of a story, e.g. a photograph with vast amounts of snow to suggest recent cooling; and computer generated data visualizations, primarily time series graphs. Examples for these types of deceptive images and graphs are discussed in this paper.

To discuss the subject of the article, we will use examples from different US-media[25]: blogs like "wattsupwiththat.com" written by former television meteorologist Anthony Watts; the website "icecap.us" ("International Climate and Environmental Change Assessment Project") founded in 2006 by weather forecaster Joseph D'Aleo, which promotes the views of climate skeptics; the 2012 book "The Mad, Mad, Mad World of Climatism" which is related to the website "climatism.net" published by engineer and business executive Steve Goreham; the book "Cold Sun. Why the Climate Catastrophe is not Taking Place" from Fritz Vahrenholdt and Sebastian Lüning, which is the most prominent climate skeptical publication in Germany; and from the online 2011 report of the "Non-governmental International Panel on Climate Change" (NIPCC), a second order assessment which examines the assessment of the IPCC, set up in 2004 by one of the most prominent climate skeptics, Fred Singer, and published by "The Heartland Institute."

Although we certainly cannot claim to paint a comprehensive picture on the general usage of images concerning climate skeptical media, we were able to observe widespread patterns of visual argumentation that can also be found in the climate skeptical media of other countries. To answer our questions we needed to rely on an idealized assumption of "correct" and "incorrect" science, and

24 Harvey, "BBC coverage criticised for favouring climate change sceptics," 2013.

25 The choice of US American sources only does not mean that such media do not exist in other countries. We could have chosen examples from Germany, Australia, Canada or elsewhere. For an international comparison see Painter, *Poles Apart,* 2011.

on the distinction between experts and laypeople, which we know—from a constructivist point of view—can only be a heuristic device.

GENERAL CONSIDERATIONS ABOUT STATISTICAL GRAPHS

"In the space of one hundred and seventy-six years the Lower Mississippi has shortened itself two hundred and forty-two miles. This is an average of a trifle over one mile and a third per year. Therefore, any calm person, who is not blind or idiotic, can see that in the Old Oolitic Silurian Period, just a million years ago next November, the Lower Mississippi River was upward of one million three hundred thousand miles long, and stuck out over the Gulf of Mexico like a fishing-rod. And by the same token any person can see that seven hundred and forty-two years from now thc Lower Mississippi will be only a mile and three-quarters long, and Cairo and New Orleans will have joined their streets together, and be plodding comfortably along under a single mayor and a mutual board of aldermen. There is something fascinating about science. One gets such wholesale returns of conjecture out of such a trifling investment of fact."[26]

This quote from Mark Twain, in which he takes the pitfalls of statistical knowledge production to absurd extremes, indicates that from their inception statistical statements have been given the blame for distortion, manipulation and, as Mark Twain called them, "damned lies." This is particularly true of graphs, an important tool for visualizing statistical data. Therefore, before analyzing examples of climate skeptical graphs, we want to look briefly at the general possibilities of charting misleading graphs independently of the topic.

When statistical graphic methods became common during the nineteenth Century, it became evident that there were many ways in which one could tamper with graphs unobtrusively. The possibilities range from distortion (by means of highlighting a certain meaning of the data without, technically speaking, falsifying it), to misrepresentation and manipulation (by means of fraud). Typical methods of tampering, for example, are doctoring the data sample or using outdated data, manipulating the mean value, distorting the course of a curve by selectively trimming or compressing the axis, incorrect extrapolation of trend lines, or shortened forecast horizons.[27] Such methods can be easily employed to put preferred arguments graphically in exaggerated or more drastic terms. Alt-

26 Twain, *Life on the Mississippi*, 1883, 176.

27 See Huff, *How to Lie with Statistics*, 1954.

hough people today should be very familiar with reading charts, the problem is still that such graphic adjustments are extremely difficult to recognize for laypeople. At the same time, visual media are an extremely powerful way to convey messages.[28] This is why visual strategies to pretty up or manipulate results can be found in many charts such with political relevance as market shares, state expenses, polling behavior or unemployment figures.

On the other hand, it can be asked what features do make a data visualization scientifically correct. Data visualization in a scientific sense is not arbitrary and must fulfill specific quality criteria. The most important criteria are expressiveness ("A visualization is said to be expressive if and only if it encodes all the data relations intended and no other data relations."[29]) and effectiveness ("Effectiveness criteria identify which of these graphical languages [that are expressive], in a given situation, is the most effective at exploiting the capabilities of the output medium and the human visual system."[30]). In theory, therefore, if climate skeptics use visualization in scientific argument, it must fulfill these criteria. So, for our investigation, we had to see (1) if the data on which a visualization is based might be already distorted, (2) if the data has been correctly transformed into the graphic and (3) if the graphic was easy to understand.

In contrast to the general vulnerability of statistical graphics to a wide spectrum of manipulations, they have, since their inception, suggested general precision and accuracy. When, in the nineteenth century, statistics became the ultimate discipline with which to produce scientific knowledge, data graphs became imbued with credibility and persuasiveness because of what Porter called the general "trust in numbers."[31] Since then, mathematics has served not only as a source for new knowledge but also as a strategy for communicating scientific objectivity and authority, leading to "prestige and power of quantitative methods in the modern world."[32] Rigorous quantification was developed as a strategy of impersonality in political and administrative affairs in response to pressures from outside.[33] Any person who refers to statistical graphs can rely on the impersonal authority of the curves and numbers s/he is pointing to, whether it be the scien-

28 See Tufte, *The Visual Display*, 1983; Brasseur, "Florence Nightingale's Visual Rhetoric," 2003; Brasseur, *Visualizing Technical Information*, 2003; Kostelnick and Hassett, Shaping Information, 2003; Latour, "Drawing Things Together," 1990.

29 Card, "Information Visualization," 2003, 523.

30 Mackinlay, "Automating the Design," 1986.

31 Porter, *Trust in Numbers*, 1995.

32 Ibid., viii.

33 Ibid., xi.

tists who actually practice statistics, or anyone communicating scientific knowledge to the public by means of charts. Pointing to quantitative expert graphs has become the modern way to claim interpretational sovereignty. This does not signify a general mistrust in statistics, but the general necessity of carefully utilizing and reading them, because they are extremely abstract tools of knowledge production and not at all easy to digest.

Yet it is important to stress the nature of charts as a heuristic tool. Often, statistics are generated with the aim of functioning as "decision-making-tools"[34]—for example, for the purpose of satisfying political interests. They are not only deployed for better scientific understanding but to create arguments as a basis for changing conditions at the same time. For example, the British social reformer and founder of modern nursing, Florence Nightingale, was only able to make her observations politically effective when, in 1858, she converted her findings on the reasons for the mortality rate in military hospitals into the famous "rose diagram."[35] The diagram made evident what decision makers had previously overlooked; only when graphic methods were put into action did they tackle the problem and change hygiene conditions in hospitals.

This effective power of graphs to cause and establish action is also the reason why statistical climate graphs evolved into fundamentally *political pictures*: they bear the potential to extract from data findings and discoveries, which can trigger consequent actions in cases of risk management. It is in this function as 'decision-making-tools' that the most prominent climate expert graphs—those of human-induced global warming—have been communicated. Among these is the "hockey stick," the reconstructed climate records derived from ice core analysis, or the scenarios of possible climate futures.

When such images are included in the *Summary for Policy Makers* of the IPCC reports, they inevitably acquire a political edge, as they suggest also political action (in simplified terms: "burn less fossils"). One could go so far as to say: when the role of graphs is switched from scientific exploration to public communication, the graphs obtain a new framing or staging—they 'want' to change the circumstances they present. The graphs deployed in the political frame of the IPCC cannot help but contain a rhetoric layer: they highlight the interpretation of the data in order to be more understandable for policy makers who have to decide upon action. This is the very essence of generating graphs for purposes of presentation and communication: to help comprehension by reducing complexity and highlighting the most important meanings.

---

34 Rosentrater, "Representing and Using Scenarios," 2010, 253.

35 Brasseur, "Florence Nightingale's Visual Rhetoric," 2009.

In summary, it can be said that producing graphs for communication has to be accompanied by a reduction in complexity. At this stage, it is correct to elide research that has been carried out to produce the graph, because graphs show only the results and not the process of how researchers arrived there—this can be followed up in the related scientific articles. When, for example, the color red is used to emphasize the detected risk, to communicate a central statement, the color implicitly contains a political layer. To deny this political function of climate graphs in order to defend the possibility of "pure science" would miss the point.[36] The general rhetorical and political layer of statistical graphs is not exclusive to climate skeptics, it is deployed throughout the whole field. There is a fundamental difference, however, when climate skeptics employ cherry picking by using graphs while at the same time hiding the full weight of evidence. This is when general rhetoric overshoots into manipulation.

CLASSIFICATION OF IMAGES IN SKEPTICAL CLIMATE DISCOURSE

In this section we will have a closer look at types and usage strategies of visual elements in skeptical climate media. On the most fundamental level, images in skeptical climate discourse can be divided into photographs and graphics, mostly computer generated data visualizations. For the photo-based approaches we found the following three usage strategies:

1. Apparent experimental proofs,
2. Conclusions by analogy, or
3. Photographs showing climate scientists with the purpose of discrediting them.

Note that some photographs can fall into more than one of these categories of usage strategies.

An example of the first usage of photographs as apparent experimental proofs would be images of the few glaciers which are not melting, but growing, or the widespread 2.14 min time-lapse video comparing the effect of a 180% increase in CO_2 on the growth of a single plant over 42 days (see Figure 1). Brought down to the convincing and simple rationality of an every day experience, view-

36 How scientists constructively relate to politics and policy beyond the most difficult role as pure scientists is the subject of Pielke, *The Honest Broker*, 2007.

ers see two plants next to each other under atmospheric CO_2 concentrations of 450 ppm (expected for around 2030) and 1270 ppm (representative for the end of the century under business-as-usual scenarios).[37] With increased CO_2 the plant on the right grows faster and taller suggesting higher future harvests—ignoring the fact that global warming does not only mean an increase in CO_2, but is accompanied by increasing aridity and weather extremes, which could annihilate the growing effect of CO_2 on photosynthesis (not to mention other negative impacts of climate change under this marked CO_2 increase). Many related arguments can be attached to this movie following the "gas of life" argument.[38]

Figure 1: Video shot comparing plan growth under two different CO_2 concentration conditions.

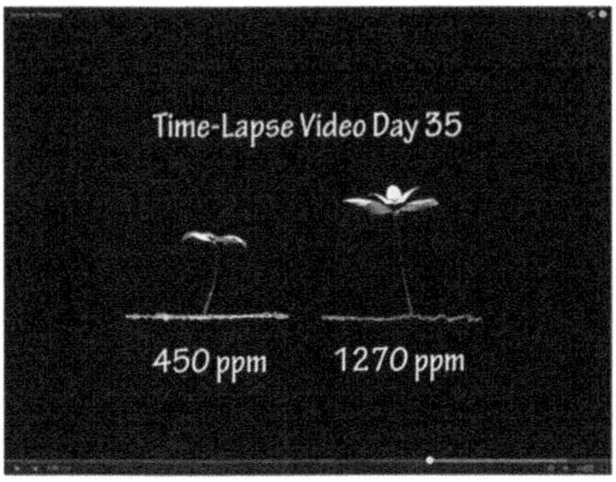

Examples for conclusions by analogy (second usage) are mainly photographs that function as more or less atmospheric illustrations of arguments in the literal sense, a technique common in many media. In climate skeptical media, for example, photographs from recent cold or, in many cases more appropriately, snow-rich winters in Europe and North America or ice floes in the Arctic Ocean are used to suggest recent cooling and, with it, a contradiction to global warming.[39] The reports fail to make the distinction between weather and climate, and between regional and global effects.

37 See Watts, Anthony (blog), posted Sept 29, 2011, http://wattsupwiththat.com/2011/09/29/plants-gobbling-up-co2-45-more-than-thought/.

38 See for example "CFACT", last modified August 18, 2013, accessed Oct 25, 2013, http://www.cfact.org/2013/08/18/time-to-praise-co2-the-miracle-gas/.

39 A very nice example is a story on "Foxnews", last modified 3 Feb, 2011, accessed Oct 25, 2013, http://nation.foxnews.com/global-warming/2011/02/03/america-crushed-snow-and-ice-global-warming with the ironic headline "America Crushed by the Snow and Ice of Global Warming."

A particularly questionable example for the third type of image usage—discrediting climate scientists—are blog entries about a recent scientific paper published by climatologist James E. Hansen illustrated by a press photo of him being arrested in the context of political protest against surface mining in 2010, completely unrelated to the science in the paper.[40] The main purpose here is to cast doubt on the impartiality of a very prominent climate scientist and thus discredit his scientific findings without bothering to study the scientific details. The strategy follows the established tactic of "attacking the messenger not the message."[41] It is also founded upon the unquestioned fact-value distinction, which has become fragile in the context of risk research that touches policy questions.[42]

An extreme usage of a photograph (in the sense of both concluding by analogy and discrediting climate scientists) was a Chicago billboard campaign[43] run by the "The Heartland Institute" in 2012.[44] The campaigners placed a mug shot of the Neo-Luddite serial killer Ted Kaczynski (the "Unabomber") next to the sentence "I still believe in global warming. Do you?," comparing those who believe in climate change to serial killers. On their web-posted press release Heartland's president, Joseph Bast, stated: "The most prominent advocates of global warming aren't scientists. They are Charles Manson, a mass murderer; Fidel Castro, a tyrant; and Ted Kaczynski, the Unabomber. Global warming alarmists include Osama bin Laden and James J. Lee […]."[45] The billboard cam-

40 For example see an article on the website of Heartland Institute, posted March 29, 2013, http://news.heartland.org/newspaper-article/2013/03/29/alarmist-hansen-admits-temperatures-not-following-predictions.

41 Oreskes and Conway, *Merchants of Doubt*, 2011, 2–3.

42 This distinction cannot be adhered to strictly in risk-relevant science, because to make scientific statements about risks for society necessarily has to be based on value-statements. See Walsh, "Visual Strategies", 2007.

43 The digital billboards were first placed along the inbound Eisenhower Expressway in Maywood.

44 The Heartland Institute is an American conservative think tank, which states that it advocates free market policies. It has a full-time staff of currently around 30 staff members and more than 200 unpaid policy advisors. The Institute has financed several meetings by climate change skeptics and has been a major factor in pushing doubt about climate change science.

45 "Heartland Institute", last modified May 3, 2012, accessed Oct 25, 2013, http://heartland.org/press-releases/2012/05/03/do-you-still-believe-global-warming-billboards-hit-chicago.

paign created an uproar; when many of Heartland's financial supporters backed out as a consequence, the campaign was immediately pulled with the comment: "This provocative billboard was always intended to be an experiment. And after just 24 hours the results are in: It got people's attention. This billboard was deliberately provocative, an attempt to turn the tables on the climate alarmists by using their own tactics but with the opposite message [...]."[46] The whole outer scientific climate change dispute is brought down to an activity of brokering which brings out the big propaganda guns.

Typically, photographs in climate skeptical media set the context of a story; they add an often emotionally charged frame to the message. As with other news media, images are sought that draw the readers' attention. But the main aim in climate skeptical media is—apart from grabbing the attention of the audience—either to prominently display an apparent contradiction to global warming or to cast suspicion on prominent climate scientists.

The Degree of Change in Skeptical Data Visualization

The main focus of our review of skeptical climate images is on the second class of images used for climate skeptical argumentation—the computer-generated data visualizations which constitute a widespread tool for reasoning climate skeptical issues. In the group of graphs, we found the following five classes that we list here in order of the increasing degree of data/visualization manipulation:

1. **Correct reproduction of up-to-date scientific graphs:** exact reproduction of scientific findings with a correct interpretation of these results. Usually used to highlight ongoing discussions and uncertainties in climate science ("there is no consensus"/"uncertainties are large" or even "there appears to be too much consensus, therefore the science must be biased or based on a conspiracy").
2. **Misinterpretation of correct graphs:** use of scientifically correct graphs, which are deceiving because they are taken out of context or interpreted in a way that contradicts the original findings.
3. **Cherry picking of obsolete research results:** deliberate selection of older and obsolete graphs, and interpretation in a way that contradicts the general

46 Carol Felsenthal, "Chicago's Heartland Institute, the Group Behind the Unabomber Billboard," in *Chicago Magazine*, last modified June 14, 2012, accessed Oct 25, 2013, http://www.chicagomag.com/Chicago-Magazine/Felsenthal-Files/June-2012/Chicagos-Heartland-Institute-and-its-Unabomber-Billboards/.

consensus. Note that classes 2 and 3 usually involve a form of cherry picking, in this case of entire graphs.

4. **Classic cherry picking:** misleading viewers by producing and displaying graphs based on carefully selected data specifically aimed to apparently contradict global warming. Classic examples include cherry picking of places or regions which show cooling despite a general global warming trend and cherry picking of periods of time with limited warming (or even cooling); visualization techniques include *exaggeration* (for example, of past climate change), *graphical understatement* (for example, of modern warming as compared to past changes) or *hiding parts* of original graphs which do not serve the argument. Often augmented by *visual enhancement techniques* (e.g., by adding highlighting curves and colors); includes comparison of quantities that cannot be directly compared (*comparing apples with oranges*, e.g., "weather" vs. "climate") to mislead viewers.
5. **Data / image manipulation:** manipulating data shown in the graph or the visualization itself (for example by changing axis labels) to deceive the viewer.

*Table 1: Phases of manipulation for the five climate skeptical image types; * in parentheses we indicate an optional incorrect image handling, however not mandatory to fall into this class.*

Class \ Manipulation phase	**Original graph selection**	**Data and graph**	**Ad hoc reaction**	**Interpretation**
Correct reproduction of up-to-date scientific graphs	Unbiased	Correct	Correct	Correct (deceiving)*
Misinterpretation of correct graphs	Unbiased (cherry picked)*	Correct	Correct (deceiving)*	Deceiving
Cherry picking of obsolete research results	Cherry picked	Correct	Correct (deceiving)*	Deceiving
Classic cherry picking	Unbiased (cherry picked)*	Cherry picked	Deceiving	Deceiving
Data/image manipulation	Unbiased (cherry picked)*	Manipulated	Deceiving	Deceiving

The five classes can be arranged in a table to show on which level they are incorrect or deceiving. Here, we distinguish the levels of the original graph that was selected, accuracy of the graph in question and the underlying data, the ad hoc reaction it invokes in a casual viewer and the more detailed interpretation in the accompanying text. The table shows the phases of manipulation levels in climate skeptical media graphs ranging from unbiased and correct to manipulated and deceiving.

In addition to the investigations discussed below, skeptical climate images can be categorized by their argumentation.[47] Here, for our systematic image investigation, we use the following classification:

1. **There is no warming/global warming has stopped/it is getting colder**: e.g. temperature measurements are biased by the urban heat island effect; there has been no warming since 1998; we are approaching a new ice age.
2. **There is warming, but it is mostly natural**: e.g. the climate has changed in the past as well; global warming is caused by the sun/cosmic rays; human greenhouse-gas emissions are tiny compared to natural fluxes.
3. **Only part of the warming is caused by our emissions/future warming will be small**: e.g. the hockey stick is wrong, it was warmer during the medieval climate anomaly; CO_2 is only a weak greenhouse gas; climate sensitivity is too high because negative feedbacks are not considered.
4. **There is warming and we cause it, but it will have good, not bad impacts**: e.g. plants will grow better under higher CO_2 levels; increases in storms/extreme events/sea level are exaggerated.

Within our analysis, we found many data visualization examples for these four classes of argumentation. For class 4, we counted in addition many photographic image argumentation examples.

IMAGE ANALYSIS—THE BANDWIDTH OF CHERRY PICKING

In this section, we will discuss examples from the image classes from the previous section in more detail. Apart from the image background and genesis, we investigate particular methods used to support the climate skeptical argumentation. Often, those visualizations are redrawn versions of climate sciences images, with additional data included. Aspects for analyzing such images are the type of data averaging, the selection of colors, the applied changes with respect to the original images, the characteristics of added labels and others.

By far the most common type of deceptive graphs in climate skeptical media are graphs which are not actually wrong in the scientific sense, but still paint a

47 For a more detailed discussion of categories see Washington and Cook, *Climate Change Denial*, 2011, chapter 3.

misleading picture of climate change research. It is not surprising that this is very common since these graphs pretend to be scientific and show a result that most instantly by means of "evidently" contradicts global warming. Obsolete or cherry picked graphs can thus appeal successfully to the general public[48] which is easy to convince by expert graphs and numbers, because without background knowledge it is impossible to decide at first glance if research was acquired by sound science. Furthermore, for graphs taken from scientific papers, the reference to a peer-reviewed article enhances the credibility of the argument presented.

Example 1: Misinterpretation of correct graphs in combination with classic cherry picking. The first example is taken from the end of the 1980s, when climate change prominently appeared on the political agenda. As early as 1989, the Marshall Institute circulated a "white paper" that strategically attacked climate scientists to weaken their message. This report got published in 1990 as a small brochure with the title "Global Warming: What does the Science Tell us?" Although the authors claimed to present an unbiased scientific assessment of global warming, their report actually presented a distorted view of climate research. They were very successful in distributing the report, directly addressing the White House and presenting it widely to members of the Office of Cabinet Affairs and other political staff.[49] Oreskes and Conway, who researched the case, found that "the briefing had a big impact, stopping the positive momentum [about global warming] that had been building in the Bush administration."[50] The main argument of the report went like this: there is no one-to-one correlation between the global warming trend (represented by the temperature curve) and the curve of CO_2 concentrations. For this reason, they claimed, it must have been caused by the sun, which had increased its output over the twentieth century.

48 Opinion surveys on the public understanding of climate change and its causes in the US and UK have suggested that a general confusion about climate change knowledge and increasing climate change skepticism is related to journalistic norms of news coverage. See Painter, *Poles Apart*, 2011, 12; see the results of a survey for BBC News in February 2011 posted on the BBC website, last modified Feb 7, 2010, accessed Oct 25, 2013, http://news.bbc.co.uk/1/hi/8500443.stm

49 Oreskes and Conway, *Merchants of Doubt*, 2011, 186.

50 Ibid.

Figure 2: Misinterpreted and cherry-picked curve of an early climate skeptical report circulated among politicians by Jastrow et al. It was singled out to incorrectly disprove influence of carbon dioxide on Earth's climate.

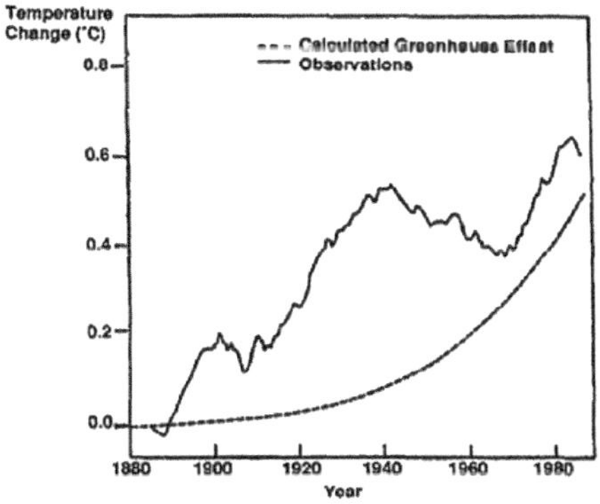

Figure 3: The original graph by J. Hansen[51], which was one of the first main pieces of evidence for human-induced global warming, and on which Jastrow et al. based their curve on (top left).

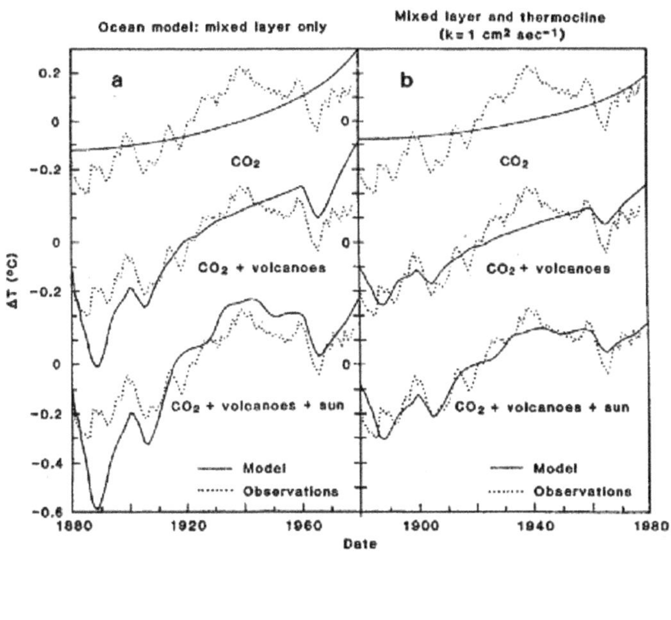

51 Hansen et al., "Climate Impact," 1981, 963.

The illustration that was employed to disprove the correlation between carbon dioxide and global temperatures (Figure 2) was an abstract scientific graph contrasting the ups and downs of the global temperature curve with the smoothly rising warming curve expected from anthropogenic greenhouse-gas emissions. Ironically, the graph originated from James Hansen's article published in Science Magazine (Figure 3), which started the whole political concern about global warming. But the Marshall report deliberately misrepresented the Hansen report by singling out only one panel from the original graph with only minor changes. By progressively adding the effects of CO_2, volcanic cooling and changes in solar irradiance, Hansen et al. could show that the overall warming trend is indeed caused by anthropogenic emissions, while the fluctuations observed in the temperature curve result from natural causes. By cutting out this piece of the graph, the authors of the Marshall report were able to turn the findings of Hansen et al. upside down, which actually concluded that changes in solar output alone could not explain recent warming.

From a visualization point of view, in contrast to the original graph from Hansen et al., the temperature change axis and the CO_2 curve were shifted such that temperature change and CO_2 curve start at the same level (zero) in the 1880s. This image adaptation enforces the ad-hoc impression that temperature rise seems to be decoupled from CO_2 increase and to be much stronger. This alteration thus appears to be intentional and reinforces the skepical interpretation.[52]

In addition, the strategy to cut away all other relevant pieces of the graph was very suggestive: "It looked like they were relying on peer-reviewed science."[53] But the authors had actually practiced a strategy that is common in unbalanced, suggestive rhetoric: to quote-mine/take out of context by cherry-picking part of the graph. This works not only for verbal statements, but also with images. At the same time, this early example shows the key function of time series graphs in the climate change debate: in the complex field of scientific climate change knowledge if one can point at a scientific curve—or better a correlation of curves, suggesting causation, the argument is likely to be convincing.

Example 2: Cherry picking obsolete graphs. The next example is also related to solar output. In 1991, Knud Lassen and Eigil Friis-Christensen published an article claiming "strikingly good agreement" between solar cycle lengths and

52 Hansen et. al.'s y-axis curve placement is more neutral because the two curves are layed out such that their mean values are placed to the y axis mean (zero) value.

53 Oreskes and Conway, *Merchants of Doubt*, 187.

northern hemisphere land temperatures over the period from 1860 to mid-1980 (Figure 4 left). Cycle lengths are the varying lengths of the about 11-year long solar activity cycles which can be traced, e.g., by sunspot numbers. The similar curve shapes, they concluded, are evidence for a major influence of the sun on the climate. The finding was challenged by other scientists for several reasons (such as inconsistencies in the smoothing of the solar cycle curve) and finally revoked in 1999 by Lassen himself. Figure 4 right shows the updated version, with a clear decoupling of solar cycle lengths and temperature after 1970.[54] As in the previous example, the technical character of the original black-and-white graph has not been changed to preserve its scientific appearance.

Figure 4: Original version of the graph claiming a correlation between solar cycle length and temperature[55] (left) and updated version, which shows a clear decoupling after 1980[56] (right).

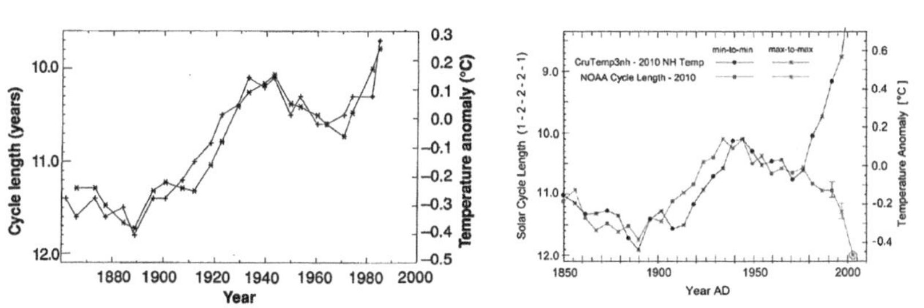

Despite the well-known facts under discussion, the original graph, based on a single outdated study published in 1991, continues to reappear again and again in climate skeptical media, trying to prove that the sun, not anthropogenic greenhouse gas emissions, is causing global warming. The original curve appears, for example, on Anthony Watts' climate skeptical blog "Watts Up With That?" in an article posted in 2011.[57] The example shows once more the general potential of graphs arising from their powerful quality of providing "immutable mobiles"—what holds for "paper-work" also applies to digital image files.[58] Expert visuali-

54 Theijl and Lassen, "Solar Forcing of the Northern Hemisphere," 1999.

55 Christensen and Lassen, "Length of the Solar Cycle," 1991.

56 Stauning, "Solar Activity—Climate Relations," 2011.

57 Watts, Anthony, blog published Jan 2, 2011, http://wattsupwiththat.com/2011/01/02/do-solar-scientists-still-think-that-recent-warming-is-too-large-to-explain-by-solar-activity

58 Latour, "Drawing Things Together," 1990, 26–35.

zations are used as an essential vehicle with which to communicate the message of climate change skeptics; via images knowledge can be successfully carried from one actor to another, even from one network to another, by at the same time retaining their significant "gestalt." The simple process of "copy and paste" can easily multiply graphic arguments—even if they are outdated. In the blog sphere, such images can live on like ghosts detached from more recent scientific developments.

Example 3: Classic cherry picking—hiding parts. Recently, a class of time series visualizations emerged in climate skeptical media, related to the argument that global warming stopped 15 years ago. This is an excellent example for cherry picking of short time periods. Neglecting the long-term trend, starting around the extreme El Niño year 1998, and not displaying other data sets that incorporate the strong warming in the Arctic, these graphs are typical examples that drop relevant parts by incorrectly reducing data and timespans. Figure 5 shows an early, widespread version of such kinds of time series images by Joseph D'Aleo.[59] An extreme example, it selects exactly the year 1998 as a starting point of the time series in order to change the significant progress of the curve. Therby a clear negative trend is shown by the graphic, whereas CO_2 is constantly rising (green curve).

Figure 5: "Global warming has stopped since 1998, and is not correlated with the CO_2 development," Joseph D'Aleo. (→ color plate 9).

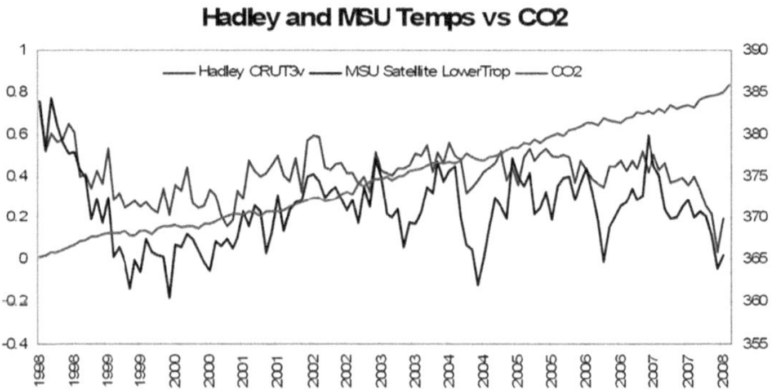

Regarding the applied visual mapping, the designer used subjective and intentional color coding of the background to highlight his argument, with 0.2 °C

59 "Icecap", accessed October 25, 2013, http://icecap.us/images/uploads/Correlation_Last_Decade.pdf.

(instead of 0°C) being the changing area between light blue and yellow, and the normalization of the green CO_2 curve such that it exactly crosses the temperature axis at 0°C. Thus, a large part of the curves lies in the light blue and the intermediate image part, giving the impression of "cold" conditions during the past 15 years.

Figure 6: a) Billboard campaign by CFACT (Committee for a Constructive Tomorrow), USA; b) Daily Mail Online. October 16th, 2012: "Global warming stopped 16 years ago [...] and here is the chart to prove it".

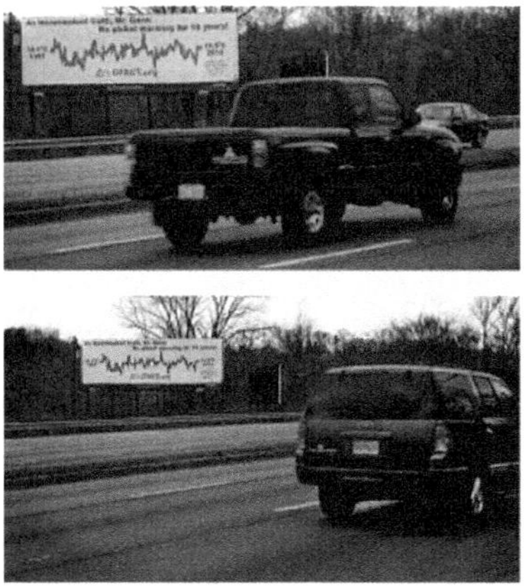

a)

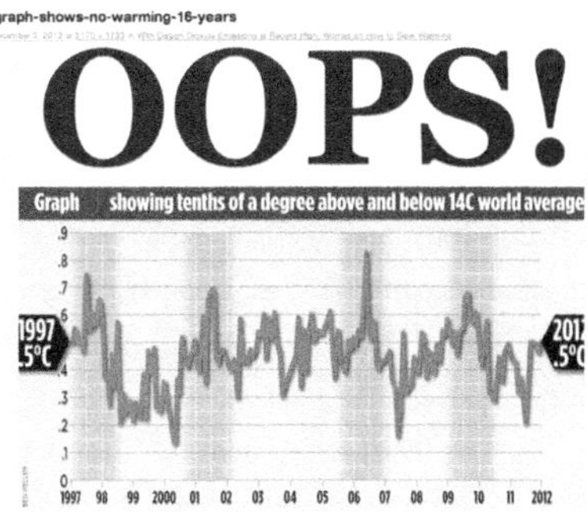

b)

Figure 7: Official version published by the Met Office (2013), integrating the NOAA NCDC and the NASA GISS temperature data sets.

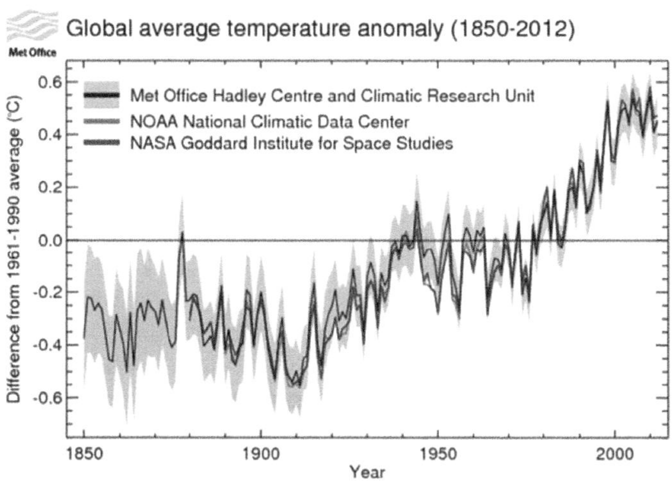

In 2013, a billboard campaign based on such short-term temperature time series visualizations was run by the Committee for a Constructive Tomorrow (CFACT), USA (Figure 6a). The billboards were placed next to highways; they presented the temperature curve commented by the sentence "An inconvenient truth, Mr. Gore. No global warming for 16 years!" The graph correctly represents the Hadley Center HadCrut4 data set, however starting with the year 1997 only and thus hiding the long-term increase in temperature, in this way again altering its meaning. The billboard campaigners stated in the press release: "If you can't count on the media to report, place the facts where they must be seen. Fight propaganda with facts!"

A similar version was published by the UK Daily Mail in 2012 (Figure 6b). The newspaper had printed the graph taken from a report by the Met Office, the United Kingdom's national weather service. The curve shows global mean temperature for the short time frame of 16 years from 1997 to 2012. By marking the similar temperatures of 14.5°C at the beginning and end of this period, the desired interpretation of the graph is highlighted graphically: global warming has actually "plateaued in 1997 and remained stable for 16 years", as the Daily Mail author writes.[60]

In the billboard design, the curve is freed from all features that normally make a chart reasonable, such as axis, reference or title. It only works with the expectation of a general public understanding of climate charts by an established collective symbolism of "normalistic curve-landscapes"[61] (Jürgen Link). A red colored curve is immediately understood as global temperature record.

Again, this is an example of misinterpretation of cherry-picked data. To illustrate this, Figure 7 depicts the official curve by the Met Office, representing a longer time period (since 1850) and additional data sets, thus showing the whole picture. In addition to this cherry picking of a time frame with an insignificant warming trend, these graphs usually use the Met Office global temperature data that are known to show a smaller warming trend than other datasets, because they omit temperatures from the polar regions (which exhibit particularly strong warming) to avoid extrapolation in regions with no weather stations.

60 David Rose, "Global warming stopped 16 years ago, reveals Met Office report quietly released [...] and here is the graph to prove it," Daily Mail, October 16, 2012, http://www.dailymail.co.uk/sciencetech/article-2217286/Global-warming-stopped-16-years-ago-reveals-Met-Office-report-quietly-released--chart-prove-it.html. The author is well-known for his climate skeptical news coverage.

61 Link and Hall, "The Normalistic Subject and Its Curves," 2004.

Example 4: Classic cherry picking—graphical understatement & hiding parts. In this class, images support the argument that climate change is not a human problem, typically by the argumentation that it has changed for long periods without human influence, displayed in time series charts of several hundred up to ten thousand years. For this kind of relativism, a popular version is the temperature curve shown in Figure 8, based on oxygen isotope ratios recorded in Greenland ice cores. Since these can be related to temperature, they can be used to reconstruct northern hemispheric temperatures. This was done by Schönwiese[62] for example, whose graphic seems to have been the source of many similar curves in climate skeptical media. There are four issues with most of the versions of this graph in those media, relating to sloppy references to the original source, concealment of the fact that a local record is not representative of global temperatures, omission of the temperature rise in recent decades, and subjective visual enhancement techniques.

First, the origin of this graph is difficult to trace based on the generally sloppy references in skeptical media. Goreham[63] references Dansgaard et al., (1984),[64] and a presentation by Avery (2009).[65] However, in the paper by Dansgaard et al., 1984, no data based on "ice cores from Crete, Greenland" (as suggested by Goreham (2012)) are shown. The slides used by Avery from a skeptical climate change conference can be inspected in a youtube video,[66] however there is no mention of the origin or the production of the graph. Vahrenholt and Lüning[67] who also use this graph, just refer to a German web site.[68] Interestingly, this web page refers to another Dansgaard publication[69] and to Schönwiese (1995) who appears to be the original image source. In these examples, the improper references of climate skeptics to their sources violate the basic rules and skills of scientific working, such as accountability and transparency.

62 Schönwiese, *Klimaänderungen, 1995.*

63 Goreham, *The Mad, Mad, Mad World of Climatism*, 2012, 63.

64 Dansgaard et al., "North Atlantic Climate Oscillations," 1984.

65 Dennis Avery, Presentation at the Int. Conference on Climate Change, Mar. 9, 2009.

66 Talk by Dennis Avery, uploaded July 10, 2013, accessed December 2, 2013, http://www.youtube.com/watch?v=zjElAxounew.

67 Vahrenholt and Lüning, *Die kalte Sonne*, 2012, 29.

68 "Das zyklische Auftreten von Optima und Pessima im Holozän," last modified Jan 15, 2013, last accessed Oct 25, 2013, http://lv-twk.oekosys.tu-berlin.de/project/lv-twk /002-holozaene-optima-und-pessima.htm.

69 Dansgaard and Johnsen, "A flow model and a time scale for the ice core from Camp Century," 1969.

Second, many authors of skeptical media (for example Avery, 2009, and Vahrenholt und Lüning, 2012) fail to mention that this temperature estimate is based on an ice-core record from Greenland and may thus not be representative of global temperatures.

Figure 8: Temperature reproduction of the last 12000 years for the Northern hemisphere, from Goreham based on Schönwiese. (→ color plate 10).

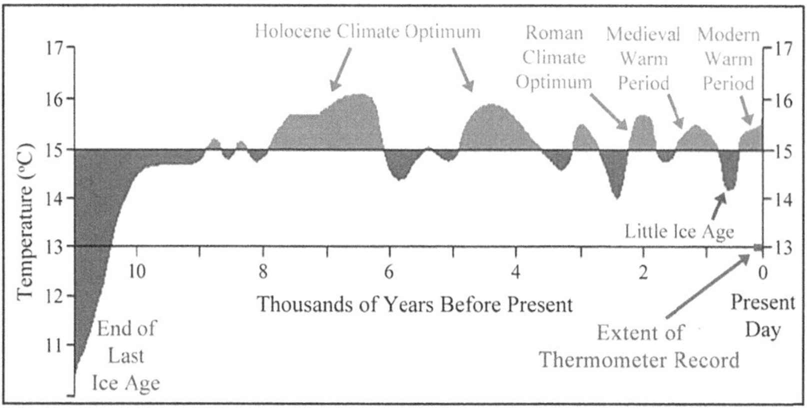

Most importantly, in Schönwiese's 1995 version the current and near future temperature changes are included. The recent warming goes far beyond the historic warm periods of the last 12000 years and should therefore have been included in the graph. In Goreham's version (Figure 8), part of the current warming is included, however, it is nearly invisible due to its thinness near the axis. In Vahrenholt and Lüning (2012) this warming is not visible at all, ending with a different curve for the last century (which might be based on Schönwiese's, 1994[70] obsolete curve, which he changed in Schönwiese (1995)). The missing temperature increase in the second half of the twentieth century in these graphs could be due to the fact that "present day" within the paleo-climate research community generally refers to the year 1950. Thus, if climate skeptics use or adapt such time series data or visualizations from paleoclimate science publications, they tend to conceal (or are unaware of) this. Resulting images leave out the part of recent warming without a clear illustration of this fact, and the graphical understatement-effect is intensified. As an extreme version of the graph in Figure 8, Vahrenholt and Lüning (2012) changed the axis ticks from "present

70 Schönwiese et al., *Das "Treibhaus"-Problem*, 1994.

day" to the year "2000" (and the other ticks accordingly), which in this case is either ignorance of the science or deliberate deceit.

From a visualization point of view, the main visual property of the climate skeptic versions of such graphs is the color filling of areas above or below a baseline. In this case, above the baseline of 15°C, areas are colored in red (highlighting warm conditions), whereas areas below this value are colored in blue (cold conditions). In Figure 8, this highlighting suggests that there were many warm periods before the present. The selection of the baseline value is often not statistically justified, rather it is used arbitrarily to support the author's arguments visually, or to simply "spice up" the graphics. Vahrenholt and Lüning, 2012[71] extensively use this method, sometimes even by taking multiple trend curves per chart or by arbitrarily leaving out the color filling.[72]

In summary, this class of long-term time series with suggestive, color-coded highlighting of warm and cold conditions is a technique very frequently used to support the argumentation that "climate has changed naturally in the past" and that the small extent of current warming is no problem. The example in Figure 6 combines several kinds of scientifically problematic issues, ranging from obscure or even false figure or data origin to suggestive highlighting of time series integrals and the missing illustration of the meaning of the time series end step.

Example 5: Cherry-picking—comparing apples with oranges & exaggerating. Figure 9, from Goreham (2010) exemplifies these two climate skeptical strategies: displaying *global* temperature change in comparison to *local* temperature variations based on data from one single weather station.[73]

From a visualization point of view, at first glance, the image seems to provide different (profound) bandwidth information for the climate debate. At second glance, within the graphics and within the text, the global mean temperature change (instead of the change of the global extremes) is set in relation to the total yearly temperature spread of a single station, thus intermixing weather (or even weather extremes) with climate: "The annual temperature variation in Chicago is typically 100 degrees, changing from about -5°F to about 95°F. In comparison, the extent of mean global temperature variation over the last 150 years is tiny [...]" (Goreham 2010, 381–82). Goreham possibly identified this problem

71 Vahrenholt and Lüning, *Die kalte Sonne*, 2012.

72 Ibid., 58.

73 Goreham, *Climatism!*, 2010, 381.

himself, because the graph is no longer to be found in his recent book (Goreham, 2012). It is, however, still available on his web site.[74]

Figure 9: Comparing apples with oranges & exaggerate: Mean global temperature increase (black horizontal line) is set in comparison with Chicago weather data (daily statistical values for 1872–2008) including daily record extremes (from Goreham 2010).

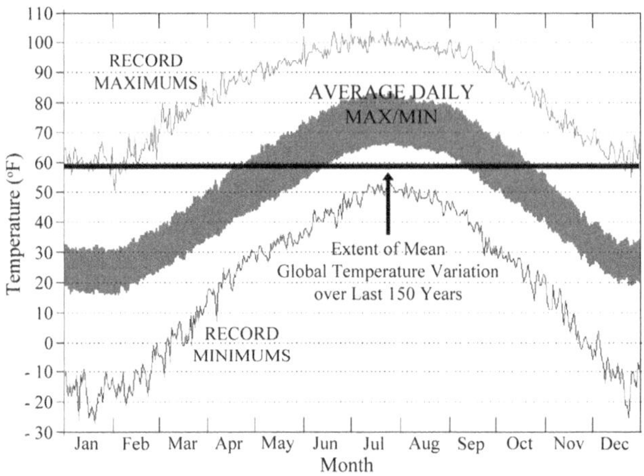

IMAGE DISCUSSION

Due to space reasons, we could not present examples for all image classes within this chapter. An important visualization category used in skeptical media and not represented in the examples above are non-adapted images from the IPCC or other climate research publications (see also Table 2, far right column). Here, we found examples of the first two image classes in Table 1 which tend to merge with each other: correct reproduction of a fact agreed within climate science that fits into the skeptical argumentation (e.g. the spread of climate model projections, climate sensitivity estimates or other uncertainties in climate science) and the more tendentious misinterpretation or isolating of facts in image captions or in the textual discussion.

The other extreme—with respect to a correct scientific visualization—is direct deceit. As an example, we discussed Figure 6 and its different variants with-

74 See Goreham, Steve, "Climatism.Net," date of last modification unknown, accessed Oct 25, 2013, http://www.climatism.net/facts-about-global-warming/#temperature.

in the skeptical media, including several properties with outright deceptive tendencies. When considering graphs used in climate change skeptical media, one might expect to find a significant number of manipulated graphs, too; however, such graphs are actually quite rare. We argue that this is to be expected, since it is easier for others to show that a clearly manipulated graph is just that, whereas it is far more difficult to explain to the general public why graphs that appear to be scientific are actually misleading and do not show the true picture of mainstream climate change science.

Within our investigation, we analyzed in detail a multitude of images within skeptical context. At the data level, showing obsolete data without providing this fact is a frequently applied procedure. At the visualization level, most climate skeptical images are very simple to interpret (show high effectiveness). This is achieved by:

- the very frequent use of time series charts,
- avoiding uncertainty/bandwidth representation,
- a limited representation of physical quantities (temperature and CO_2 are the most frequent, other quantities such as precipitation and further greenhouse gases, or interactions of climate factors, and the wide range of impact variables are visualized only rarely).[75]

Skeptical visualizations are easy to produce (high appropriateness). However, in many cases the most important expressiveness criterion is violated, mostly by not representing all the data within the graphic and by providing misleading image captions or interpretations.

As indicated above, the main visualization techniques within climate skeptical media are time series graphs, which map the time dimension onto the x-axis and another variable onto the y-axis. To quantitatively underline this subjective impression, we systematically counted the frequency of different types of visualization within selected climate skeptical books and reports (see Table 1).

We subdivided time series graphs of climate history and future into 6 subcategories: 1 to 15 years before present, 15 to 60 years before present, 60 to 200 years before present, time series with future projections and time series representing the course of a year. In addition, we counted bar or pie charts, scatter

75 The NIPCC report is an exception to this observation.

plots,[76] spatial representations[77] and others. Furthermore, we gathered the number of directly reproduced IPCC graphics (last column).

This investigation shows that time series graphs for 15–60, 60–200 and more than 200 years before present are most frequently used in the skeptical media analyzed here. Future projections and spatial presentations are less common. We argue that this has its reasons in the typical arguments in climate skeptic discourse, and in the relatively easy production and high degree of understandability of such time series representations. In particular, discussions regarding past temperature proxies are omnipresent. As an exception, the NIPCC report has a very low number of data visualizations, and many scatter plots with a higher degree of complexity.

Table 2: Quantification of visualization techniques used in climate skeptical publications.

Visualization type \ Source		**NIPCC-IR (2011)[78]**	**Goreham (2012)**	**Vahrenholt & Lüning (2012)**
Time series	**1–15 years b.p.**	1	2	2
	15–60 years b.p.	8	14	29
	60–200 years b.p.	6	7	30
	>200 years b.p.	8	19	39
	with future	3	1	10
	course of the year	0	2	0
Bar or pie chart		1	10	3
Scatter-plot		9	1	1
Spatial / map		3	4	3
Others		0	2	12
IPCC charts		1	8	6

76 In a scatter plot, arbitrary variables are mapped onto the x- and y-axes, with points or other graphical primitives mapped to the x and y position with respect to their value.

77 Typically, geographical maps.

78 Idso, Carter, and Singer, *Climate Change Reconsidered (NIPCC)*, 2011.

CONCLUSION AND FUTURE WORK

To summarize, within skeptical media, photographs and expert graphics are both major tactics for casting doubt on climate science. They are employed to present basic findings about climate change as an open discussion. The visual language of skeptics' graphs is comparable to the images used by climate researchers, however the number of applied types of visualization techniques and their complexity is considerably lower. For readers of such graphics, indicators for misleading skeptics' images include the use of time series spanning an interval which is too short to show long-term climate trends (say three decades or less), the display of time series which end long before the current date, or of series that are specific for a limited geographical range only (e.g. single measurement stations) within a global argumentation. However, it is not a straightforward matter of generalizing visual indicators for detecting misleading images. Probably the best way is to judge the context of the image and the reliability of the source, and to compare it to the arguments presented from more reliable sources. This is made difficult, however, by the inherent tendency of the Internet to link to content presenting a similar point of view, and by a general "illiteracy" many people have when it comes to interpreting and judging scientific information. But it is not about literacy only: the understanding of scientific graphs is also influenced by already existing beliefs, values and ideologies that might effect judgements about credibility even more strongly—an issue that could not be adequately adressed in this paper by looking at the images only.

We stated at the beginning that such expert graphics of climate change in general have become political images. But while such images also play a fundamental role in the communication of mainstream climate science as an object of socio-political concern, we found that there is a distinction in the usage of graphics between the two groups: when climate researchers highlight their findings graphically they need to stick to the correct frame of scientific working. Otherwise their results get rejected in the strict process of peer-reviewing; by contrast, climate change skeptics do not expose themselves to the peer-reviewing process within climate science. This is why they can use graphics according to the strategies of cherry picking: they manipulate the data, single out arguments or distort findings in order to intentionally misinform, mislead and deceive readers.

Mark Twain's reservations about statistics as a means of deception and being open to misinterpretation is still strongly valid. In the struggle to convince climate skeptical graphs often even assume a role as means of propaganda in the literal sense—selectively presenting only one side of an argument in order to effectively influence people's awareness of the topic.

Data graphics are very dominant evidence in the context of climate science in general. But because they are full of preconditions they are not at all easy to interpret for people not involved in climate research. By presenting mainly time series curves of temperature and CO_2 they decided to use the most simple and straightforward pictures to communicate doubt. Timeline images seem to be understandable without previous knowledge. At the same time, by selectively showing only time series graphs, skeptics are unnoticeably leaving out the complexities of interactions, geographical distributions and other weight of evidence, all of which need to be considered to understand climate change. The simplicity of the graphs compliments the publishing conditions of the web: it is actually playing into the skeptics' hands.

We also observed a sharp difference in the targeted reader groups. While climate researchers address both inner and outer scientific communities, skeptics primarily aim to convince people with no background in the field. They mainly address the outer scientific communities and therefore the credibility of the broader public and lay people. Arguments and visual evidence often only seem scientific but are actually detached from the scientific process of scrutinizing and peer reviewing. The result is a paradox: while strongly relying on scientific graphs, climate change skeptics present reviewed climate science as a process that is not trustworthy at all. Either there is too much of a consensus among climate researchers, in which case it is conspiracy or biased, or climate researchers contradict each other with certain results—then the diagnosis is uncertainty. In both cases, climate science is given a vote of no confidence.

We see future work taking different directions. Firstly, it would be interesting to test what effect both skeptical and scientifically reviewed expert graphics have on lay people with regard to credibility and understandability. Secondly, it would be helpful to research ways in which the scientific understanding of the general public and the way it deals with information in modern media could be enhanced—resp. how media users could be encouraged to not only scrutinize the message but also the messenger. Thirdly—and this would be the issue for further research in theory—the skeptical debate proves that the "case of climate change shows the difficulties of distinguishing strictly scientific questions for experts alone to decide from associated value-laden and often politicized matters of concern for the lay public to debate."[79] Here it becomes important to do further research on the ideologies, beliefs and interests conntected to the contradictory perceptions of climate change and to the consequences to be taken today.

79 Demeritt, "Science Studies, Climate Change," 2006, 474.

BIBLIOGRAPHY

Beck, Ulrich. *Risk Society: Towards a New Modernity*, trans. M. Ritter, London: Sage 1992.

Boykoff, Maxwell and Jules Boykoff. "Balance as Bias: Global Warming and the US Prestige Press." *Global Environmental Change* 15, no. 4 (2004): 125–36.

Brasseur, Lee. "Florence Nightingale's Visual Rhetoric in the Rose Diagrams." *Technical Communication Quarterly* 14, no. 2 (2005): 161–82.

—. *Visualizing Technical Information: A Cultural Critique*. Amityville, NY: Baywood Pub., 2003

Card, Stuart. "Information Visualization." In *The Human-Computer Interaction Handbook: Fundamentals, Evolving Technologies, and Emerging Applications*, edited by Andrew Sears and Julie A. Jacko, Lawrence Erlbaum, 2003, 544–82.

Cook, John. The Scientific Guide to Global Warming Skepticism, posted December 2010, http://www.skepticalscience.com/The-Scientific-Guide-to-Global-Warming-Skepticism.html

Dansgaard, Willi and S. J. Johnsen. "A flow model and a time scale for the ice core from Camp Century, Greenland." *Journal of Glaciology* 8, no. 53 (1969): 215–23.

Dansgaard, Willi et al., "North Atlantic Climate Oscillations Revealed by Deep Greenland Ice Cores." In *Climate Processes and Climatic Sensitivity*, Geophysical Monograph Series 29, edited by J. E. Hansen and T. Takahashi, 288–298. Washington D.C.: American Geophysical Union, 1984.

Demeritt, David. "Science Studies, Climate Change and the Prospects for Constructivist Critique" *Economy and Society* 35, no. 3 (2006): 453–479.

Friis-Christensen, Eigil and Knud Lassen. "*Length of the Solar Cycle. An Indicator of Solar Activity Closely Associated with Climate.*" *Science, New Series* 254, no. 5032 (1991): 698–700.

Goreham, Steve. *Climatism!: Science, Common Sense, and the 21st Century's Hottest Topic*. New Lenox: New Lenox Books, 2010.

Goreham, Steve. *The Mad, Mad, Mad World of Climatism: Mankind and Climate Change Mania*. New Lenox: New Lenox Books, 2012.

Greitmeyer, Tobias. "Beware of Climate Skeptic Films." *Journal of Environmental Psychology* 35, (2013): 105–109.

Hansen, James E. et al., "Climate Impact of Increasing Atmospheric Carbon Dioxide," *Science* 28, (1981).

Huff, Darrell. *How to Lie with Statistics*. New York: Norton, 1954.

Hulme, Michael. *Why we Disagree About Climate Change. Understanding Controversy, Inaction and Opportunity*. Cambridge, UK: Cambridge University Press, 2009.

Idso, Craig D., Robert M. Carter, and S. Fred Singer, editors. *Climate Change Reconsidered: 2011 Interim Report of the Nongovernmental Panel on Climate Change (NIPCC)*. Chicago, IL: The Heartland Institute, 2011.

Jastrow, Robert, Frederick Seitz, and William Nierenberg: "Global Warming: What Does the Science Tell Us?", George Marshall Institute, August, 1991.

Jaeger, Carlo and Julia Jaeger. „Warum zwei Grad?" *Politik und Zeitgeschichte* 32–33, (2010): 7–15.

Kostelnick, Charles and Michael Hassett. *Shaping Information. The Rhetoric of Visual Conventions*. Carbondale: Southern Illinois University Press, 2003.

Latour, Bruno. "Drawing Things Together." In *Representation in Scientific Practice*, edited by Michael Lynch and Steve Woolgar, 19–68, Cambridge, MA: MIT Press, 1990.

Leuschner, Anna. *Die Glaubwürdigkeit der Wissenschaft. Eine wissenschafts- und erkenntnistheoretische Analyse am Beispiel der Klimaforschung*. Bielefeld: transcript, 2011.

Link, Jürgen and Mirko M. Hall. "The Normalistic Subject and Its Curves: On the Symbolic Visualization of Orienteering Data." *Cultural Critique*, no. 57 (Spring, 2004): 47–67, University of Minnesota Press.

Mackinlay, Jock. "Automating the Design of Graphical Presentations of Relational Information." *ACM Transactions on Graphics* 5, no. 2 (1986): 110–141.

McCright, Aaron. "Dealing with Climate Change Contrarians." In *Creating a Climate for Change: Communicating Climate Change and Facilitating Social Change,* edited by Susanne Moser and Lisa Dilling, 200–212. Cambridge, UK: Cambridge University Press, 2007.

Norgaard, Kari. *Living in Denial: Climate Change, Emotions and Everyday Life*. Boston, MA: MIT Press, 2011.

Oreskes, Naomi and Erik M. Conway. *Merchants of Doubt. How a Handful of Scientists Obscured the Truth on Issues from Tobacco Smoke to Global Warming*. New York: Bloomsbury Press, 2011.

Painter, James. *Poles Apart: The International Reporting of Climate Scepticism*. Oxford: Reuters Inst. for the Study of Journalism, 2011.

Pielke, Roger A. *The Honest Broker: Making Sense of Science in Policy and Politics*. Cambridge, UK: Cambridge University Press, 2007.

Porter, Theodore M. *Trust in Numbers: The Pursuit of Objectivity in Science and Public Life*. Princeton, N.J.: Princeton University Press, 1995.

Rosentrater, Lynn D. "Representing and Using Scenarios for Responding to Climate Change". In: *WIREs Climate Change* 1, no. 2 (2010): 253–59.

Schönwiese, Christian D. *Klimaänderungen. Daten, Analysen, Prognosen*. Berlin: Springer, 1995.

—. et al., *Das "Treibhaus"-Problem: Emissionen und Klimaeffekte. Eine aktuelle wissenschaftliche Bestandsaufnahme*. Berichte des Instituts für Meteorologie und Geophysik der Universität Frankfurt/Main, no. 96. Frankfurt/Main: Institut für Meteorologie, 1994.

Stauning, Peter. "Solar Activity—Climate Relations: A Different Approach," *Journal of Atmospheric and Solar-Terrestrial Physics* 73, no. 13 (2011): 1999–2012.

Thejll, Peter and Knud Lassen. "Solar Forcing of the Northern Hemisphere Land Air Temperature: new data," Scientific Report 99–9, *Copenhagen: Danish Meteorological Institute*, 1999.

Tufte, Edward R. *The Visual Display of Quantitative Information*. Cheshire, CT.: Graphics Press, 1983.

Twain, Mark. *Life on the Mississippi*, Vol. 1. Leipzig: Tauchnitz, 1883.

Vahrenholt, Fritz and Sebastian Lüning. *Die kalte Sonne: Warum die Klimakatastrophe nicht stattfindet*. Hamburg: Hoffmann und Campe, 2012.

Walsh, Lynda. "Visual Strategies to Integrate Ethos Across the "is/ought" Divide in the IPCC's Climate Change 2007: Summary for Policy Makers." *Poroi, Issues in the Rhetoric of Science and Technology* 6, no. 2 (2009): 33–61.

Washington, Haydn and John Cook. *Climate Change Denial: Heads in the Sand*, London and New York: Earthscan, 2011.

Watts, Anthony, blog published Jan 2, 2011, http://wattsupwiththat.com/2011/01/02/do-solar-scientists-still-think-that-recent-warming-is-too-large-to-explain-by-solar-activity/.

Weingart, Peter, Anita Engels, and Petra Pansegrau. *Von der Hypothese zur Katastrophe: Der anthropogene Klimawandel im Diskurs zwischen Wissenschaft, Politik und Massenmedien*. Opladen: Barbara Budrich, 2008.

IMAGE COPYRIGHTS & SOURCES

Figure 1: CO_2science. “Comparison of Plant Growth under two Different CO_2 Concentration Conditions”. Online video. 2010. wattsupwiththat.com/2011/09/29/plants-gobbling-up-co2-45-more-than-thought/.

Figure 2: Jastrow, Robert, William A. Nierenberg and Frederik Seitz. “Comparision between observed global average temperature and calculations by Hansen et al.,” *Global warming: What does the science tell us?* Graph. 1991. www.sciencedirect.com/science/article/pii/0360544291900068.

Figure 3: Hansen, J., D. Johnson, A. Lacis, S. Lebedeff, P. Lee, D. Rind, and G. Russell. “Climate Impact of Increasing Atmospheric Carbon Dioxide,” *Science,* 213: 963. Graph. 1981. www.atmos.washington.edu/~davidc/ATMS211/articles_optional/Hansen81_CO2_Impact.pdf.

Figure 4a: Friis-Christensen, Eigil, and Knud Lassen. „Length of the Solar Cylce. An Indicator of Solar Activity Closely Associated with Climate,“ *Science,* New Series, Vol 254, Nr. 5032, 698–700. Graph. 1991. errortheory.blogspot.de/2013/05/is-john-cook-planning-to-use_3465.html.

Figure 4b: Stauning, P. “Solar activity—climate relations: a different approach,” *Journal of Atmospheric and Solar-Terrestrial Physics,* Volume 73, Issue 13, August 2011. Graph. 2011. www.sciencedirect.com/science/article/pii/S1364682611001866.

Figure 5: D’Aleo, Joseph. „Correlation Last Decade and This Century CO2 and Global temperatures Not There.“ Paper published by *SPPI* (Science and Public Policy Institute). Graph. 2008. http://icecap.us/images/uploads/Correlation_Last_Decade.pdf.

Figure 6a: CFACT (Committee for a Constructive Tomorrow). „No global warming for 16 years.“ Photo. 2013. www.cfact.org/2013/03/03/pictures-of-cfacts-d-c-billboard-campaign-in-the-field-no-global-warming-for-16-years

Figure 6b: http://www.dailymail.co.uk/sciencetech/article-2217286/Global-warming-stopped-16-years-ago-reveals-Met-Office-report-quietly-released--chart prove-it.html. Accessed December 3, 2013.

Figure 7: Met Office. “Global annual average temperature anomaly time series.” Graph. 2013. http://www.metoffice.gov.uk/research/monitoring/climate/surface-temperature.

Figure 8: Goreham, Steve. *The Mad, Mad, Mad World of Climatism*. Graph. New Lenox: New Lenox Books, 2012: 63.

Figure 9: Goreham, Steve. *Climatism!*, Graph. New Lenox: New Lenox Books, 2010: 381.

Towards an Interactive Visual Understanding of Climate Change Findings on the Net: Promises and Challenges

MARKUS WROBEL, DOMINIK REUSSER

Web-based applications can facilitate the dissemination of scientific information and reduce existing access barriers to climate change research findings. In addition, interactive visualization offers increased flexibility in representing data compared to static pictures, and allows the user to adapt the visual representation to her or his personal interest to some extent. The focus of this article is on benefits and challenges emerging from combining these two information technological strands in order to improve the dissemination of scientific results on climate change to a wider audience. Web-based interactive visual representations of climate change findings (WIVCC) have the potential to improve the communication of research results by supporting users in better comprehending the inherent complex, multi-dimensional data of this field. However, applying WIVCC for effective climate change communication is no trivial task. We analyze current challenges by integrating findings from various disciplines, including climate change research, climate change communication, computer-based visualization and user interface design. After outlining general benefits and challenges and introducing two examples of existing applications, we address two issues in greater detail and present results relating to (i) the lack of established conventions in designing user interfaces for WIVCC applications and (ii) appropriate visual representation of spatially heterogeneous uncertainty. We discuss these findings together with a set of further open questions related to designing effective interactive visual representations of scientific results to foster web-based climate change communication.

1 INTRODUCTION

Building awareness is essential for the shift in behavior necessary to avoid dangerous climate change and to advance adaptation. Efficient communication of the scientific understanding of anthropogenic climate change can support such awareness building, and visual representations play a central role within this communication process. Two broad categories can be distinguished:

(1) The first category relates to (often iconic) images intended to evoke emotions towards climate change. Such images typically aim to illustrate negative effects of climate change in a concrete manner, e.g., by depicting a polar bear in an environment affected by increasing temperature. It has been recognized that images can be influential in determining peoples' perceptions of climate change.[1] Hulme[2] discusses the combined role of pictures, (narrative) scenarios, and probabilities in climate change communication and suggests using additionally a set of icons (including both natural systems like coral reefs, cities and individual species) focusing on a relatively small number of highly visible climate change impacts for communicating dangerous climate change. Images with emotional connotations play a prominent role in climate change campaigns;[3] typical examples include pictures showing polar bears[4] or melting glaciers.[5] The persuasive power of images falling into this first category raises , among other things, ethical concerns that need to be carefully considered.[6]

(2) The second category of visual representations in the context of climate change communication relates to the visualization of scientific results. Visual representations of this category are typically more abstract than those of the first category. They are, for example, applied to depict observed or modelled climate variables in their spatial distribution or their development over time, often in the form of maps and diagrams (various examples include the figures provided with the 4th IPCC report[7]). Importantly, as distinct from category one, visual

1 e.g. Nicholson-Cole, "Representing Climate Change Futures," 2005.

2 Hulme, Pictures, *Scenarios or Probabilities*, 2004.

3 e.g. Manzo, "Imaging Vulnerability: The Iconography of Climate Change," 2010.

4 Slocum, *Polar bears and Energy-efficient Lightbulbs*, 2004.

5 Doyle, "Picturing the Clima(c)tic: Greenpeace," 2007.

6 Nicholson-Cole, ibid.; Sheppard, "Landscape Visualisation and Climate Change," 2005.

7 IPCC, "Synthesis Report: Climate Change 2007. An Assessment of the Intergovernmental Panel on Climate Change," 2007.

representations of the second category need to adhere strictly to the requirements of scientific visualization[8]: their role and duty is to truly and solely represent the information contained in scientific data rather than to primarily evoke emotions. Note that categories one and two cannot always be strictly separated: although the requirements of scientific accuracy need to be strictly fulfilled for images of the second category, this does not imply that such images cannot trigger emotions towards climate change.[9]

The focus of this article is on a specific subtype of the second category: web-based interactive visual representations aiming to communicate scientific results on climate change (hereafter referred to as WIVCC). *Interactivity* relates to the fact that the user is to some degree enabled to interact with the visual representation, for instance, by selecting the kind of data to be depicted, by determining location and temporal reference for the data to be displayed, or by obtaining feedback on specific data values by moving the mouse cursor over their visual representation. *Web-based* means that such functionality is provided remotely over the Internet and can be accessed via a typical web browser. An appropriate combination of these two information technological strands can aid science in the necessary popularization of climate change results.

While static images have been made available via the web since its early days, current web technologies additionally facilitate developing *interactive* visual representations. As compared to static publication of information, these allow for increased interaction with the underlying data by enabling the user to influence and adjust the data's visual representation. Through WIVCC, it is thus possible to provide the user with quantitative information that is less aggregated, and to offer more freedom in exploring, selecting and determining the degree of detail in the data to be displayed. In particular, WIVCC extends a passive reception of climate change research results by enabling the wider audience a limited degree of data exploration. We illustrate this advantage with a typical example from the climate change research context: if model-based projections are represented using a static map, the data that can be depicted typically needs to be restricted to a single variable (e.g. temperature), to a fixed temporal reference and aggregation (e.g. a 30-year average), and either to a single model and scenario or to an aggregation over several models and / or scenarios. Using an interactive map, on the other hand, the user can be enabled to dynamically change both content and temporal reference to be depicted, and additionally to

8 e.g. Mackinlay, "Automating the Design of Graphical Presentations of Relational Information," 1986, 110-141.

9 e.g. Schneider, "Image Politics," 2011; Walsh, "Visual Strategies," 2009.

use zooming to obtain either on a general overview or to view locations of interest in greater detail. Accordingly, such visual representations are increasingly utilized as part of web-based applications addressing the dissemination of climate change related research results.

To date, neither the visual communication as aspect of climate change communication, nor climate change communication using new media have been sufficiently researched and understood.[10] There is also little specific and integrated information about the benefits and challenges of communicating results from climate change research with WIVCC available in the literature. There are several fields of research that can contribute to such an understanding. On the one hand, the general discussion of communication of scientific information on climate change (e.g. Moser and Dilling 2007) also applies here. On the other hand, findings from interactive visualization in general, as well as from Human-Computer Interaction (HCI), contribute to the body of understanding.

In the remainder, we first give a brief introduction to the challenge of climate change communication and its relation to interactive visualization and web-based information access. Subsequently, we narrow down our focus to two out of the many points of importance in the context of WIVCC, the user interface and the communication of uncertainty. To give some practical examples, we will discuss both aspects briefly against two out of the range of established web applications designed for access to climate change related information (sec. 2). By presenting empirical investigations, a more detailed view on challenges related to two relevant issues is given, the design of suitable user interfaces for WIVCC and the visual communication of uncertainty (sec. 3). Finally, we briefly discuss further challenges in the context of WIVCC (sec. 4).

2 VISUALIZATION AND THE COMMUNICATION OF CLIMATE CHANGE FINDINGS

In general, reconciling the supply of scientific information with user demands needs to be improved. Large research gaps exist, for instance, related to the nature of decision processes and to the selection of relevant supportive information (intelligence phase) as well as to appropriate aggregation and presentation (promotion phase).[11] Sufficient understanding of the target audience

10 e.g. Moser, "Communicating Climate Change," 2010, 31–53.

11 McNie, "Reconciling the Supply of Scientific Information with User Demands: An Analysis of the Problem and Review of the Literature," 2007, 17–38.

is crucial to enhance the supply of relevant information. The challenge of efficient communication of climate change to policy and decision-makers as well as to a broader audience is neither trivial nor has it been solved.[12] Further, various criteria contribute to specific difficulties in communicating climate change, including inherent complexity, uncertainty and a lack of immediacy which render it difficult to perceive causes and impacts.[13]

2.1 The Role of Interactive Visualization for Climate Change Research

Visualization can support comprehension of large amounts of data as well as perception of data properties that have not been anticipated.[14] Due to such benefits, visualization is an important instrument for aiding research: scientific visualization[15] supports the understanding of scientific data and is of high relevance for research conducted in various disciplines. It can be applied for explorative and confirmative analyses as well as for the presentation and communication of results.[16] This also holds true for climate change research: besides playing a key role in supporting research, visual data representations are additionally of importance in the communication of scientific results from this field.[17] Interactive computer-based visualization allows users to create and adjust visual artefacts on the fly, and to decouple the visualization designer from the data to be represented.[18] The benefits of fast interactive computer-based visualization can greatly improve the understanding of data,[19] for instance, by allowing for comparison of different views, the posing of queries and receipt of immediate answers.[20] While the advantages of interactive visualization are crucial for the scientific analysis of data, they can also be helpful for popularizing climate change results to a wider audience.

12 e.g., Moser, ibid.; Patt, "Communicating Uncertainty to Policy Makers," 2009.

13 For a discussion of these and other aspects that add to the difficulties in communicating climate change see Moser, ibid.

14 e.g. Ware, Information Visualization, 1999.

15 McCormick et al., "Visualization in Scientific Computing," 1987.

16 Schumann and Müller, Visualisierung, 2000.

17 Nocke et al., "Visualization of Climate and Climate Change Data," 2008.

18 Card et al., *Readings in Information Visualization*, 1999.

19 Ibid.

20 Cook and Swayne, *Interactive and Dynamic Graphics for Data Analysis*, 2007.

Figure 1: Schematic relation between the amounts of scientific data included in scientific analysis, in interactive and in static result presentation (own illustration)

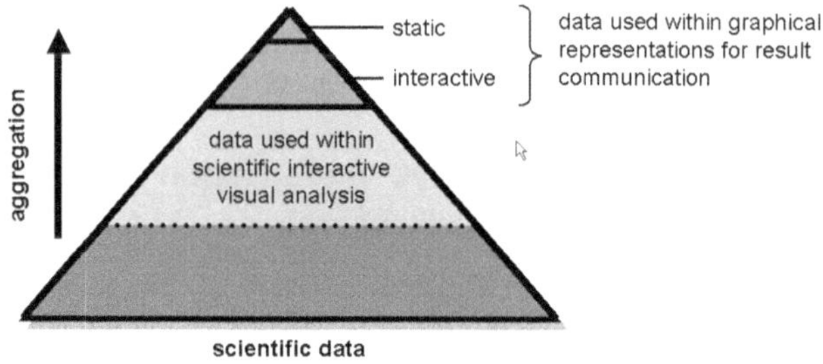

In the context of communicating results from climate change research, interactive visualization has several advantages over static, e.g. print-based visual representations. Data in this context is typically spatio-temporal and inherently multi-dimensional. Since interactive visualizations allow the user to influence the visual representation of data, she/he can inspect more data than with static representations and adapt the representation to content of personal interest. By enabling the user to zoom in and out of a visual representation, it is possible to look both at larger patterns and smaller features of interest, hereby following the basic principle of Shneiderman's often quoted 'Visual Information Seeking Mantra': "Overview first, zoom and filter, then details-on-demand."[21]

WIVCC increases the amount of data made accessible to the end user as compared to static visual representations. This situation is depicted in Figure 1. While static visual representations typically present data that is highly aggregated along multiple dimensions, interactive visual representations can allow the user to look at more data, e.g., by offering to select between various predefined temporal aggregations (including long-term annual, monthly and seasonal aggregations). Note, however, that this extended data representation is not to be confused with a free exploration of the "raw" data[22] underlying these aggregations. Analyses of raw data do require a sufficient level of expertise to ensure correct data handling and processing and to avoid creating inappropriate, misleading visual representations and misinterpretation of the data involved. Thus, despite supporting an extended degree of freedom, WIVCC also needs to guide and carefully restrict the user. This includes, for example, constraining

21 Shneiderman, "The Eyes Have It," 1996.

22 Such data is provided for example by the IPCC Data Distribution Centre (http://www.ipcc-data.org/).

access to meaningful aggregations pre-defined on the supply side as well as sensibly limiting the supported degree of freedom in mapping the data to visual variables.

2.2 Climate Change Information Applications

In future, science will be increasingly required to provide access to results and related datasets via the web. Such deployment can improve the link between society and science, since selection of information can, to some extent, be determined on the demand side.[23] Visual representations can play an important role in this process where a broad audience with different levels of expertise and information demands can be reached, not least since web sites which are too text-heavy are often not well perceived.[24] At present, WIVCC is offered by an increasing number of applications provided by various institutions. We briefly present two examples to depict the current state of the art and to outline how interactive visual representations are used. The first application is the web-based visualization tool of the IPCC Data Distribution Centre, the second the Climate Impacts: Global and Regional Adaptation Support Platform (ci:grasp). Both applications use interactive maps to present information about projected future climate data such as temperature and precipitation. We highlight basic characteristics of the user interface, the treatment of uncertainty and differences in the information on offer.

IPCC Data Distribution Centre. The assessment reports of the IPCC, (e.g. IPCC 2007) build on a multitude of Global/General Circulation Model (GCM) projections generated by a pool of different models and for a set of emission scenarios. The web-based visualization tool of the IPCC Data Distribution Centre[25] (Figure 2a) allows the interactive generation of maps of GCM projections that have been used for the 4th (AR4) or the 3rd (AR3) IPCC assessment report. In addition, maps of observational climatologies from the Climate Research Unit (CRU) can be displayed. The application's user interface consists of an interactive map that can be zoomed and panned plus the necessary controls for data selection. These controls, organized in three different panels arranged below and to the right of the map, are used in sequential order to specify the data to be visualized. To display a GCM projection, the user selects

23 McNie, ibid.

24 For special requirements on using text in the web see, e.g., Krug, *Don't Make me Think*, 1989.

25 http://www.ipcc-data.org/maps/

(i) between AR4 or AR3, (ii) an emission scenario, (iii) either mean climatologies or mean anomalies for 20 or for 30 years, (iv) a GCM, and (v) one of the available variables generated by the GCM (e.g. precipitation anomaly). After a variable has been selected, the map is updated accordingly and (vi) a different temporal reference (long-term monthly aggregations) can be specified subsequently. Uncertainties are treated implicitly by providing data for various scenarios and models along with short descriptions. Since this application focuses only on the data display we just outlined, the complete user interface is made available on a single web page. This page is provided together with a set of other visualization applications for IPCC related data (maps and diagrams) via an entry page[26] from which the various applications can be reached.

Ci:grasp. The range of information offered on the Climate Impacts: Global and Regional Adaptation Support Platform ci:grasp-2[27] is broad (Figures 2b and 2c). Besides climate projections, derived expected impacts of climate change and a database of climate adaptation projects with detailed information are presented. In addition to interactive map representations, projected climate is also depicted with Walter diagrams (Figure 2c) and further time series plots which the user can select by picking[28] on an interactive map. The ci:grasp-2 user interface allows access to the information through multiple ways, e.g., through a set of thematic modules related to topics like stimuli, impacts, cities or transitions, and through a graphical navigation tool based on climate impact chains. The latter provides links to available maps and adaptation projects related to a given climate stimulus or impact. Similar to the IPCC data visualization tool, uncertainties are mainly treated implicitly by providing results (about climate change and its impacts) for multiple scenarios and multiple model runs.

The selected examples outline only two out of a multitude of available applications that target web-based access to scientific results on climate change and, to that end, make use of interactive visual representations. Nevertheless, these examples indicate typical tasks addressed in the context of WIVCC and current approaches followed to support the user in achieving them. WIVCC makes wide use of interactive maps to display the spatial distribution of values from a set of user-selectable climate or impact variables. Another common approach is to use maps to allow selection of a spatial reference in order to

26 http://www.ipcc-data.org/ddc_visualisation.html.

27 http://www.cigrasp.org; http://www.pik-potsdam.de/cigrasp-2/; Wrobel et al., "Novel Approaches for Web-Based Access," 2013.

28 Picking refers to selecting an object by clicking its visual representation with the mouse.

Figure 2: Examples of interactive web-based visual representations in applications presenting climate change related information: (a) IPCC Data Distribution Centre Visualization Tool and (b) and (c) Climate Impacts: Global and Regional Adaptation Support Platform (ci:grasp).

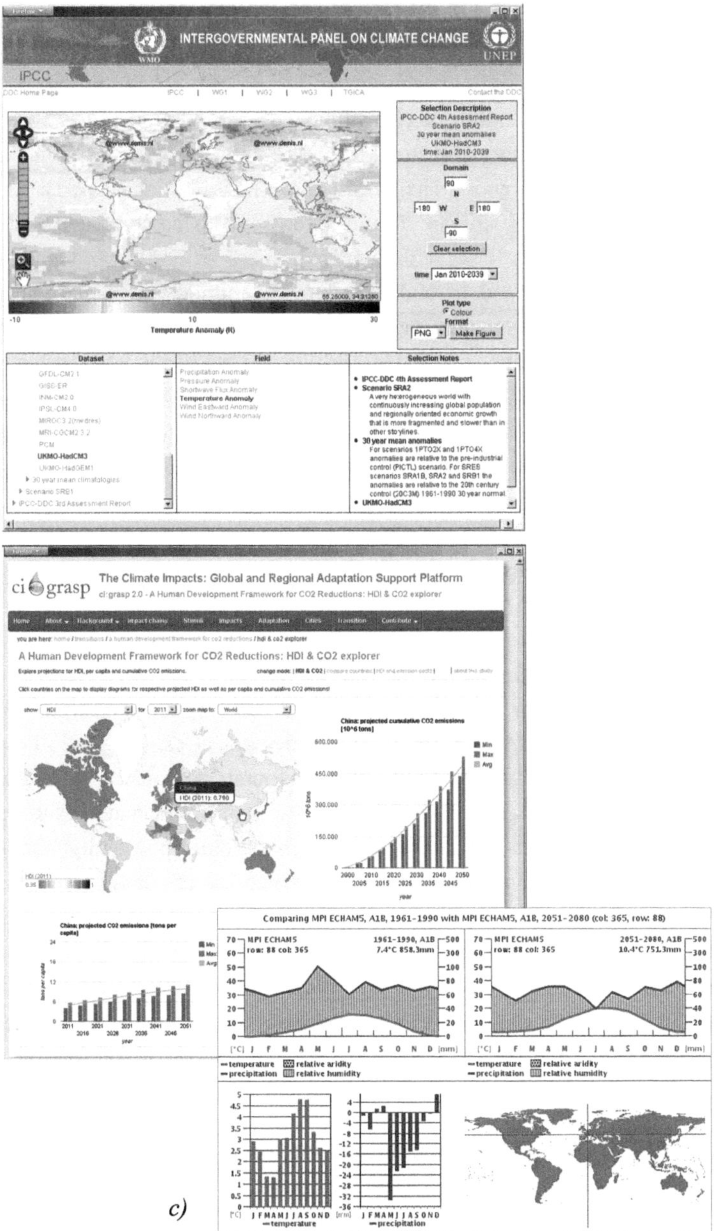

a)

b)

c)

subsequently access location-specific information (which can be both visual or textual, e.g., time series diagrams or descriptions of adaptation case studies).

3 Challenges of Interactive Web-Based Visualization of Climate Change Results

Appropriate presentation of climate change information is *per se* challenging. Interactive, web-based platforms for communicating climate change results are affected by these general challenges as well as additional issues. According to Moser[29] and others, climate change is difficult to perceive and understand for most lay audiences. Thus, communicators need to find clear, simple metaphors and imagery adjusted to and suited to extending the mental models of the audience. Further, suitable framings are required to lay the foundation for more appropriate cognitive processing. While generally not trivial, this also poses demanding challenges to the design of interactive, web-based platforms where users are typically not known in advance. Assisting multiple user roles[30] with an accordingly adjusted treatment of relevant information and its complexity and uncertainty could enhance the support for different users. However, available web-based applications providing scientific results on climate change information often require at least a basic level of background knowledge on underlying concepts and typically lack support for different user roles. Visualization in communicating climate change results has to meet the challenge of providing a dual message. Information must be objective, trustworthy and do justice to existing complexity and uncertainty. At the same time, the information should be appealing, persuasive and clear in its message. Designing web-based climate change related information applications and appropriate interactive visual representations as part of these has to also take into account that typical users will explore them autonomously, i.e., without opportunity for direct, bi-directional communication with an expert to clarify questions on content and functionality.

In the following, we focus on two out of the many challenges related to effective WIVCC. First, enabling efficient interaction of the user with visual

29 Moser, ibid.

30 Typical user roles in this context can be determined, e.g., by differences in information needs, climate change background knowledge, familiarity with interactive visualization of climate change data, etc.

representations demands *supportive user interfaces.*[31] We explore whether today, conventions in designing user interfaces for WIVCC have emerged (sec 3.1). Second, although intuitive perception of large and complex data can be supported and facilitated by visual representations, it is not sufficiently known how well such representations are suited to addressing *uncertainties*. Against this background we look more closely at the challenge of visual representation of spatially heterogeneous uncertainty (sec. 3.2).

3.1 Missing Conventions for User Interfaces to Support WIVCC

The ultimate goal to be achieved in designing computer-based applications is user acceptance. To this end, careful consideration of human factors is crucial. The user interface plays a key role in this context; it should support intuitive and fearless exploration of the application, e.g., by minimizing cognitive overload and reducing the probability of usage errors. Users' trust in applications is easily lost and it is crucial to avoid frustrating or even scaring away potential users. These non-trivial tasks become even more challenging if users with different backgrounds and preferences are to be supported.[32] The specifics inherent to the web and web-based applications pose additional usability challenges,[33] as does the increasing degree of interactivity of web sites.[34]

Central user interface challenges in the context of WIVCC relate to an appropriate use of *interactive maps* (see also sec. 2.2). Web mapping has been rated as a step change in the evolution of geographic applications on the Internet,[35] and we can observe a generally increasing role of interactive mapping in the web[36] as well as a close relation between geo-spatial web and visual methods.[37] A large amount of the data relevant for climate change research is spatially explicit; it is thus not surprising that interactive maps form a typical mean to present results from this field to a wider audience in the web.

31 The challenge of designing appropriate user interfaces for computer based applications is researched in the field of Human-Computer Interaction (e.g., Shneiderman, ibid.).

32 e.g., Shneiderman, ibid.; Norman, *The Psychology of Everyday Things*, 1988.

33 e.g. Krug, ibid.

34 Petrie and Power, "What do Users Really Care About?" 2012.

35 Haklay et al., *Web Mapping 2.0*, 2008.

36 e.g. Schmidt and Weiser, "Web Mapping Services", 2012; Haklay et al., ibid.; van Elzakker, "Use and Users of Maps on the Web," 2000; Skarlatidou and Haklay, "Public Web Mapping: Preliminary Usability Evaluation," 2006.

37 Elwood, "Geographic Information Science," 2010.

Designing appropriate map applications for WIVCC is no trivial task. First, (static) climate maps as a communication medium already come with their own challenges. McKendry and Machlis,[38] for instance, point out several cartographic flaws in a prominent map from the 4th IPCC report.[39] The ability of end users to comprehend scientific information depicted on climate maps cannot be taken for granted: in using qualified students as peers for prospective decision-makers in a usability test, Ishikawa et al.[40] found that the test persons had difficulties in correctly interpreting probability-related climate forecast maps. Second, there are usability issues related specifically to web-based interactive maps.[41] Nivala et al.[42] found in total over 400 usability problems of varying severity on four prominent general web-mapping sites. Such map-related usability problems can affect the usage of the overall application.[43] Not least, it needs to be carefully determined under which conditions interactive maps actually add to increased usability.[44] Third, specific challenges inherent to WIVCC relate to enabling the user to intuitively select the actual content to be displayed on such a map. Since climate change related content is multi-dimensional and potentially complex, maps in WIVCC need to be embedded in an appropriate user interface. It is thus important to understand to what end the design of such user interfaces supports or hampers WIVCC.

Due to the inherent complexity of climate change related information, one single WIVCC application can hardly cover all aspects of climate change driven processes. Consequently, information relevant for a given user may be dispersed across several applications, each providing a different user interface. Conventions followed in designing these various user interfaces can contribute to facilitated usage, since usage patterns could be transferred more easily from one WIVCC application to another (referred to as "transfer of training"). A lack of such user interface conventions, however, indicates the need for usability improvements. Complying with conventions when designing web sites can reduce users' learning effort and increase familiarity.[45] In this sense, the notion

38 McKendry and Machlis, "Cartographic Design," 2009.

39 IPCC: *IPCC AR4: Synthesis Report*, 2007.

40 Ishikawa, T. et al. "Climate Forecast Maps," 2005.

41 Nivala et al., "Usability Methods' Familiarity," 2007.

42 Nivala et al., "Usability Evaluation of Web Mapping Sites," 2008.

43 e.g. Meng et Malczewski, "Web-PPGIS Usability and Public Engagement," 2010.

44 see, e.g. Shneiderman, "The Eyes Have It," 1996; Wrobel and Costa, "Web Based Applications," 2012.

45 Krug, *Don't Make me Think*, 1989.

of conventions relates closely to the concept of consistency.[46] Consistency across applications can be described as *external consistency* as opposed to the internal consistency of a system or its metaphorical consistency to real world features.[47] If the same actions of the user have the same effect independent of the context, we can talk of *action-effect consistency*; if similar goals or tasks require similar sets of actions, *task-action consistency* will be the term given.[48]

We investigated whether we could observe external consistency across the user interfaces of a set of 32 different web-based climate change related information applications.[49] All applications in the sample utilized maps to provide access to information on climate change. The maps in the sample are used to (1) display the spatial distribution of values of climate change related variables, e.g., past, present or future climate, climate impacts or socio-economic information (22 applications), and/or (2) to allow a selection of spatial entities to access further information, for example, in the form of time series diagrams or text on location-specific impacts (20 applications).

The review revealed a remarkably high level of external user interface inconsistency, even for fairly simple and common tasks like zooming a map or selecting content to be displayed. The observed facets include both action-effect and task-action inconsistencies. For example, scrolling the mouse wheel might trigger a map zooming in one application but a web page scroll in another, and clicking on a map might result in map zooming in some applications while it may lead to the selection of a location in others (action-effect inconsistency). Then again, different applications require the user to perform different actions in order to achieve the same task, e.g., zooming, selecting content to be displayed on a map or updating the map display (task-action inconsistency). Note that further research is required to estimate to what extent the observed external user interface inconsistencies actually add to the cognitive workload and learning efforts of a user. However, we can assume that they are likely to contribute to usage errors while trying to interact with different WIVCC applications and to limit transfer of training. Furthermore, unexpected or unfamiliar behaviour of

46 For benefits and limits of consistency relating to the usability of user interfaces see, e.g. Grudin, "The Case against User Interface Consistency," 1989; Nielsen, "Coordinating User Interfaces," 1989; Kellogg, "The Dimensions of Consistency," 1989; Chimera and Shneiderman, "User Interface Consistency," 1993; Shneiderman, *Designing the User Interface,* 1998; Pyla et al., "Multiple User Interfaces," 2006.

47 Grudin, ibid.

48 Monk, *Noddy's Guide to Consistency*, 2000.

49 Wrobel et al. "A Review of User Interface Conventions," 2010.

interactive visualizations in the web can irritate the user and affect cognitive processes, inadvertently building up an additional barrier to the communication of climate change.

3.2 Visualizing Uncertainty

Uncertainties about climate change and its impacts are in some cases substantial, and spatially heterogeneous. While current information applications treat uncertainties often as rather implicit, this section discusses challenges for a more explicit visual representation of uncertainties. Uncertainty in our understanding of climate change and its effects has multiple sources.[50] Some processes are random and non-linear in nature, making predictions about the development inherently uncertain (stochastic uncertainty). Also, the current understanding and ways of representation of the relation between factors may be limited, thus improving knowledge may reduce uncertainty (epistemic uncertainty). Finally, in the case of humans and societies, the discussion about scenarios of the future may alter behaviour and policies. In this way, scenarios will affect the very future they attempt to envision, which causes additional uncertainty about human-nature interactions (human-reflexive uncertainty).

The existence of these uncertainties has an effect on decision-making. In some cases, even if the uncertainties are well known and clearly represented, humans will fail to make the right decisions, for example due to bias with respect to small probabilities.[51] Strategies used to present uncertainty for decision-making not only depend on the decision issue itself but also on the scientific theory used to describe the decision process.[52] Patt[53] distinguishes three categories of decision theory: economic, psychological and political, and discusses decision failures in view of each of these theories. He concludes that the presentation of uncertain information should be matched to the thought process of the target audience. Recommended information includes the range of potential outcomes, a verbal description of the likelihood of important outcomes, the speed of the process of uncertainty resolution, the source of uncertainty and some guidance for exemplary problems indicating if they are sensitive to the uncertainty at hand. These considerations are challenging in themselves and do not explicitly address the visual presentation of uncertainties, for instance, in maps.

50 e.g. Dessai and Hulme, "Does Climate Adaptation Policy Need Probabilities?" 2004.

51 Kahneman et al., "Judgment under Uncertainty: Heuristics and Biases," 1994.

52 Patt, "Communicating Uncertainty to Policy Makers," 2009.

53 Patt, ibid.

Spatially distributed uncertainties add to the general challenge of presenting uncertainties. Pang[54] discusses the visualization of uncertainty in natural hazards and reviews a range of techniques used to visualize spatially heterogeneous uncertainties. Focusing on spatial information and maps, Senaratne and Gerharz[55] provide a categorization system for uncertainty visualization techniques. MacEachren et al.[56] identify seven (still) open research questions with respect to visualization of information uncertainty on maps, especially in the context of decisionmaking. One of these open questions is about assessing the usability of uncertainty representation techniques.

Common techniques used to present uncertainties include adjacent maps, contouring, symbols and transparency as illustrated in Figure 3. Senaratne et al. (2012) observed that user performance in interpreting uncertainties varies depending on technique and task. In the worst case, participants may produce a wrong answer, while having the impression that the visualization is easy to understand. Thus, they may be overconfident in their judgement. Against this background, it seems appropriate to offer the user multiple uncertainty visualization techniques; on the downside this would also increase application complexity and thus add to existing user interface challenges (sec. 3.1).

Flexibility may improve efficiency of the visual presentation of uncertain information. However, getting the visualization technique right is not sufficient; appropriate communication of uncertainties requires a well-founded understanding of the target audience's thought and decision processes.[57] This is also in agreement with problems reported in understanding probabilistic climate forecast maps.[58] Print-based publications of climate change results like IPCC (2007) already apply a multitude of techniques to depict uncertainties. By contrast, the number of approaches for explicit uncertainty visualization in current WIVCC is as yet limited, and often restricted to representing different model projections and scenarios using maps or time series plots.[59] Further research is needed to better understand this crucial aspect of WIVCC based communication.

---

54 Pang, "Visualizing Uncertainty in Natural Hazards," 2008.

55 Senaratne et al., "Usability of Spatio-Temporal Uncertainty Visualisation," 2012.

56 MacEachren et al., "Visualizing Geospatial Information Uncertainty," 2005.

57 Patt, ibid.

58 Ishikawa et al., "Climate Forecast Maps," 2005.

59 This information is often offered by allowing a user defined selection of model and scenario to be depicted on a map or in aggregated manner, e.g., in form of quantiles (for an example of the latter see http://www.climatewizard.org).

Figure 3: Commonly used visualization techniques to represent uncertainties:

a) Using transparency: information is represented by a fully saturated color scale between yellow and red when the information is certain. If data is uncertain in a region, the color saturation is reduced, giving the impression that the map is transparent;

b) Adjacent maps, the left one shows the data using a color scale between yellow and red. Uncertainty is presented in the right-hand map with white areas showing low uncertainty and blue areas showing high uncertainty;

c) Contours, with thick contours signifying high uncertainties; and

d) Texture of symbols (dots) of increasing size with increasing uncertainty.

The yellow-red color scale is used to represent synthetic data while blue is used to represent synthetic uncertainties. Since synthetic data is used, scales and coordinates are omitted. (→ color plate 11).

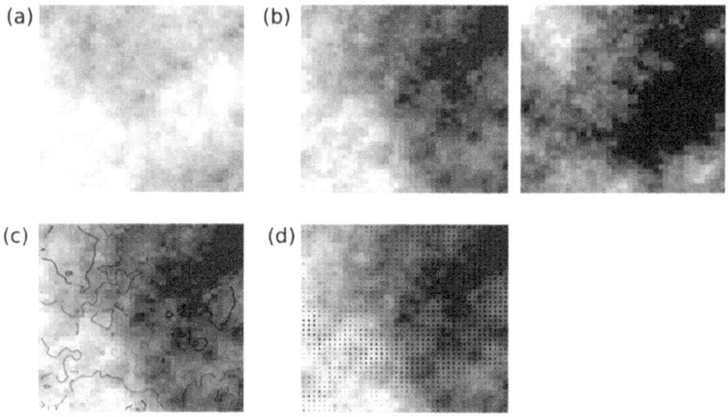

4 DISCUSSION

Applying WIVCC is an interdisciplinary task that requires building upon a range of existing and ongoing research. The integration of these different evolving sources of knowledge is still a great challenge. We have illustrated the nature of existing open questions in section 3 by focusing on the user interface and the representation of uncertainty as two representative challenges out of a large number of issues to be addressed in this context. In the following, we broaden the discussion and briefly highlight further unresolved challenges in designing and applying WIVCC to improve access to scientific information for a wider audience. Issues include fostering a more localized experience of climate change, target audience specific communication, ethical aspects, and complementing the communication with climate change experts.

Hulme has observed and criticized the fact that the cultural, local dimensions of climate tend to get lost in an understanding of climate change dominated by natural sciences. He states that this absence might contribute to the explanation of why progress "*towards meeting stated climate change policy goals is so tortuous*" and calls for a stronger involvement of geographers in climate change research to remedy this situation.[60] According to Nicholson-Cole, visualization of climate change designed to engage people should be easy to relate to and applicable both spatially (locally recognisable) and temporall (not too far in future).[61] We would argue that appropriate WIVCC bears the potential to contribute to a more 'personalized' perception of climate change research results. For example, interactive maps that combine the display of climate data projections with a base map showing satellite terrain images and allow zooming to recognizable locations might add to such a localized experience. This may also be achieved by applications that allow picking diagrams and other climate change related information for a location that is familiar to the user. The actual effectiveness of such approaches has yet to be researched. An additional, yet crucial challenge lies in data availability: while the spatial resolution of available global GCM projections is quite coarse, appropriate downscaled information of climate drivers and related impacts is not at hand for all locations and subjects of current research efforts.[62]

To be effective, designing climate change communication requires the identification of the target audience to be addressed as well as the appropriate framing.[63] Similarly, effective communication of uncertainties to avoid biased decisions requires a match to the underlying thought processes.[64] For visual climate change communication activities, Nicholson-Cole[65] stresses that characteristics of the target audience need to be considered and that careless presentation can also trigger defensive psychological responses, leading either to a feeling of powerlessness or to desensitization and "issue fatigue." Notably,

60 Hulme, *Pictures, Scenarios or Probabilities*, 2004.

61 Nicholson-Cole, "Representing Climate Change Futures," 2005.

62 see, e.g. the Coordinated Regional climate Downscaling Experiment (CORDEX), http://www.meteo.unican.es/en/projects/CORDEX

63 Moser, "Communicating Climate Change," 2010.

64 Patt, "Communicating Uncertainty to Policy Makers," 2009.

65 Nicholson-Cole, ibid.

existing web-based climate change information applications are often quite reluctant to clearly indicate their targeted user group(s).[66] While offering the user the option to interactively select visual representations of data and locations of interest might form a first step, more research is needed to come to a better understanding of how WIVCC can contribute to a target audience specific climate change communication.

Visual communication of climate change also has ethical implications.[67] Visualizations are generalized interpretations and as such are essentially subjective.[68] This is obviously the case for the iconographic images, like those of melting glaciers,[69] used in climate change campaigns. Climate change iconography itself is inherently political, and might even harbor hidden, old Western colonial thinking.[70] But ethical issues also arise in using rather abstract representations in the form of scientific diagrams or maps. Avoiding misleading visual representations of data is crucial for all scientifically sound visualizations. This necessity also inherently restricts the degree of freedom in interactively manipulating visual representations of scientific data that should be provided to a wider (lay) audience.[71] In addition, cultural differences of the intended target audience need to be considered, for example, by taking into account that colors may convey different meanings in different disciplines and cultures[72] or by using appropriate (equal-area) map projections that do not distort the size of one part of the world as compared to others. The ongoing progress of information technology will offer new opportunities for communicating climate change visually, but will pose additional ethical challenges as well; see, e.g., Sheppard[73] for a discussion of ethical questions in the context of climate change and photo-realistic landscape visualization.

In any event, it should be clear that WIVCC does not replace access to human experts. As climate change and its impacts is such a broad and multi-layered topic, two-way communication may often be necessary.[74] Ideally, stakeholders, decision-makers and the general public would be able to directly

---

66 Wrobel et al., ibid.

67 e.g. Moser, ibid.

68 Nicholson-Cole, ibid.

69 Doyle, ibid.

70 Manzo, ibid.

71 In this context it should also not be forgotten that visual representations can be of limited use—or even completely useless—for colour blind or blind users.

72 e.g. Schumann and Müller, *Visualisierung*, 2000.

73 Sheppard, "Landscape Visualisation and Climate Change," 2005.

74 Patt, ibid.

communicate with the respective experts whenever specific information on climate change is required. Obviously, this is not a realistic scenario, and web-based applications like WIVCC hold some promise in complementing this access to human experts to a certain degree. Thus, it needs to be carefully assessed under which conditions and to what extent WIVCC can meaningfully augment information access via human experts.[75]

5 CONCLUSIONS

In this article, we have addressed web-based interactive visual representations in the context of climate change (WIVCC) as means of improving the communication of scientific results from this field. We have highlighted the capability of WIVCC to engage with a wider range of scientific information about climate change as compared to using static visual representations. While interactive visualization is already a well-established scientific tool for experts, enabling a meaningful interaction with visual representations of scientific data for a wider audience has the potential to contribute to improved research result communication.

However, we have also seen that, irrespective of information technological progress over recent decades, appropriate utilization of WIVCC is far from easy and requires interdisciplinary efforts and integration to address existing challenges. Since target audiences of web-based information applications are often not sufficiently known and accessible in advance, designing effective communication strategies using these channels remains difficult. Moreover, there exists the need to carefully balance guidance and flexibility. There is no one-size-fits-all solution in sight: in some cases, less flexibility and highly condensed data will be more appropriate while in others, a higher degree of flexibility and possibilities for exploration may be desirable. Further, the information technological ingredients involved in WIVCC, both visual and general web technologies, are under constant development and will bring new opportunities as well as new challenges.

75 An interesting additional option, besides direct communication with an expert or web-based access to selected results, consists in designing tailored interactive visualization applications for use by climate change experts in the context of face-to-face result communication. For an example, see Nocke et al., "Visualization of Biosphere Changes in the Context of Climate Change," 2009.

Implementing web-based interactive visual representations is today technically less demanding than it used to be a decade ago. However, turning them into intuitive and powerful tools that efficiently provide access to useful information, rather than adding to information overload, requires careful design, especially of the images used for communication. While an increasing number of web-based applications embedding interactive visualization of climate change related data has been developed in recent years, a comprehensive and systematic evaluation of the effectiveness of such tools is still lacking. Additional research is required to derive clear guidelines on applying existing techniques and to explore new methods for interactive visual representations that effectively support the pressing task of climate change communication.

BIBLIOGRAPHY

Card, Stuart K., Jock D. Mackinlay, and Ben Shneiderman. *Readings in Information Visualization: Using Vision to Think.* Burlington: Morgan Kaufmann Publishers, 1999.

Chernoff, Herman "The Use of Faces to Represent Points in K-Dimensional Space Graphically." *Journal of the American Statistical Association,* vol. 68.

Chimera, Richard and Ben Shneiderman. "User Interface Consistency: An Evaluation of Original and Revised Interfaces for a Videodisk Library." In *Sparks of Innovation in Human-Computer Interaction,* edited by Ben Shneiderman. Norwood: Ablex Publ., 1993.

Cook, Dianne and Deborah F. Swayne. *Interactive and Dynamic Graphics for Data Analysis*. Berlin: Springer Science + Business Media, LLC, 2007.

Dessai, Suraje and Mike Hulme. "Does Climate Adaptation Policy Need Probabilities?," *Climate Policy* 4 (2) (2004): 107-128.

Doyle, Julie "Picturing the Clima(c)tic: Greenpeace and the Representational Politics of Climate Change Communication," *Science as Culture 16* (2) (June 2007): 129–50.

Elwood, Sarah "Geographic Information Science: Visualization, Visual Methods, and the Geoweb." *Progress in Human Geography 35* (3) (Jul. 2010): 401–08.

Grudin, Jonathan "The Case against User Interface Consistency," *Communications of the ACM* 32 (10) (Oct 1989): 1164–73.

Haklay, Muki, Alex Singleton, and Chris Parker. "Web Mapping 2.0: The Neogeography of the GeoWeb." *Geography Compass* 2/6 (2008): 2011–39.

Hulme, Mike *Pictures, Scenarios or Probabilities: How Should we Portray Dangerous Climate Change, Conference on 'Perspectives on Dangerous Climate Change',* 2004, University of East Anglia, Norwich, UK.

Hulme, Mike "Geographical Work at the Boundaries of Climate Change." Transactions of the Institute of British Geographers 33 (1) (Jan 2008): 5–11.

IPCC. *IPCC AR4. Synthesis Report: Climate Change 2007—An Assessment of the Intergovernmental Panel on Climate Change.* Technical Report November 2007, Intergovernmental Panel on Climate Change.

Ishikawa, Toru, Anthony G. Barnston, Kim A. Kastens, Patrick Louchouarn, and Chester F. Ropelewski "Climate Forecast Maps as a Communication and Decision-Support Tool: An Empirical Test with Prospective Policy Makers," *Cartography and Geographic Information Science* 32 (1) (Jan 2005): 3–16.

Kahneman, Daniel, Paul Slovic, and Amos Tversky (1982). "Judgment under Uncertainty: Heuristics and Biases," *Science, New Series* 185, no. 4157 (Sep. 27, 1974):1124–31.

Kellogg, Wendy A. "The Dimensions of Consistency." In *Coordinating User Interfaces for Consistency,* edited by Jakob Nielsen, 1989, Academic Press Professional

Krug, Steve. *Don't Make me Think: A Common Sense Approach to Web Usability*. San Francisco: New Riders, 1989.

MacEachren, Alan M., Anthony Robinson, Susan Hopper, Steven Gardner, Robert Murray, Mark Gahegan, and Elisabeth Hetzler. "Visualizing Geospatial Information Uncertainty: What we Know and What we Need to Know." *Cartography and Geographic Information Science 32* (3) (2005): 139–60.

Mackinlay, Jock D. "Automating the Design of Graphical Presentations of Relational Information." ACM Transactions on Graphics (TOG) 5 (2) (1986): 110–41.

Manzo, Kate "Imaging Vulnerability: the Iconography of Climate Change." In *Area* 42 (1) (Mar 2010): 96–107.

McCormick, Bruce H., Thomas A. DeFanti, and Maxine D. Brown. "Visualization in Scientific Computing," *Computer Graphics 21* (6), 1987.

McKendry, Jean E. and Gary E. Machlis. "Cartographic Design and the Quality of Climate Change Maps," *Climatic Change 95* (1-2) (Nov 2009): 219–30.

McNie, Elizabeth. "Reconciling the Supply of Scientific Information with User Demands: An Analysis of the Problem and Review of the Literature," *Environmental Science & Policy 10* (1) (Feb 2007): 17-38.

Meng, Yunlian and Jacek Malczewski. "Web-PPGIS Usability and Public Engagement: A Case Study in Canmore, Alberta, Canada," *URISA Journal 22* (1) (2010): 55–64.

Monk, Andrew. *Noddy's Guide to Consistency*. Interfaces 45 (2000): 4–7.

Moser, Susanne C. "Communicating Climate Change: History, Challenges, Process and Future Directions," *Wiley Interdisciplinary Reviews: Climate Change* 1 (1) (2010): 31–53.

Moser, Susanne C. and Lisa Drilling. *Creating a Climate for Change: Communicating Climate Change and Facilitating Social Change.* Cambridge: Cambridge University Press, 2007.

Nicholson-Cole, Sophie A. "Representing Climate Change Futures: A Critique on the Use of Images for Visual Communication." *Computers, Environment and Urban Systems* 29 (3) (May 2005): 255–273.

Nielsen, Jakob. "Coordinating User Interfaces for Consistency." *ACM SIGCHI Bulletin* 20 (3) (Jan 1989): 63–65.

Nivala, Annu-Maaria, Stephen Brewster, and L. Tiina Sarjakoski. "Usability Evaluation of Web Mapping Sites." *The Cartographic Journal,* 45 (2) (May 2008): 129–138.

Nivala, Annu-Maaria, L. Tiina Sarjakoski, and Tapani Sarjakoski. "Usability Methods' Familiarity among Map Application Developers." *International Journal of Human–Computer Studies*, 65 (9) (2007): 784–795.

Nocke, Thomas, Till Sterzel, and Markus Wrobel. "Visualization of Climate and Climate Change Data: An Overview." *Digital Earth Summit on Geoinformatics 2008: Tools for Global Change Research (ISDE'08),* Volume 1999, edited by Ehlers.

Nocke, Thomas, Ursula Heyder, Stefan Petri, Katrin Vohland, Markus Wrobel, and Wolfgang Lucht (2009). Visualization of Biosphere Changes in the Context of Climate Change. *Information Technology and Climate Change,* 2nd International Conference IT for empowerment (ITCC'08), 29–36, edited by V. Wohlgemuth. Berlin, Germany: Trafo Wissenschaftsverlag, 2009.

Norman, Donald A. (1988). *The Psychology of Everyday Things*. New York: Basic Books, 1988.

Pang, Alex "Visualizing Uncertainty in Natural Hazards." *Risk Assessment, Modeling and Decision Support* 14, 2008, 261–94.

Patt, Anthony. "Communicating Uncertainty to Policy Makers." In *Uncertainties in Environmental Modelling and Consequences for Policy Making.* Berlin: Springer, 2009.

Petrie, Helen and Christopher Power. "What Do Users Really Care About?," *Proceedings of the 2012 ACM Annual Conference on Human Factors in Computing Systems—CHI '12,* New York, New York, USA, pp. 2107. ACM Press.

Pyla, Pardha S., Manas Tungare, and Manuel A. Pérez-Quinones. "Multiple User Interfaces: Why Consistency is not Everything, and Seamless Task Migration is Key." *Proceedings of the CHI 2006 Workshop on The Many Faces of Consistency in Cross-Platform Design,* 2006.

Schmidt, Manuela and Paul Weiser. "Web Mapping Services: Development and Trends." *Online Maps with APIs and WebServices,* Chapter 2. Berlin: Springer, 2012: 13–21.

Schneider, Birgit. "Image Politics: Picturing Uncertainty. The Role of Images in Climatology and Climate Policy." In *Climate Change and Policy. The Calculability of Climate Change and the Challenge of Uncertainty,* edited by Gabriele Gramelsberger and Johann Feichter, 191–09. Berlin/Heidelberg: Springer, 2011.

Schumann, Heidrun and Wolfgang Müller. *Visualisierung: Grundlagen und Allgemeine Methoden.* Berlin: Springer, 2000.

Senaratne, Hansi and Lydia Gerharz. "An Assessment and Categorisation of Quantitative Uncertainty Visualisation Methods for Geospatial Data." In *Advancing Geoinformation Science for a Changing World,* edited by S. Geertman, W. Reinhardt, and F. Toppen. Berlin: Springer, 2011.

Senaratne, Hansi, Lydia Gerharz, Edzer J. Pebesma, and Angela Schwering. "Usability of Spatio-Temporal Uncertainty Visualisation Methods." *Bridging the Geographic Information Sciences: Lecture Notes in Geoinformation and Cartography,* 3–23. Berlin: Springer, 2012.

Sheppard, Stephen L.J. "Landscape Visualisation and Climate Change: The Potential for Influencing Perceptions and Behaviour," *Environmental Science & Policy* 8 (6), Dec 2005, 637–54.

Shneiderman, Ben *Designing the User Interface: Strategies for Effective Human-computer Interaction.* Reading: Addison-Wesley, 1998.

Shneiderman, Ben "The Eyes Have It: A Task by Data Type Taxonomy for Information Visualizations." (Proceedings) *IEEE Symposium on Visual Languages,* 1996.

Skarlatidou, Artemis and Muki Haklay. "Public Web Mapping: Preliminary Usability evaluation." (Proceedings) *GIS Research UK 2006,* Nottingham, UK.

Slocum, Rachel. *Polar Bears and Energy-efficient Lightbulbs: Strategies to Bring Climate Change Home.* Environment and Planning D: Society and Space 2004, vol. 22.

van Elzakker, *Corné P.J.M.* "Use and Users of Maps on the Web." *Cartographic Perspectives 37* (2000): 34–50.

Walsh, Lynda "Visual Strategies to Integrate Ethos Across the 'is/ought' Divide in the IPCC's Climate Change 2007: Summary for Policy Makers." *Poroi, Issues in the Rhetoric of Science and Technology,* 6 (2) (2009): 33–61.

Ware, Colin. *Information Visualization.* Amsterdam: Elsevier, 1999.

Wrobel, Markus, Alexander Bisaro, Dominik Reusser, and Jürgen. P. Kropp. "Novel Approaches for Web-Based Access to Climate Change Adaptation Information—MEDIATION Adaptation Platform and ci:grasp-2." In *Environmental Software Systems. Fostering Information Sharing IFIP Advances in Information and Communication Technology* 413 (2013): 489–99

Wrobel, Markus and Luis Costa. "Web Based Applications for Climate Change Related Information: A Usability Test." edited by R. Seppelt, A. Voinov, S. Lange, and D. Bankamp, *International Congress on Environmental Modelling and Software. Managing Resources of a Limited Planet.* Leipzig, Germany: Sixth Biennial Meeting, 2012.

Wrobel, Markus, Luis Costa, Tabea Lissner, Marta Moneo Laín, Tobias Weiß, and Jürgen P. Kropp. "A Review of User Interface Conventions in Web Applications for Climate Change Information." International Congress on Environmental Modelling and Software: Ottawa, 2010.

IMAGE COPYRIGHTS & SOURCES

Figure 1: © Wrobel and Reusser.
Figure 2a: © IPCC.
Figure 2b and 2c: © Potsdam Institute for Climate Impact Research.
Figure 3: © Wrobel and Reusser.

Color Plates

Plate 1: Satellite map of total ozone above the Antarctic region.

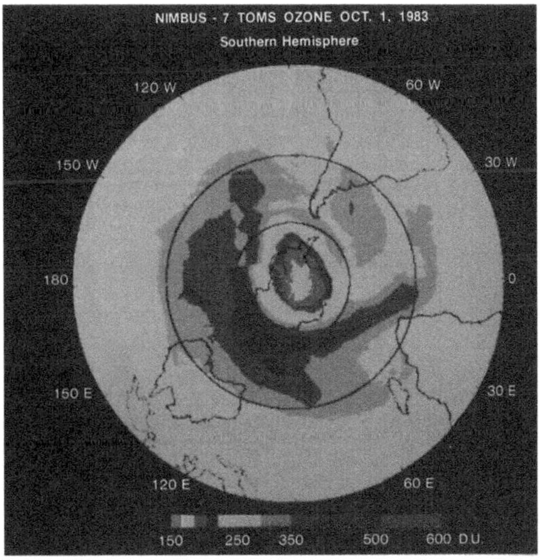

Plate 2: Visualization technique parallel coordinates.

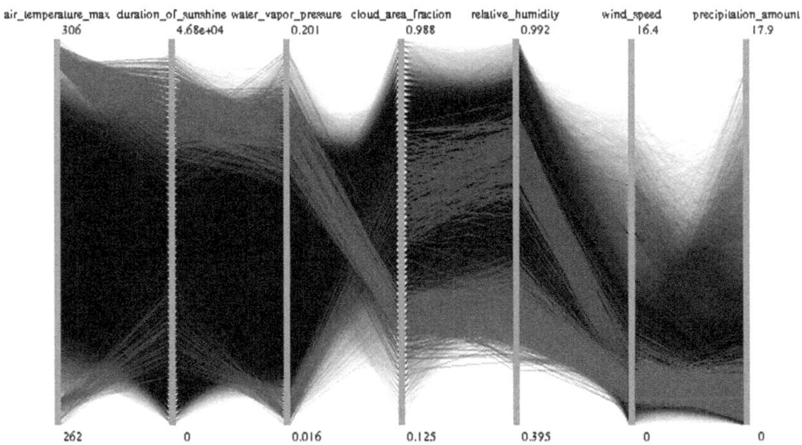

Plate 3: Visualization Design Wizard for Climate Data.

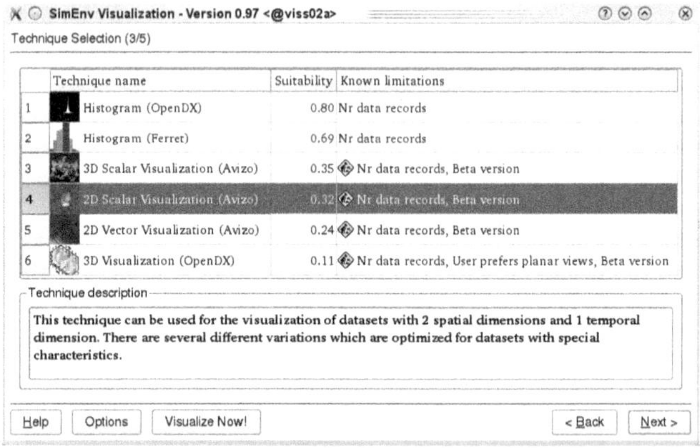

Plate 4: The infamous "hockey stick" graph from the Third Assessment Report.

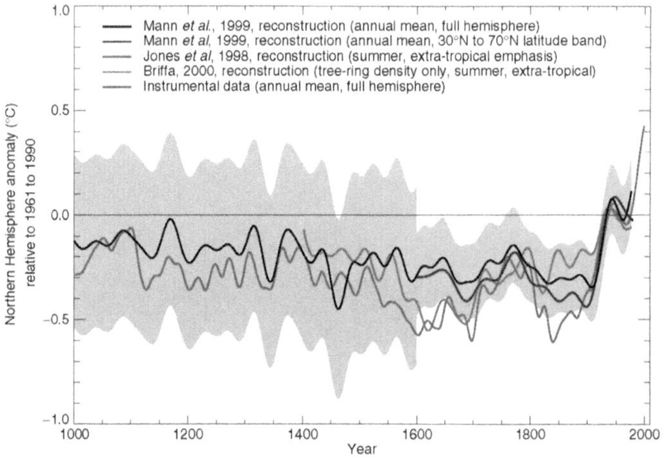

Figure 2.21: Comparison of warm-season (Jones *et al.*, 1998) and annual mean (Mann *et al.*, 1998, 1999) multi-proxy-based and warm season tree-ring-based (Briffa, 2000) millennial Northern Hemisphere temperature reconstructions. The recent instrumental annual mean Northern Hemisphere temperature record to 1999 is shown for comparison. Also shown is an extra-tropical sampling of the Mann *et al.* (1999) temperature pattern reconstructions more directly comparable in its latitudinal sampling to the Jones *et al.* series. The self-consistently estimated two standard error limits (shaded region) for the smoothed Mann *et al.* (1999) series are shown. The horizontal zero line denotes the 1961 to 1990 reference period mean temperature. All series were smoothed with a 40-year Hamming-weights lowpass filter, with boundary constraints imposed by padding the series with its mean values during the first and last 25 years.

Plate 5: The cover image of the World Meteorological Organization Statement.

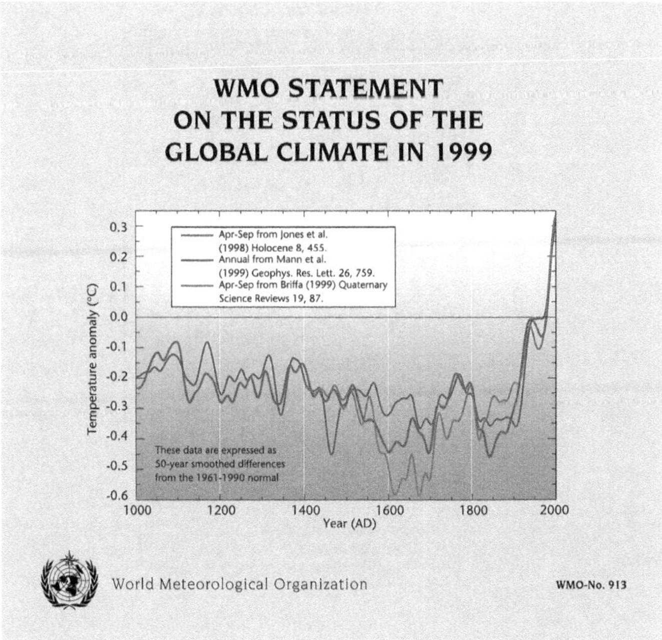

Plate 6: Projected temperature changes under different emissions scenarios (left) and "reasons for concern" or "burning embers" (right).

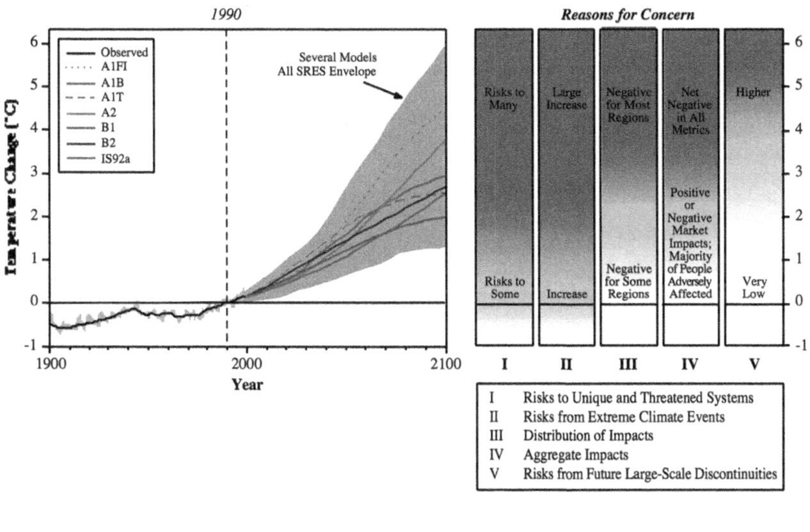

Plate 7: Adaptation of the "burning embers" diagram.

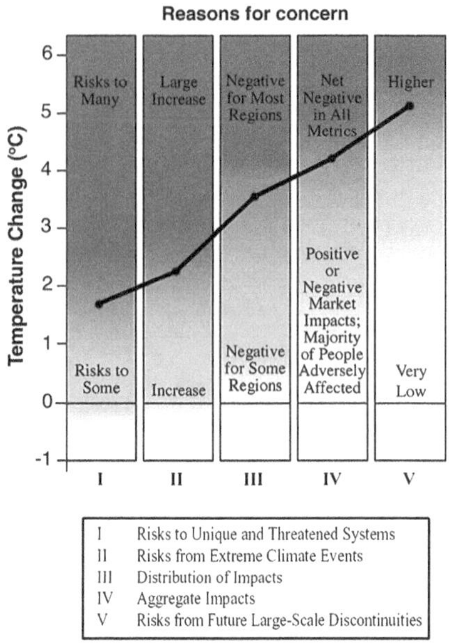

Plate 8: Nature under threat: spectacular photographs by Bernhard Edmaier.

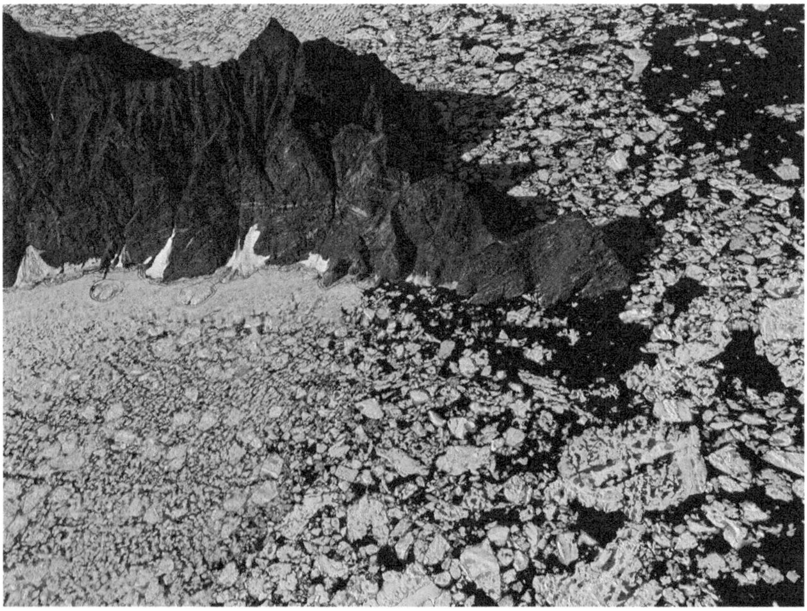

Plate 9: Global warming has stopped in the last 15 years? Joseph D'Aleo.

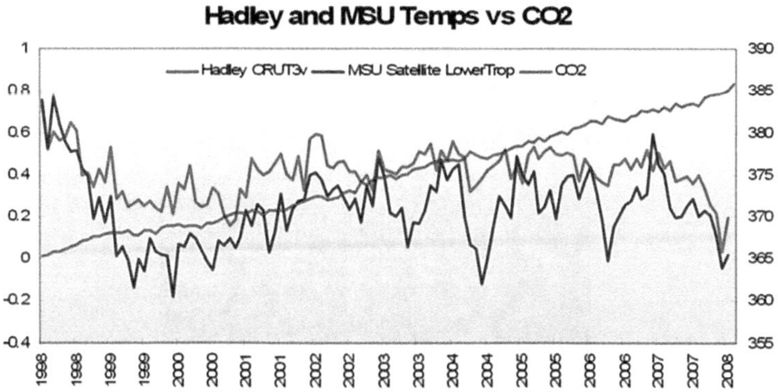

Plate 10: Temperature reproduction of the last 12000 years, from Goreham.

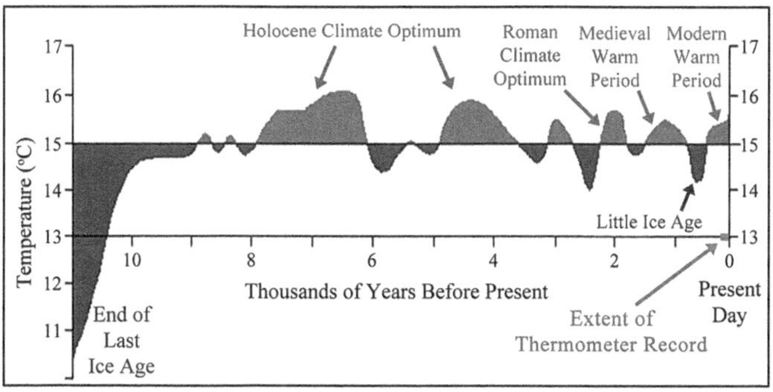

Plate 11: Visualization techniques commonly used to represent uncertainties.

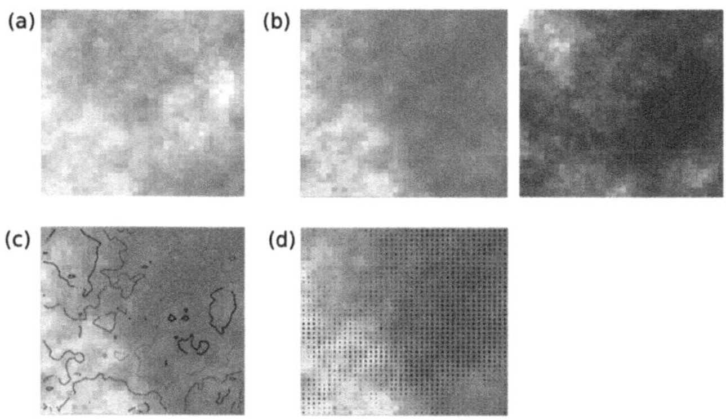

Plate 12:Climate Time Bomb: Signs of Climate Change.

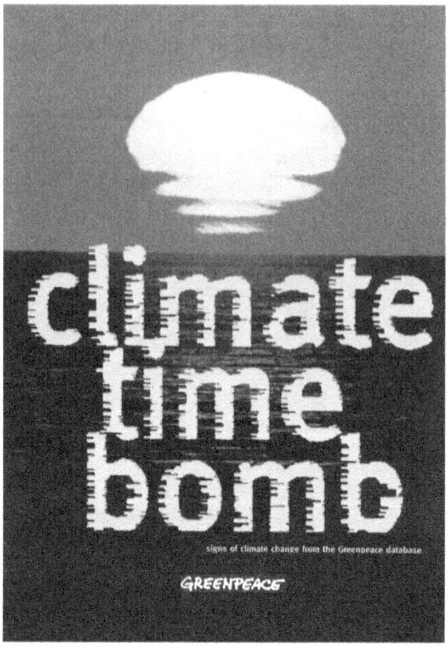

Plate 13: Braschler and Fischer, The Human Face of Climate Change.

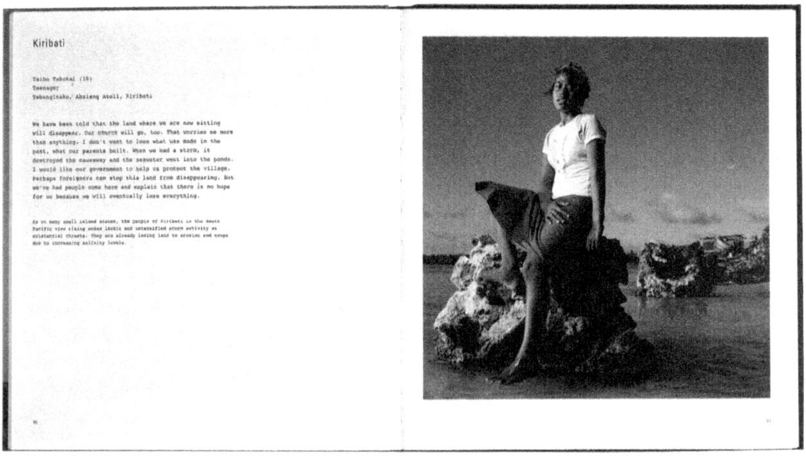
Kiribati

Taibo Tabokai (15)
Teenager
Tebunginako, Abaiang Atoll, Kiribati

We have been told that the land where we are now sitting will disappear. Our church will go, too. That worries me more than anything. I don't want to lose what was made in the past, what our parents built. When we had a storm, it destroyed the causeway and the seawater went into the ponds. I would like our government to help us protect the village. Perhaps foreigners can stop this land from disappearing. But we've had people come here and explain that there is no hope for us because we will eventually lose everything.

As on many small island states, the people of Kiribati in the South Pacific view rising ocean levels and intensified storm activity as existential threats. They are already losing land to erosion and crops due to increasing salinity levels.

Plate 14: Climate Change. Picturing the Science, photo by Joshua Wolfe.

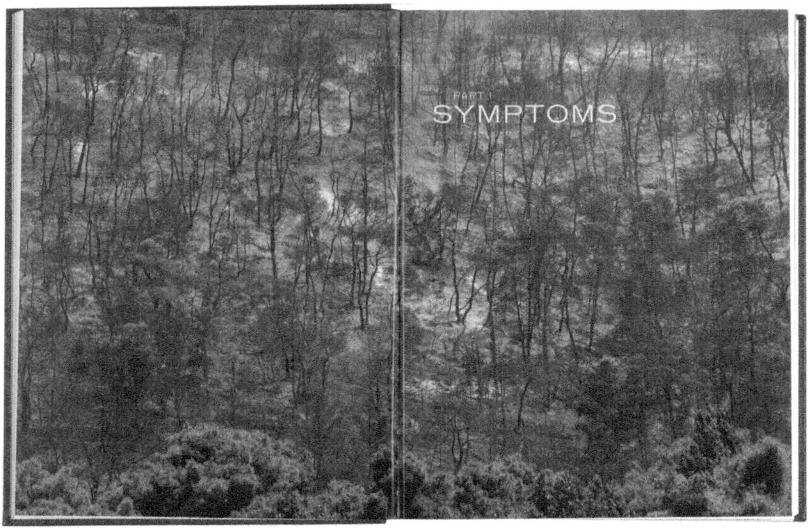

Plate 15: Al Gore, An Inconvenient Truth. Photos by Bruno Messerli.

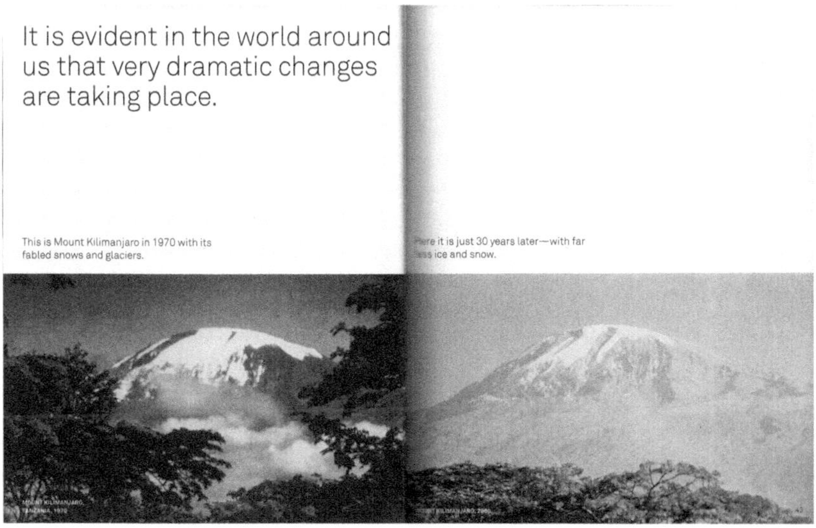

Plate 16: Sayler/Morris: Extreme Weather Events XV, Louisiana, 2005.

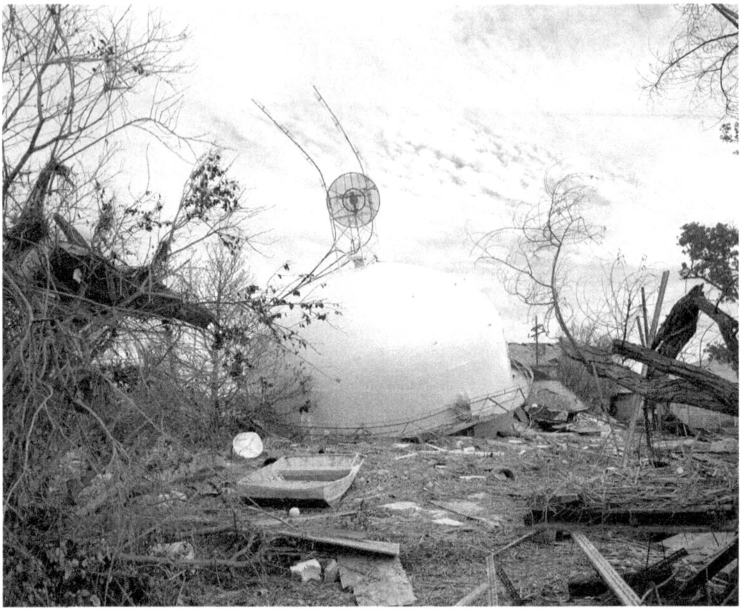

Plate 17: Extreme Weather Events XV: XXIII: Coney Island, NY, 2012.

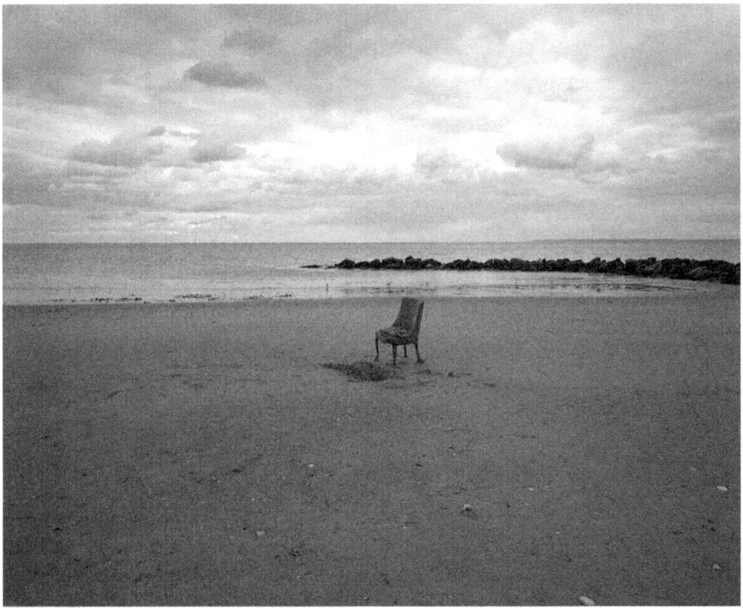

Plate 18: Schmidt and Wolfe, Climate Change, 2009.

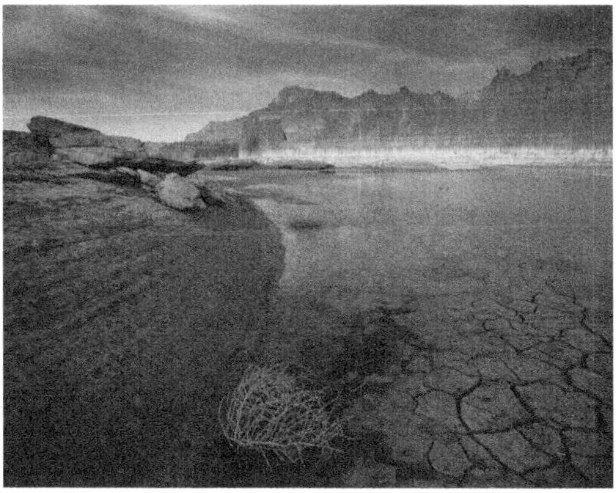

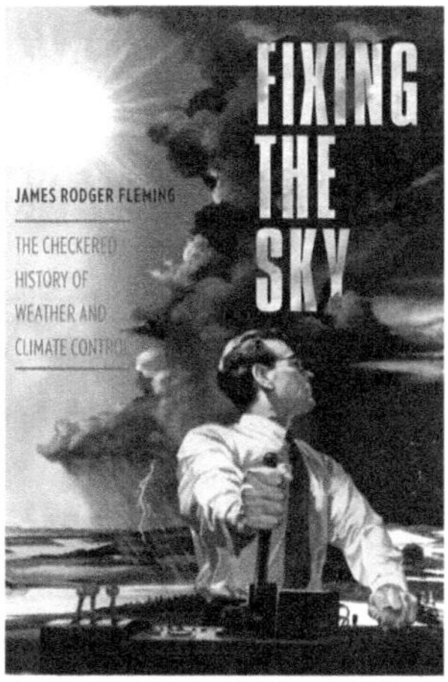

Plate 19: Technocrat pulling the levers of weather control.

Plate 20: Cartoon on diagnosing climate change, in Nature 455.

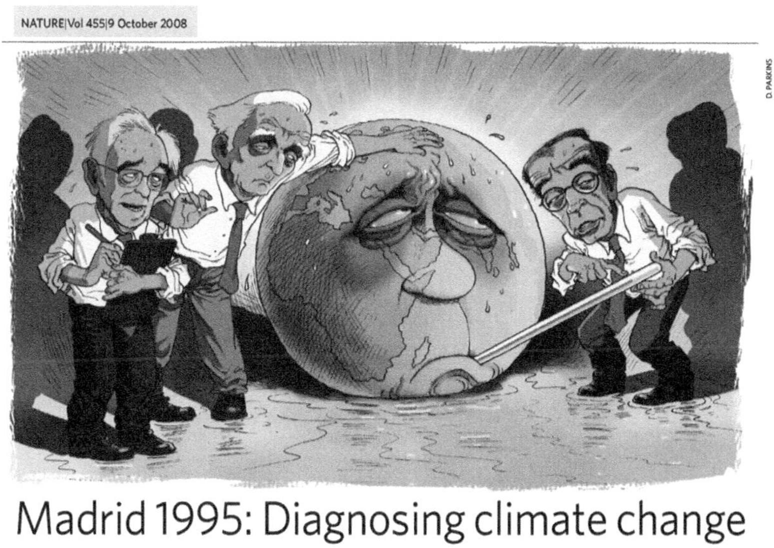

Chapter 4
From Vision to Action? Making the Invisible Imaginable through Art and Photography

Picturing the Clima(c)tic: Greenpeace and the Representational Politics of Climate Change Communication*

JULIE DOYLE

Images of melting glaciers have come to dominate the pictorial language of climate change. This paper argues that such images of melting glaciers engender a representational problem in the communication of climate change as they depict the already seen effects of climate change. Given the dominance of the photograph within Greenpeace campaigns, the paper examines this adherence to visual immediacy by analyzing Greenpeace climate change campaign literature since 1994. Identifying five representational phases over the last decade, the analysis shows how a symbolic pictorial language of climate change was being created, and the ways in which risks were communicated as actual rather than potential. Understood retrospectively however, this visualization calls attention to the problems of communicating environmental issues of a temporal (long term) and unseen nature through a medium of representation which privileges the 'here and now'.

INTRODUCTION

Images of melting glaciers dominate the pictorial language of climate change; powerful symbols of a fragile earth at risk from climate impacts (Figure 1). Environmental campaign group Greenpeace has been instrumental in documenting these impacts through film and photography. Since 1997,

* A version of this chapter was originally published as: Doyle, Julie. "Picturing the Clima(c)tic: Greenpeace and the Representational Politics of Climate Change Communication", *Science as Culture* 16, no. 2 (2007): 129–150.

Greenpeace has led expeditions to the Arctic and Antarctic to record climate change effects witnessed through melting glaciers. The discourse of visual 'truth' invested in the photograph functions here in a powerful way when considered in the context of struggles by Greenpeace, and other environmental NGOs, to communicate the reality of climate change to governments and publics since the Intergovernmental Panel on Climate Change (IPCC) published its First Assessment Report in 1990.[1] Yet photographs of melting glaciers engender a representational problem concerning the visual communication of environmental issues. Photographs of glaciers represent temporally the *already seen* effects of climate change; a distinct problem within historical efforts by environmental NGOs to bring attention to the climate change, *before its impacts could be seen.*

Figure 1: Comparative photographs of Patagonian glaciers taken in 1928 and January 2004.

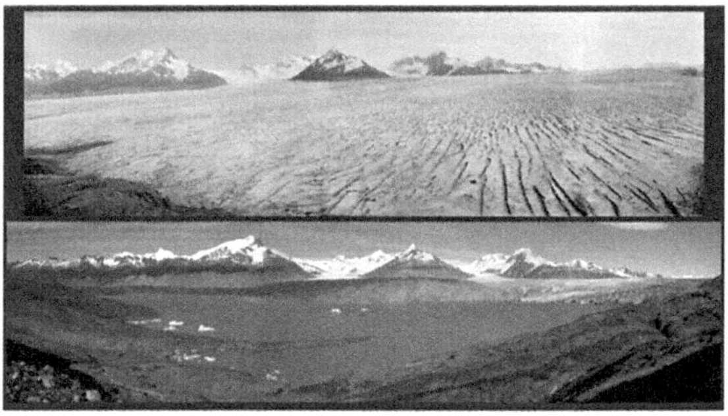

Within the discourse of photography, the comparative photographs of melting glaciers signify as the certainty of what is, or "what has been."[2] The potential effects of climate change upon the landscape have become actual and made evident by the photograph.[3] The photographic visualization of climate change thus calls attention to the difficulties of communicating environmental issues that are temporal (long term and developmental) and unseen (not always visible), through a medium which privileges the 'here and now' of the visual.

1 The IPCC was established in 1988 "to provide independent scientific advice on the complex and important issue of climate change" (Houghton, *Global Warming,* Foreword, 2004). Conducting no new research, the IPCC assesses existing scientific reports to provide a comprehensive overview of the scientific basis of human-induced climate change.

2 Barthes, *Camera Lucida,* 2000, 85.

3 Doyle, "Seeing the Climate?" 2009.

In this chapter, I examine the problems associated with communicating a temporal issue like climate change by analyzing the history of climate change communication produced by Greenpeace from 1994 to 2007.[4] My intention is to consider how Greenpeace has sought to make 'real' the potential and often-invisible risks of climate change, and to consider the limitations this mode of representation has posed for climate change communication. As climate change has been dependent upon science for its legitimation, my analysis also takes into account how climate science has been framed and represented by Greenpeace during this period. Furthermore, as environmental problems do not gain legitimacy as problems outside of social, economic and political conditions, my analysis of the campaigns takes these changing conditions into account and considers how these may have impacted upon the communication strategies of Greenpeace.

Before I engage directly with Greenpeace's climate change campaign materials, I will discuss the historical constraints that have affected the communication of climate change. Firstly, I will be looking at the role of the visual within Greenpeace campaigns and environmental politics more generally, and secondly, at the socio-political role of science within the validation and communication of climate change.[5]

GREENPEACE AND THE PROBLEM OF IMAGING CLIMATE CHANGE

"Greenpeace can be regarded as an organization with photography as its vital medium [...] Greenpeace's strategy is implicitly based upon the premise that it can and must counter images of power with forceful images of its own."[6]

The image is a well-known communication strategy for Greenpeace, used to document environmental destruction and the visual beauty of nature at risk. Photography is also central to the non-violent direct action, which forms the basis of Greenpeace's philosophy of 'bearing witness' to environmental damage,

---

4 1994 was the year that Greenpeace published its first public campaign report on climate change (Greenpeace International, *Climate Time Bomb, 1994.*)

5 For a visual analysis of Greenpeace climate campaigns in historical relation to the visualization of climate change by science and the mass media see Doyle, "Seeing the Climate?," 2009.

6 Boettger, "The Role of Photography in Greenpeace's Strategy," 2001, 12.

where the photographic framing of an action is a constitutive aspect of that action.[7] Greenpeace is honest about the power of images to communicate and to persuade. A former Greenpeace picture editor comments, "it has been said that the messages of Greenpeace's actions would have been difficult to understand had they not been translated into the language of photography [...] which is also used by the mass media."[8] The visual mediation of protest is thus a key communication strategy.

Greenpeace's claim for the transparency of photography as internationally comprehensible corresponds to the global nature of the organization and its campaign messages. Greenpeace consists of "Greenpeace International [...] in Amsterdam" with "27 national and regional offices around the world, providing a presence in 41 countries."[9] Co-ordinated by Greenpeace International, the global campaigns encompass seven key messages, aiming to: "stop climate change, protect ancient forests, save the oceans, say no to genetic engineering, stop the nuclear threat, eliminate toxic chemicals and encourage sustainable trade."[10] The national/regional offices rework the global campaign messages within a specific geographical and political context, making the campaign message relevant to a national/regional audience, which is communicated further by local campaign groups to a local audience.[11]

With 2.8 million supporters worldwide, Greenpeace is funded solely by supporter subscriptions and donations, and does not accept funding from industry or government.[12] Its dependence upon public donations makes its media presence an important aspect of its campaigning and public profile. Boettger claims "one of the reasons for Greenpeace's great popularity is that the organization does provide material for the spectacular, not to say—by aesthetic standards where politics are involved—beautiful pictures."[13] Such aesthetics corresponds to the discourse of landscape and environment since the Enlightenment, as images of the landscape have come to define what the

7 Brown and May, *The Greenpeace Story,* 1989.

8 Boettger, ibid.

9 Greenpeace International, "How is Greenpeace Structured," 2006.

10 Greenpeace International, "About Greenpeace," 2007.

11 This communication structure differs from other (international) environmental NGOs who vary their campaign messages according to a local context.

12 Greenpeace International, ibid.

13 Boettger, "The Role of Photography in Greenpeace's Strategy," 2001, 20.

environment is and how it is to be understood.[14] Yet this commitment to the beautiful and spectacular becomes problematic in the context of climate change communication. The persuasive power of comparative photographs of melting glaciers as evidence of climate change gain part of their legitimacy through an aestheticization of the landscape, which arguably renders its loss more shocking. Such shock also corresponds to the events-based tactics of news media coverage of the environment, which forecloses the communication of long term and accumulative (non visible) environmental problems.[15]

The visual communication strategies of Greenpeace correspond with the limitations of events-based reportage. Investment in photographing the beautiful and spectacular is constrained by the temporal nature of climate change as an ongoing and accumulative problem, not always visible. Yet, environmental groups do play an important role in framing environmental issues. Marc Mormont and Christine Dasnoy (1995) have acknowledged the significant role that Greenpeace played in first identifying the problem and science of climate change. Like Hansen,[16] they argue that that frameworks of interpretation need to be established.

> "When it comes to climate change, the facts completely escape common experience for it is only by communication that the issue is given meaning [...] The taking account of these facts (or hypotheses) by public opinion presupposes an interpretative context, one that may designate the risks and the victims, in short, the social context which gives these facts meaning."[17]

The framing of climate change science is crucial to bringing attention to, and legitimating, the issue.

The legitimacy of environmental issues and perceived risks can be linked to "questions about the ease with which some issues link into powerful, historically established, symbolic imagery."[18] Alison Anderson also states that issues gain public legitimacy through their capacity to signify symbolically: when "they

---

14 Urry, *The Tourist Gaze,* 1990; Wilson, *The Culture of Nature,* 1992; Macnaghten and Urry, *Contested Natures,* 1998.

15 Hansen, "The Media and the Social Construction of the Environment," 1991; Anderson, *Media, Culture and the Environment,* 1997; Alan et al., *Environmental Risks and the Media,* 2000.

16 Hansen, "The Media and the Social Construction of the Environment," 1991.

17 Mormont and Dasnoy, "Source Strategies," 1995, 61.

18 Hansen, ibid., 451.

become icons or symbols for a wide range of concerns that people can easily identify with."[19] In the case of climate change predictions, I would argue that the lack of visible evidence of this problem has made it difficult for it to be linked to an established set of symbolic imagery, and thus to become an issue of political and public concern.

THE SOCIO-POLITICS OF CLIMATE SCIENCE

Environmental issues require articulating as problems in order to gain political and public legitimacy.[20] Yet identifying an environmental problem, or the magnitude a threat may pose, does not guarantee that the issue will gain public or political attention. Hansen states that "we must examine the wider notion of cultural resonances [...] in order to account for the very different careers of environmental problems."[21] Cultural resonances are widely held beliefs within society, for example, a belief in scientific authority to legitimate claims to truth about risk.

Since the publication of the First Assessment Report by the IPCC in 1990, scientists have been central to legitimizing climate change risks, having drawn attention to the issue of climate change before environmentalists.[22] Greenpeace has been committed to communicating and validating this science, publishing a reader's guide to the IPCC's first report.[22] Yet, the validation of climate science has been constrained by the institutional and political processes involved in the authentication of scientific knowledge. As Robert Weart has shown,[23] the slow and non-linear processes of scientific legitimation were particularly acute in the case of climatology, which remained underfunded and isolated from other disciplines for years. Climate science has also been negatively affected by its reliance upon computer models to predict the rate of increasing global temperatures, by the political management of the IPCC, and by the oil industry's financing of global warming skeptics.[24]

Given the institutional, political and economic complexities of legitimizing scientific knowledge, Greenpeace maintains the view that "science is crucial to

19 Anderson, *Media, Culture and the Environment,* 5–6.

20 Hansen, "Claims-making and Framing in British Newspaper Coverage," 2000.

21 Hansen, "The Media and the Social Construction of the Environment," 1991, 451.

22 Greenpeace UK, *Climate Change,* 1990.

23 Weart, *The Discovery of Global Warming,* 2003.

24 Weart, ibid.; Carter, *The Politics of the Environment,* 2001.

environmental protection."[25] Indeed, climate change "can only be detected and understood through science."[26] Greenpeace also demonstrates awareness that science "is used to justify the existence and deployment of environmental threats, such as nuclear power," and characterizes this as a "double-edged relationship with science."[27] This continual negotiation of scientific knowledge is not unusual for environmental NGOs, who often use science to legitimate their arguments.[28] For Greenpeace, this negotiation is aided by the specific production of its 'own' science through work carried out at the Greenpeace Science Laboratory at University of Exeter, UK.[29] Greenpeace is thus involved in the public communication of science, as well as reflexively critiquing this, for popular consumption.

GREENPEACE COMMUNICATION STRATEGIES

Greenpeace is involved in a variety of campaigning strategies: producing scientific reports, lobbying governments, working with industry, staging direct actions, and providing environmental information for the public, governments and industries. However, it is the public campaign literature, often distributed by the local Greenpeace groups, which I am directly examining in this chapter. The printed public campaign literature was sourced from visits to the Greenpeace UK offices and focuses primarily on Greenpeace UK campaigns. Greenpeace UK does not have a systematized archive of campaign materials, and campaigners often change every few years, to be replaced by new staff. Accessing the material, and constructing an historical overview of its campaigns, was thus based upon knowledge provided by interviews with current campaigners,[32] as well as my own knowledge as a volunteer for Greenpeace from 1995–2002.

This chapter now examines the strategies employed by Greenpeace to communicate climate change in light of the problems of communication already discussed. The five representational strategies identified in my analysis establish particular relations between the temporal and the visual in the very process of making climate change meaningful and relevant.

25 Greenpeace International, "Greenpeace Science Unit," 2007.

26 Ibid.

27 Ibid.

28 Eden, Donaldson and Walker, "Green Groups and Grey Areas," 2006.

29 Greenpeace International, "Greenpeace Science Unit," 2007.

PHASE ONE: IMMINENT AND INEVITABLE DESTRUCTION

"Greenpeace has been monitoring scientific and news reports of major extreme weather events and natural climate-related disasters since the IPCC's [Intergovernmental Panel on Climate Change] first warning in 1990. Greenpeace believes that the only conclusion that can be drawn from this 'catalogue of disasters' is that *global warming is now detectable* and *the first impacts* of human-induced climate change are in fact already *being felt.*"[30] [my emphasis]

It is widely acknowledged that the speech made by UK Prime Minister, Margaret Thatcher to the Royal Society in September 1988 put climate change on the UK political agenda, followed by an increase in media coverage.[31] However, whilst the IPCC's First Assessment Report in 1990 asserted that human activities were increasing the atmospheric levels of greenhouse gases, by 1992 political and media interest in the issue had waned. Uncertainties identified by the IPCC over the extent of climate change were being highlighted by some as a way of preventing policy changes.[32]

Having published a reader's guide to the IPCC's First Assessment Report in 1990, Greenpeace made an attempt to engage the public on climate change in 1994, during a period of low political and public concern. As a first step towards giving the issue symbolic meaning, *Climate Time Bomb: Signs of Climate Change from the Greenpeace Database* (Figure 2) sought to engage the public through images and text. Through explicit reference to its the monitoring of scientific and news reports since the IPCC's 1990 report, Greenpeace also positions itself as both legitimator and interpreter of scientific knowledge. As the first impacts are "being felt" rather than seen, early frameworks of interpretation are established by foregrounding the potential magnitude of impacts upon the natural and human environment through a visual and textual language of danger and threat. The temporality of climate change is presented as an imminent future anchored in the present: a time bomb ticking away. The blurred large yellow text signifies fast movement and the quick passing of time. Reminiscent of an atomic mushroom cloud, the yellow glowing sun establishes a metaphorical relation between the devastation of nuclear weaponry and the potential destruction of human-induced climate change.

30 Greenpeace International, *Climate Time Bomb,* 1994.

31 Farrow, "Communicating about Climate Change," 2000; Weart, *The Discovery of Global Warming,* 2003; Carvalho and Burgess, "Cultural Circuits," 2005.

32 Ibid.

The metaphorical relation between nuclear disaster and climate change corresponds with the temporal discourse of environmentalism: "Environmentalism, above all, links the past with the future. The present is always presented as the sum of the consequences of the past...The future and past are presented as imminent in the present."[33] In this example, the need to visually reference the past in order to communicate the present and the future is contingent upon the temporal nature of climate change as a future possibility.

Figure 2: Climate Time Bomb: Signs of Climate Change from the Greenpeace Database 1994. (→ color plate 12)

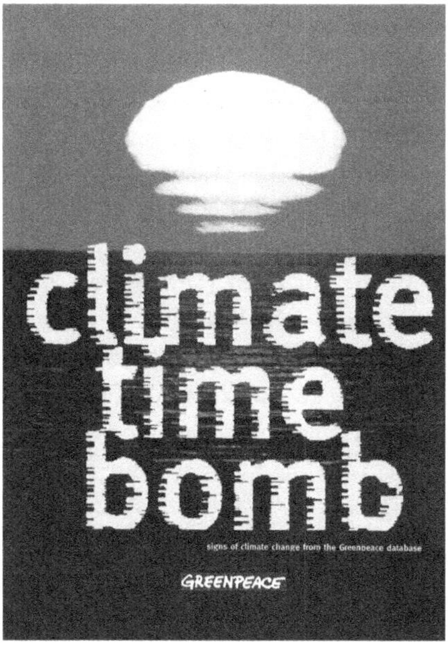

Signified on the front cover as a matter of time, inside, the future of climate change is conveyed by photographic images and text identifying twelve likely environmental and human impacts. These include rising temperatures; thinner ice; spread of disease; more pests and increasing insurance claims. Individual photographs of a wave, the close up shot of an insect, or the idyllic deserted beach convey little about the impacts of climate change. The photographs that portray destruction, such as a flooded road or a hurricane-devastated house, communicate impacts more immediately as indexical signs of the visibly

33 Harré et al., *Greenspeak,* 1999, 7.

damaging effects of weather. Yet the limits of photographic signification are also revealed, as the images need to be read together in order to signify the future of climate change through the visible authentication of present impacts.

The photographs also require explanatory text (through a timeline) to anchor their meaning as climate signs. The timelines begin in 1990 and end in 1994, recording present impacts, such as "The eight hottest years on record have all occurred since 1980," whilst implying future threats. The temporality of climate change as a long term, accumulative problem is thus negotiated by reference to both present impacts and potential future threats.

Climate Time Bomb presents climate change as a complex issue, the effects of which are wide ranging, whilst giving precedence to the language of destruction rather than solutions and hope. The visual aestheticization of nature is minimally present. The environment is already destroyed by wind, water, fire, drought and disease. Only the last page provides an indication of solutions via renewable energy and public transport.

The phrase "climate change," rather than "global warming," is used throughout, demonstrating Greenpeace's adherence to the scientific terminology used by the IPCC. However, the choice to visually frame climate change through the symbolic resonance of the color red (on the front page) frames this science through the more easily understood link between rising temperatures and global warming. Viewed in retrospect, framing climate change as an imminent catastrophe draws attention through shock tactics. This early example of "climate change discourse in the making" highlights the problems in communicating the complexity of the issue, resorting to cataclysmic images that render the viewer powerless in terms of agency.

PHASE TWO: CAUSES, IMPACTS AND SOLUTIONS—DIRTY OIL VERSUS PRISTINE HABITATS

In 1997, the UK Government's opening up of the North Atlantic for new oil exploration off the North West coast of Scotland signaled an opportunity for the issue of climate change to be linked more directly to the oil industry, making relevant to a national audience. Greenpeace UK launched its 'Atlantic Frontier Campaign,'[34] communicating a stronger message by making a causal link between the burning of fossil fuels and climate change. Whereas the discourse of *Climate Time Bomb* communicated a catastrophic event about to happen, the

34 Greenpeace UK, *The Atlantic Frontier*, 1998.

title of this report, *Putting the Lid on Fossil Fuels*,[35] shifts the temporal dimension to one where present action can be effective.

The verb "putting" places immediate emphasis upon present action. Like the temporal theme of *Climate Time Bomb*, this campaign shares a similar color scheme of red, yellow and black to signify the dangers of a warming planet, communicated by the redness of the globe on the front cover as representative of a possible future. However, as Tim Ingold has discussed, "the significance of the image of globe in the language of contemporary debate about the environment" is problematic, because it renders the world "as an object of contemplation detached from the domain of lived experience."[36] Thus, while Greenpeace has chosen the recognizable image of a globe to convey the potential impacts of global climate change, this cover also reinforces a distance between humans and the environment in the very process of establishing a popular discourse of climate change.

Climate science is given an authoritative voice throughout: translated, on the front cover, into the form of computer model predictions of rising temperatures to demonstrate and forecast rising temperatures from 1860 to 2090. It directly quotes the IPCC's Second Assessment Report of 1995 on the opening page: "'The balance of evidence suggests a discernible human influence on global climate' IPCC, 1995" although this science is somewhat simplified by Greenpeace's translation of this into an affirmation by the IPCC that "climate change had arrived."[37]

The report offers a linear, visual narrative of cause, effects and solutions, conveying a less alarming message than *Climate Time Bomb.* The inclusion of photographs of oil platforms is intended to establish an associative link between oil and climate change. The effects of climate change are conveyed by a series of four photographs placed together, depicting the devastation wrought upon buildings and the landscape by wind and water, and thus indexical signs of an already warming world. At the end of the report, a photograph of a group of children surrounding a solar panel metonymically signifies the future, and solution, to climate change.

The temporal dilemma which underpins climate change communication—the need to communicate both the present reality and future threats without extensive visual evidence—is negotiated in the report through the inclusion of

35 Greenpeace UK, *Putting the Lid on Fossil Fuels,* 1997.

36 Ingold, "Globes and Spheres," 1993, 31–32.

37 Greenpeace UK, ibid.

photographs of existing destruction wrought by weather, alongside the future possibility of a warming planet signified by the computer model predictions.

The "risks and the victims"[38] of climate change designated here are more clearly (and narrowly) identified than in the previous publication. The burning of fossil fuels as the cause of climate change is communicated through the destructive image of an oil platform producing a fiery plume, with the risk of flood and hurricane damage communicated via photographs. Yet more significantly, humans are not identified as victims, rather the loss of beautiful ocean habitats and its wildlife are victims of the dirty oil industry, with the clean technology of solar panels on the final page setting up an opposition between dirty versus clean energy.

PHASE THREE: GLACIAL IMPACTS AND RENEWABLE SOLUTIONS

The first photographic 'evidence' of climate change impacts was captured by Greenpeace in 1997 through the splitting of the Larsen B ice shelf in Antarctica (Figure 3). From existing knowledge that climate change involves rising global temperatures, the cracked ice functions indexically as evidence of climate impact. However, this image also works on a symbolic level. The visual pristineness of Antarctica as a seemingly virginal land is in stark visual contrast with the devastation signified by the split. The cropping of the image places greater emphasis upon the size of this split, further amplified by the small human figure in the frame.

This visual and epistemological opposition between human and nature stages a moral opposition, where the negative impacts of human activities are conveyed by the visible damage to ice. Yet the problem with such an image, and its subsequent circulation within the news media, is that it simplifies the impact of climate change, reducing it to a single dramatic instance, and a synecdochal sign for climate change.

38 Mormont and Dasnoy, "Source Strategies," 1995, 61.

Figure 3: Crack in the Larsen B Ice Shelf, Antarctica, 1997.

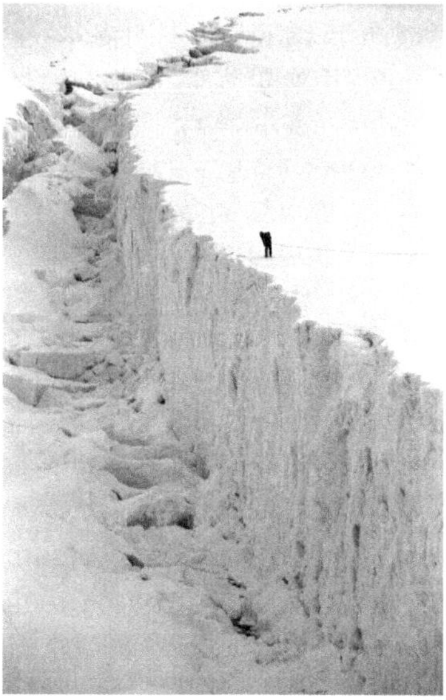

The synecdochal function of the cracked ice image facilitated a shift in the visual discourse of Greenpeace climate change communication during 1998. A visual language of glacial ice and the fragility of the polar ice caps was established. Dominating the cover of the publication, *Nature's Bottom Line: Climate Protection and the Carbon Logic,*[39] is the photograph of the Larsen B ice shelf crack. The text foregrounds the need for 'protection' of nature, a common discursive trope in environmental discourse, thus framing climate change as a consequence for the natural landscape, rather than humans.

From establishing the evidence of climate change upon Antarctic ice, *New Power For Britain: A Strategy for a Renewable Energy Industry*[40] attempts to communicate a positive message for the development of a renewable energy industry for Britain. The image of the wind turbine that dominates the cover is supported by images of the natural elements that are intended to identify the raw materials of nature used for renewable energy. The symbolic resonance[41] of images of natural energy in the form of wind, sun and waves confers meaning upon the technology of the wind turbine. A positive meaning is acquired by its

39 Greenpeace UK, *Nature's Bottom Line,* 1998.

40 Greenpeace UK, *New Power For Britain,* 1998.

41 Hansen, "The Media and the Social construction of the Environment," 1991.

referential links with the natural power of energy forms that can be harnessed and made useful for energy production, as well as for British industry.

In different ways, both publications make use of images of the natural world: one that signifies the fragility of nature, the other connoting the power of natural forces that will protect the fragile earth. Furthermore, a temporal shift is established: now that the symptoms of climate change have become visible in the present, authenticated by the photographic sign, Greenpeace appears to re-focus its message on promoting solutions. This is also consistent with requests, at the time, from NGO members and the public, for NGOs to offer more solutions to climate change rather than to present doom scenarios.[42]

The move towards solutions is enabled because of the visible evidence of climate change impact in the present. This evidence, however, means that the polar ice caps function as privileged signs within the discourse of climate change. Undoubtedly providing an important impetus to climate politics, in light of the formation of the Kyoto Protocol in 1997, at the same time, such images produce a distancing effect, relegating climate impacts to a remote and inaccessible place, where animals and habitats are affected, rather than humans.

Attempts to explain the complex science of climate change were renewed. In 1997, *Fossil Fuels and Climate Protection: The Carbon Logic*[43] outlined the science of carbon logic, which calculated the amount of carbon dioxide that could safely be released into the atmosphere without causing irreparable climate change. Public and political perception at the time was that fossil fuel reserves would run out, thereby encouraging the search for new oil. This is countered here by the prediction that fossil fuel reserves outnumber the carbon budget and that renewable forms of energy should be developed instead.

The Carbon Logic provided the scientific rationale for the next phase of climate change campaigning: a focus upon, and attempts to stop, new oil exploration. Thus, whilst critical of the distancing effect engendered in the new visual language of climate change, I also acknowledge the efforts by Greenpeace to explain the causes of, and offer solutions to, climate change, to some extent enabled by the very production of photographic evidence.

42 Farrow, "Communicating about Climate Change," 2000.

43 Greenpeace International, *Fossil Fuels and Climate Protection,* 1997.

PHASE FOUR: DIRTY OIL, DIRTY POLITICS

Greenpeace climate change campaigns during the latter years of the 1990s continued targeting oil companies and new oil exploration, whilst also promoting renewable energy. The Arctic campaign in 1999 focused upon drawing attention to the opening up of new oil reserves off the Alaskan coast. In contrast to the Atlantic Frontier Campaign of 1997, this campaign had stronger symbolic resonance due to the geographic location of the area and its associations with the pristine polar landscapes already identified as being affected by climate change. In conjunction, exploration in the Arctic was being led by British Oil company BP, thus making the area potentially more relevant to a UK audience. BP was promoting itself as a company with high environmental credentials.[47] To counter this, Greenpeace campaign literature made the specific visual link between clean ice landscapes and the dirtiness of oil. The use of images of polar bears became part of the symbolic repertoire of climate change images.

The targeting of oil companies in the campaign against climate change led to the development in 2001 of the biggest international campaign aimed at bringing global awareness to the issue of climate change. "Stop Esso" (Figure 4) highlighted the role of oil in contributing to climate change, the links between governments and oil industries demonstrated by ExxonMobil's financial backing of George Bush's US presidential election campaign, and Bush's subsequent withdrawal of the US from the Kyoto Protocol. This was the first Greenpeace public climate change campaign to not use images of climate impacts, instead using the Esso logo and images of Bush to promote its message.

Figure 4: Stop Esso Campaign 2002.

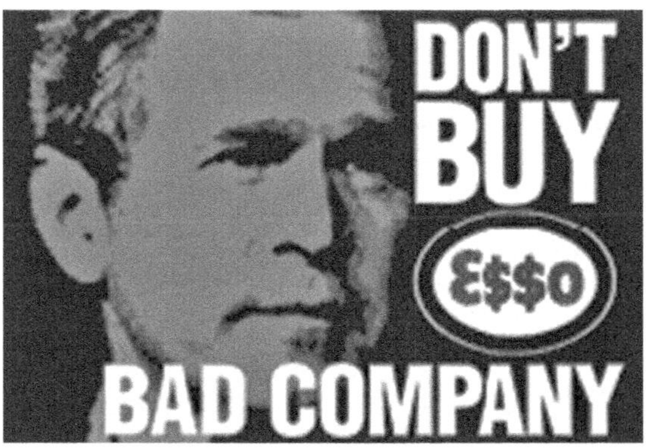

This shift in visual communication demonstrates how Greenpeace had perceived the issue of climate change as having developed sufficient cultural resonance for the public to understand without including evidence of climate impacts. Instead, climate change is framed in relation to the production and use of oil, and the economic-political role of governments in supporting the oil industry, as well as their funding of climate change skeptics,[44] who were outnumbered by the consensus of climate scientists.[45] The IPCC's Third Assessment Report in 2001 presented strong evidence that the increase in global temperatures during the last 50 years "was likely...due to the increase in greenhouse gas concentrations."[46] As Weart has argued, the uncertainty now lay not in the science but in "what people will choose to do" about it.[47]

PHASE FIVE: THE 'HERE AND NOW' OF CLIMATE CHANGE

"*New visual evidence* of the impacts of climate change was released by Greenpeace in Patagonia today. *Dramatic new photos* of Patagonian glaciers [...] show the extent to which climate change has caused the ice to melt this century, when compared to photos of the same glaciers taken in 1928."[48] [my emphasis]

Since 2002, photographic documentation of the world's melting glaciers has figured prominently in the climate change communication strategies of Greenpeace. This documentation relies upon recreating in the present, historic photographs taken during early twentieth century geological expeditions. The visual evidence of current climate impacts upon melting glaciers is thus authenticated by the photographic evidence of the past.

Yet the "dramatic new photos"[49] of climate impacts simultaneously assigns climate change to the past. This is achieved through the use of comparative "before and after" (climate change) shots that present a linear perspective, identified by Adam as representative of an "abstraction from context" which underpins scientific knowledge and its emphasis upon visible materiality as a

44 Greenpeace International, "Exxon's Weapons of Mass Deception," 2002.

45 Weart, *The Discovery of Global Warming,* 2003.

46 Ibid., 187.

47 Ibid., 192.

48 Greenpeace International, "Pictures of Climate Change," 2004.

49 Ibid.

sign of "the real."[50] They also signify the beautiful and the spectacular; a key strategy of Greenpeace's success.[51] In the context of climate change communication, such beautiful devastation could quickly be interpreted as inevitable and unavoidable loss, where nostalgia for the irretrievable past is fixed by the passing of time within the photograph.

Since 2000, there has been "a new sense of urgency attached to risks from climate change"[52] due to the increase in everyday experiences of extreme weather events. The heatwaves that swept across Europe in 2002, and occurrences of local flooding, have meant that climate change evidence has been brought closer to many European homes. Alongside the documentation of climate impacts upon the world's glaciers, since 2002, Greenpeace UK climate campaigns have also been increasingly focused upon the human impacts of a changing climate and the need for action on a UK governmental and personal level. Public campaigns have focused upon promoting renewable energy through wind power.[53]

The need for a political commitment to renewables has developed in the context of the political framing of nuclear as a safe option for addressing climate change. This discursive reframing of nuclear by the UK government has occurred over the last few years, with debates first appearing in the UK media during 2003–2004.[54,55] Greenpeace's response to this has been to dispel the myths surrounding the safety of nuclear by foregrounding the problems of radioactive waste, critiquing its economic viability, and by (problematically) linking nuclear to concerns about national security through discourses of terrorism.

Given the increasing global awareness and media coverage of climate change, Greenpeace UK has shifted some of its attention to the solutions available on a local/regional level. Decentralizing energy production has been a key message of Greenpeace UK climate change campaigns since 2006.[56] The

50 Adam, *Timescapes of Modernity*, 1998, 39.

51 Boettger, "The Role of Photography in Greenpeace's Strategy," 2001.

52 Carvalho and Burgess, "Cultural Circuits of Climate Change," 2005, 1466.

53 Greenpeace UK, *Speak Out!* 2002; Greenpeace UK, *Sea, Wind, Europe,* 2004.

54 Carvalho and Burgess, ibid.

55 This signals a u-turn on the UK Labour Government's commitment to decommissioning all nuclear power stations by 2025 and is in opposition to public opinion. See Bickerstaff et al., "Reframing Nuclear Power," 2006 and Doyle, "Acclimatizing Nuclear?" 2011.

56 Greenpeace UK, "Decentralised Energy," 2007.

promotion of more localized energy production through the generation of heat and power together has been communicated as a common sense option that is already happening in other countries (Denmark and Finland) and, as such, presents a positive message about tackling climate change.

CONCLUSION—LIMITATIONS OF THE VISUAL WITHIN ENVIRONMENTAL DISCOURSE

The privileging of the visual/visible within environmental and scientific discourse has proven problematic for the representation of climate change. Environmental problems develop over a long period of time and are often invisible.[57] A prioritization of the visible, at the expense of that which is hidden, thus reinforces the view that 'seeing is believing,' based upon the empiricism of scientific knowledge. This belief in the visible has contributed since the early 1990s to the difficulties encountered in the communication of climate change as a 'real' threat.

Adherence to the visible as a form of (scientific) knowledge is reinforced further by media coverage of the environment, which routinely employs events-based tactics,[58] rendering climate change difficult to cover. Prioritizing the visual is also more generally characteristic of environmental discourse. A visual aestheticization of the landscape, as representative of the environment, has come to define environmental discourse since the Enlightenment, placing vision/sight as one its key features.[59]

Drawing upon these perspectives in order to illustrate the difficulties encountered in the representation of climate change, I have also called more specific attention to the role of photography within environmental discourse and the communication strategies of Greenpeace. Where photographic discourse places emphasis upon the indexical nature of the photographic sign as a form of truth,[60] I have argued that this proves particularly problematic in the context of

57 Adam, *Timescapes of Modernity,* 1998.

58 Hansen, "The Media and the Social Construction of the Environment," 1991, "Greenpeace and Press Coverage of Environmental Issues," 1993, "Claims-making and Framing in British Newspaper Coverage," 2000; Anderson, *Media, Culture and the Environment,* 1997.

59 Urry, *The Tourist Gaze, 1990*; Wilson, *The Culture of Nature, 1992;* Mcnaghten and Urry, *Contested Natures, 1998*; Ingold, "Globes and Spheres," 1993.

60 Barthes, *Camera Lucida,* 2000.

climate change communication. The moment climate change can be photographed is the moment it becomes visible as a symptom, and thus too late for preventative action. This temporal dilemma is particularly problematic for Greenpeace's investment in photography as "internationally comprehensible."[61]

Greenpeace has been committed to communicating the reality of climate change since 1990[62] and to using photography as a privileged form of environmental communication. By identifying the representational problems encountered in communicating a temporal, and not always visible, environmental risk such as climate change, attention has been focused in this chapter upon how Greenpeace sought to make 'real' the potential and invisible risks of climate change, in light of its commitment to photographic communication. Through historical analysis of Greenpeace public campaign literature on climate change since 1994, I have identified five representational phases that illustrate how Greenpeace has framed climate change in its attempt to make it meaningful to the public.

Phase one (1994–1997) established the dangers of a warming planet, presenting the future as an imminent and catastrophic present, communicated through metaphors of heat and nuclear catastrophe. Phase two (1997) shifted attention to the causes (both present and past) and solutions to climate change, making the issue meaningful by designating the victims as the habitats and wildlife of the Atlantic Ocean. The first photographic evidence of climate change impacts in Antarctica in 1997 helped create a synecdochal sign for climate change in the present, with phase three (1997–1999) focusing upon the fragility of the polar regions, as well as offering renewable solutions.

With photographic evidence of climate impact upon polar ice, and increasing public consciousness, phase four (2001–2003) concentrated upon the links between dirty oil and dirty politics, identifying Esso and George Bush as the climate change villains. Phase five (2002–2007) returned to the visual trope of melting glaciers, explicitly using the indexical power of the photograph as a sign of climate change truth. At the same time, this photographic 'truth' has enabled Greenpeace to shift its campaign messages to solutions on a national/regional level, through decentralized energy production and the promotion of non-nuclear energy sources.

The temporality of all environmental discourse that makes present environmental problems relevant through reference to the past and the future has been present in all phases of Greenpeace climate change communication. Yet in

61 Boettger, "The Role of Photography in Greenpeace's Strategy," 2001, 12.

62 Mormont and Dasnoy, "Source Strategies," 1995.

the case of climate change, this has been shown to be complicated by its non-visible nature. An examination of the history of Greenpeace climate communication shows that Greenpeace has always sought to present climate change as a present threat.

So what of the role of photography? Photographic evidence of climate impacts upon the Antarctic ice in 1997 proved a significant moment in public and governmental acknowledgement of climate change (as seen in the Kyoto Protocol in 1997). This also enabled Greenpeace, to some extent, to shift the focus of its campaign to a more solutions-led approach. However, I would maintain that continued adherence to "the language of photography [...] which is also used by the mass media"[63] reinforces the very problems encountered in the communication of environmental issues, including the science upon which they are often dependent for legitimation.

Greenpeace's commitment to the photograph as a powerful tool of environmental persuasion and documentation is inscribed by the representational limitations of the visual as a discourse of seeing and truth. Glacial retreat "as one of the most visible signs that warming and climate change is real"[64] could quickly figure climate change as being observable from a distance, where fatalistic images operate to disempower human agency. However, more recent attempts by Greenpeace to offer credible solutions through the option of de-centralized energy signals a more hopeful narrative, based upon climate change solutions, rather than one of catastrophe.

It is worth thinking about how the representational limitations of the visual, in relation to the communication of environmental issues, has broader imply-cations for a world invested in the visual as a form of truth; an investment seen in this instance as having detrimental impacts upon the environment. Rather than proving that climate change is real through visible means, it might be more useful for environmental NGOs, and environmental scientists, to persuade the public that not all environmental problems can be seen.

ACKNOWLEDGEMENTS

Thanks go to the staff of Greenpeace UK for allowing me access to and use of campaign material, particularly Emma Gibson, Angela Glienicke and Stokely Webster.

---

63 Boettger, "The Role of Photography in Greenpeace's Strategy," 2001, 12.

64 Greenpeace International, "Arctic Environment Melts Before our Eyes," 2002.

BIBLIOGRAPHY

Adam, Barbara. *Timescapes of Modernity: The Environment and Invisible Hazards*. London: Routledge, 1998.

Allan, Stuart, Barbara Adam, and Cynthia Carter, eds. *Environmental Risks and the Media*. London: Routledge, 2000.

Anderson, Alison. *Media, Culture and the Environment*. London: UCL Press, 1997.

"Arctic Environment Melts Before our Eyes." *Greenpeace International*, August 7, 2002. Accessed May 17, 2006. http://www.greenpeace.org/international/news/glaciers-melt-before-our-eyes.

Barthes, Roland. *Camera Lucida: Reflections on Photography*. London: Vintage, 2000.

Boettger, Conny. "The Role of Photography in Greenpeace's Strategy." In *Greenpeace, Changing the World: The Photographic Record*, edited by Conny Boettger and Fouad Hamdan. Steinfurt: Rasch & Rohring, 2001.

Brown, Michael, and John May. *The Greenpeace Story*. London: Dorling Kindersley, 1989.

Bickerstaff, Karen, Irene Lorenzoni, Nick Pidgeon, Wouter Poortinga, and Peter Simmons. "Reframing Nuclear Power in the UK Energy Debate: Nuclear Power, Climate Change Mitigation and Radioactive Waste." *Public Understanding of Science* 15 (2006): 1–25.

Carter, Neil. *The Politics of the Environment: Ideas, Activism, Policy*. Cambridge: Cambridge University Press, 2001.

Carvalho, Anabela and Jacquelin Burgess. "Cultural Circuits of Climate Change in U.K. Broadsheet Newspapers, 1985–2003." *Risk Analysis* 25, 6 (2005): 1457–69.

Doyle, Julie. "Seeing the Climate?: The Problematic Status of Visual Evidence in Climate Change Campaigning," In *Ecosee: Image, Rhetoric, and Nature*, edited by Sidney Dobrin and Sean Morey. New York: State University of New York Press, 2009.

—. *Mediating Climate Change*. Farnham: Ashgate, 2011.

—. "Where Has all the Oil Gone? BP Branding and the Discursive Elimination of Climate Change Risk." In *Culture, Environment and Eco-Politics: History/Theory/Practice*, edited by David Wragg and Nick Heffernan. Newcastle: Cambridge Scholars Press, 2011.

—. "Acclimatizing Nuclear? Climate Change, Nuclear Power and the Reframing of Risk in the UK News Media." *International Communication Gazette* 73, 1–2 (2011): 107–25.

Eden, Sally, Andrew Donaldson and Gordon Walker. "Green Groups and Grey Areas: Scientific Boundary-work, Nongovernmental Organisations, and Environmental Knowledge." *Environment and Planning* 38, (2006): 1061–76.

Farrow, Cherry. "Communicating about Climate Change." In *The Daily Globe: Environmental Change, the Public and the Media,* edited by Joe Smith. London: Earthscan, 2000.

Greenpeace International. *Climate Time Bomb: Signs of Climate Change from the Greenpeace Database.* Greenpeace International: Amsterdam, 1994.

—. "About Greenpeace." Accessed January 16, 2007. http://www.greenpeace.org/international/about.

—. *Fossil Fuels and Climate Protection: The Carbon Logic.* Amsterdam: Greenpeace International, 1997.

—. *Exxon's Weapons of Mass Deception,* October 2002. http://www.greenpeace.org.uk/MultimediaFiles/Live/FullReport/5291.pdf.

"Greenpeace Science Unit," *Greenpeace International.* Accessed January 20, 2006. http://www.greenpeace.org/international/about/greenpeace-science-unit-2.

Greenpeace UK. *Climate Change: A Reader's Guide to the IPCC Reports.* London: Canonbury Villas, 1990.

—. *Putting the Lid on Fossil Fuels: Why the Atlantic Should be a Frontier Against Oil Exploration.* London: Canonbury Villas, 1997.

—. *Nature's Bottom Line: Climate Protection and the Carbon Logic.* London: Canonbury Villas, 1998.

—. *New Power For Britain: A Strategy for a Renewable Energy Industry.* London: Canonbury Villas, 1998.

—. *The Atlantic Frontier: Britain's Last Ocean Wilderness.* London: Canonbury Villas, 1998.

—. *Speak Out! No to Nuclear, Yes to Wind.* London: Canonbury Villas, 2002.

—. *Sea, Wind, Europe: Clean, Green Energy for Every Home in Europe.* London: Canonbury Villas, 2004.

—. "Decentralised Energy," 2007. Accessed February 3, 2007. http://www.greenpeace.org.uk/climate/solution/revolution.cfm

Harré, Rom, Jens Brockmeier, and Peter Mühlhäuser. *Greenspeak: A Study of Environmental Discourse.* London: Sage, 1999.

Hansen, Anders. "The Media and the Social Construction of the Environment." *Media, Culture and Society,* 13 (1991): 443–56.

—. "Greenpeace and Press Coverage of Environmental Issues." In *The Mass Media and Environmental Issues,* edited by Anders Hansen. Leicester: Leicester University Press, 1993.

—. "Claims-making and Framing in British Newspaper Coverage of the 'Brent Spar' Controversy." In *Environmental Risks and the Media,* edited by Allan Stuart, Barbara Adam and Cynthia Carter. London: Routledge, 2000.

Houghton, John. *Global Warming: The Complete Briefing.* Cambridge: Cambridge University Press, 2004.

"How is Greenpeace Structured," *Greenpeace International.* Accessed January 20, 2006.
http://www.greenpeace.org/international/about/how-is-greenpeace-structured#

Ingold, Tim. "Globes and Spheres: The Topology of Environmentalism." In *Environmentalism: The View from Anthropology,* edited by Kay Milton. London: Routledge, 1993.

Macnaghten, Phil and John Urry. *Contested Natures.* London: Sage, 1998.

Mormont, Marc and Christine Dasnoy. "Source Strategies and the Mediatization of Climate Change." *Media, Culture and Society,* 17 (1995): 49–64.

"Pictures of Climate Change from the Disappearing Glaciers of Patagonia." *Greenpeace International,* Press release, 10 February 2004. http://www.greenpeace.org/international/press/releases/pictures-of-climate-change-fro#.

Urry, John. *The Tourist Gaze: Leisure and Travel in Contemporary Societies.* London: Sage, 1990.

Wapner, Paul. "Greenpeace and Political Globalism." In *The Globalisation Reader,* edited by Frank Lechner and John Boli, 385–91. Oxford: Blackwell, 2000.

Weart, Spencer R. *The Discovery of Global Warming.* Cambridge, Mass: Harvard University Press, 2003.

Wilson, Alexander. *The Culture of Nature: North American Landscape from Disney to Exxon Valdez.* Oxford: Blackwell, 1992.

IMAGE COPYRIGHTS & SOURCES

Figure 1: © Archivo Museo Salesiano © Greenpeace/Daniel Beltra.
Figure 2: Greenpeace International 1994.
Fig. 3: © Greenpeace/ Steve Morgan.
Fig. 4: © Greenpeace UK 2002.

The Uncanny Polar Bear: Activists Visually Attack an Overly Emotionalized Image Clone

VERA TOLLMANN

This essay takes as its starting point an observation on popular culture: The polar bear has become an icon of climate change discourse. Against the backdrop of cultural and visual studies, the text examines the bear's semiotic shifts, from an "early warning signal" of climate change to the iconoclastic use and disruption of the image by activists, to its use as commercialized "carrier of emotions"—or even "emoticon." How does context change the polar bear's message? Following W.J.T. Mitchell's method of "iconology," which reconsiders the image as a living thing, this essay takes a look at images of polar bears across the media.

The polar bear on the ice is an icon of climate change. In the media, the motif regularly serves as an extreme example illustrating global warming and the precarious situation of the Arctic's inhabitants in the face of polar melting. The polar bear is no longer portrayed as solitary and independent, but as a plague, or an animal with "feelings." What kind of times are these, when polar bears still roam freely in northern Europe? When "grolar bears"—polar bear-grizzly bear hybrids—are already being born in Alaska? The image of the "white" polar bear is being spoiled.

What has happened? Can the polar bear still be the representative of a problem? Or is the motif worn out? In the frigid days of winter 2012–13, the front pages of various newspapers featured a snow-dusted polar bear covering its ears with its front paws. The photographer must have been drawn to the seeming humanity of the gesture, rather than the polar bear as a warning of climate change. As a "warm" image, the close-up served to make the cold outside more tolerable. "I can't bear it any longer," read the headline in the German national newspaper *Frankfurter Allgemeine Zeitung*. If the double-digit below-freezing temperatures were unbearable even for a creature of the Arctic, then it must have been really cold. And so even in the serious media the polar-bear motif, by now familiar from the discourse around climate, turned up under the heading of entertainment.

Everyone likes pictures of animals, especially young animals acting like awkward, playful children and, with their furry, chubby appearance, fitting the childlike image especially well. With polar bears, there is also a resemblance to the typical teddy-bear face. Even Walt Disney relied on this familiar likeness with Winnie the Pooh. The latest view counts for YouTube videos of clumsy animals attest to the continuing popularity of this motif. An especially pertinent indication of the popularity of polar bears and the perception of them as cute was provided by Knut, the polar bear that was born at the Berlin Zoo in 2007 and hand-raised. Within days of his first public appearance in front of the press with his zookeeper, the baby animal had become an icon, made by the German media. This intimate relationship between bear and keeper caused the image to oscillate between human child and pet.

ICONS

Since climate change was re-identified as a part of the political arena in 2007, the polar bear seems to have acquired two media identities, which have no existence independent of one another. One on hand: the polar bear in the wild, on the ice, intended to embody the drama of global climate change, a token image of nature. On the other: the denatured zoo bear aka stuffed animal, equally intended to charm visitors, but without further encoding (to create awareness of climate change, for example). Both identities briefly coincided in Berlin's polar bear Knut. Thus the distinction between "discrete icon" and "generic icon" made by communication theorist David D. Perlmutter in the context of war and crisis photography[1] must, in this case, be set aside. In the case of generic icons, we are dealing with pictures "that describe many images across time and space and events. The 'starving child of Africa' is such a commonplace—a veritable visual cliché repeated over the last century." These ordinary icons are contrasted with discrete icons, "those images that we recognize as single famous 'shots' taken at one time and place in relation to one event."[2] These latter correspond to icons that meet criteria such as celebrity, prominence, frequency and profit.

1 Perlmutter, "Hypericons," 2006, 54.

2 Ibid.

In 2007, the year of the fourth IPCC report on climate change, pictures of Knut possessed this double presence because of the way the bear was presented in politics and the media as both the image of a serious problem and an endearing entertainment figure—"Cute Knut."[3] The media made the zoo animal into a pop star, with German *Vanity Fair* calling him an "international star from Germany." The moment was opportune; climate change was a topic in global politics. Knut the polar bear even made it onto the cover of *Vanity Fair*'s international edition. That cover showed a montage depicting the three-month-old bear and celebrity climate activist Leonardo DiCaprio on an ice floe. DiCaprio was really there; Knut, of course, was not. Both had been photographed by celebrity photographer Annie Leibovitz, the icon of American portrait photography.

Animal advocates anthropomorphized the bear in their campaigns, as did photo editors and YouTube amateurs. The animal-rights group PETA, for example, described Knut's father as "a broken creature with a stunted mind" and campaigned with celebrity activists such as Stella McCartney. After Knut's death, the German animal-protection federation called for an end to the "unnatural" practice of keeping polar bears in zoos. Note: only after Knut was no longer alive.

THE WHITE BEAR

Likewise, the white polar bear functions as a metaphor: as a blank sheet of paper. Thus, it provides the ideal projection screen for all manner of interests. This is why polar-bear images cannot attain the status of "hypericons." "The new indelible images of world affairs might be called hypericons. They gain fleeting attention and then are replaced quickly by new icons. Since 9/11, a number of pictures have received saturation coverage, from Saddam's capture in his bunker through the orange-jumpsuited hostages."[4]

The "life" of polar bear images in visual culture has more in common with what W.J.T. Mitchell calls the convergence of iconic concepts, cloning and terrorism. In the context of the U.S. government's rhetorical and visual strategies

3 "The polar bear has been dubbed 'Cute Knut' by the media, and Germans have been eagerly following his progress on television, and the internet." Moore, "Baby bear becomes media star," 2007. See also Spiegel Online International, "Berlin's Cute Knut," 2007.

4 Perlmutter, "Photojournalism and Foreign Affairs," 2005.

after 9/11, he analyzes how fear of cloning and fear of terrorism became literal in speech (the "war on terror") and images (such as Abu Ghraib Man). "Cloning has been re-metaphorized as a figure of speech for all kinds of processes of copying, imitation, and reproduction—as, in other words, an 'image of image-making.'"[5] Mitchell traces transformations of image samples that reappear as "cloned images" in visual art, cartoons and activism.

From New York to Shenzhen, images of polar bears are copied and reproduced primarily in the context of environmental campaigns and advertising posters. In the more ideal scenario, it is a mascot for climate defenders, in which case the image it presents is usually harmless, without mechanisms of cause and effect. In the Arctic, the bear still seems to fulfill a longing for unspoiled nature. As long as the polar bear can survive there, nature is still in balance—this seems to be how the image is interpreted.

Summer 2012 marked another high point for the climate-political connotations of the polar-bear image. Yet the image in circulation seemed more "broken," as in a new ad from Greenpeace (Figures 1 and 2), in which a polar bear rummages unnoticed through garbage on the streets of London.[6] Our guilty conscience set loose in our everyday lives? In our city? And nonetheless unobserved, except by us viewers and a girl riding in a car, whose senses would seem to be less burdened than those of adults: a typical Hollywood motif, providing children with (extra)sensory perception or fantastic sensitivity to the (spirit) world, with which they can make contact with animals or more extraordinary nonhuman beings.

For its 2012–13 winter campaign, a traditional Swiss producer of down jackets used an image showing two polar bears in bad weather in the Arctic. The bears, a big one and a smaller one, were not even real animals or realistic images anymore; they had obviously dressed two people in polar bear costumes. The message was clear: Wearing one of their jackets means being as protected against the cold as a polar bear—or, one could even say, feeling like a polar bear. Without the recent history of the serious use of polar bear images in the context of climate-change discourse, this example of "imagineering" would not have been as emotionally affecting.

5 Mitchell, *Cloning Terror*, 2011, 22.

6 Partizan, "A Homeless Polar Bear in London," 2012.

Figure 1 and 2: Stills from "Save the Arctic," Greenpeace campaign video, 1:36, 2012.

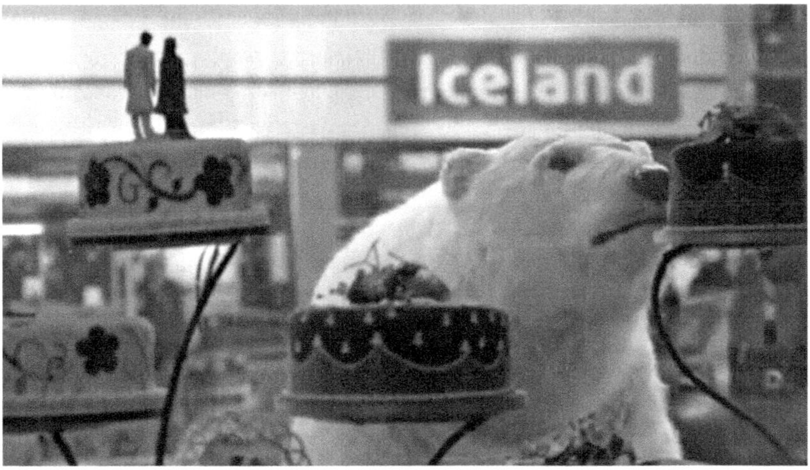

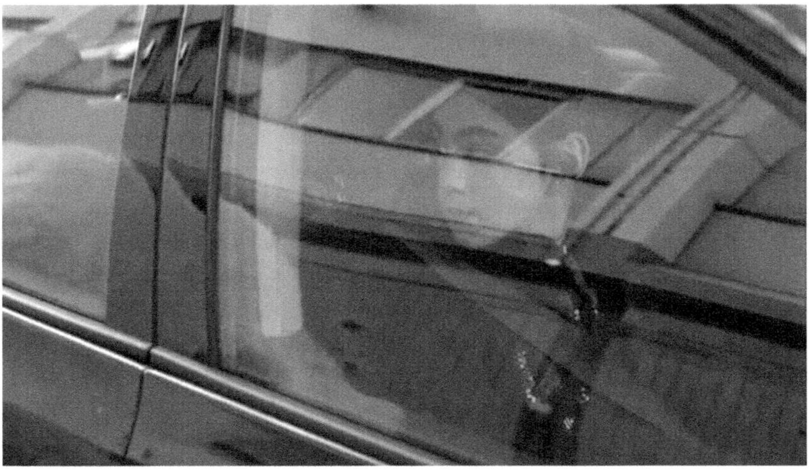

MENTAL IMAGES

As a result of our cultural experiences, we have images in our heads—media images—that can converge in unexpected combinations, as with London and the polar bear. In its polar-bear video clip, Greenpeace employs the same method Martha Rosler used for her famous photomontage series on the Vietnam War, though without her political acerbity. A video narrative directed by British artist David Harradine bears a striking resemblance to the Greenpeace clip but deconstructs the polar bear-human being metaphor as the story develops.[7] In "Bringing the War Home" (1967–72), she literally did bring the repressed home, into the living rooms of the American middle class. By combining photos from contemporary home-furnishings catalogs with media images of the war, she found a visual response to society's schizophrenic state. The conscious experience of the separation of here and there—here the American single-family home, there the violence; and by a similar token, here the hedonistic lifestyle, there the endangered polar bear in its habitat—is made clear in the pictures, and thus the repressed becomes unavoidable. In 2004, Rosler updated the work with photos from the Iraq War ("Bringing the War Home: New Series").

So Greenpeace sticks with the polar bear, the established motif in climate-change discourse, but reframes it into the realm of the fictional with a protagonist that roams the urban environment as in a fairy tale. The story ends on a moral note: the bear finds the sources of damaging carbon dioxide, sniffing first at a car's exhaust, then at a forgotten pinwheel before at last settling wearily and heavily against a tree trunk in a park. The stoical bear and its surroundings are meant to shake up viewers, but the Radiohead soundtrack casts a melancholy net. The avant-garde electronic band provided the song "Everything in Its Right Place," which appeared on their 2000 album *Kid A*, for the video. At a concert in Japan, Radiohead presented the same song in the cause of Tibetan independence,

---

7 "It's the Skin You're Living In." In a presentation at the "Image Politics" conference held in 2012 at the Potsdam Insitute for Climate Impact Research, media studies scholar Julie Doyle gave "insights into her own experiences of collaborating with the British artist David Harradine in 2010 in order to produce 'moving images of climate change.' One result of this was the critical reworking of the polar bear symbol realized within a performance video done by Harradine." (see Conference Report *Image Politics*). A manifesto coauthored by Doyle and Harradine during that process stipulates that climate-change art "must not contain images in which polar bears appear to be cute." See Doyle and Harradine, "Manifesto for Producing Artwork on Climate Change."

with the familiar flag, with its central sun and bright red and blue rays, hanging on the stage. Climate change, Tibet—all that really matters is global protest? And the music does incline more to a depressive descent into worry and poor prospects than to action.

Along with Radiohead, another British celebrity, the actor Jude Law, has lent his soft voice to this appeal from the "Save the Arctic" campaign against Arctic oil drilling. Awakening the Radiohead fans with emphatic gentleness—can this strategy work? When a song on the same album is called "How to Disappear Completely"?

Perhaps fans who like to immerse themselves in introspection and *Weltschmerz* are more open to an affective demonstration of solidarity with Greenpeace's demands. Then the celebrity strategy might work. A significantly more considered statement, though, was made by the American actor Ed Norton at the UN's Rio+20 summit in Brazil in summer 2012: "You can't protect something that people don't value. So we have to work to make sure that people value biodiversity and see how it has value in their own lives."[8] Whether these values can be conveyed by a polar bear torn from its larger context remains to be scrutinized.

Why do Western media and activists turn to animal images to represent the problem of climate change? What does the image of the polar bear evoke? What are animals in comparison to people? Why don't the media devote the same attention to people who are feeling the effects of climate change—coastal dwellers in Bangladesh, for instance, or islanders in Tuvalu or the Maldives?

THE SITUATION OF A SOCIAL PRAXIS THROUGH THE MEDIA

The media fetish that is the "event" determines what is reported. The spectacle of Greenpeace's actions serves this fetish, because most of their storyboards are based on confrontation—of nature and its industrial exploitation. In contrast to this feeding of the media spectacle, other environmental NGOs, and even some companies, downplay the complexity of the problem by providing easy steps to take. The recipient is usually offered a simple instruction that can be easily acted upon in everyday life. It used to be don't fill up at Shell; today it's don't fill up at BP, don't buy fish not certified by the Marine Stewardship Council, don't fly, don't drive; take the train, buy green electricity.

8 Barchfield, "Celebs Speak out at Rio Earth Summit," 2012.

These instructions, however, do not call the recipient's fundamental actions or lifestyle into question; rather, they allow one to think of oneself as a critical participant in political activity without criticizing the composition of society as a whole—without situating the solution in a larger context, where it would be recognizable as a minuscule adjustment. A surface treatment, as it were, when in fact an overhaul of society and its mode of production and consumption is required. The result is less a social praxis than the situation and representation (through the media) of what a social praxis could be. Thus a potential day-to-day praxis is replaced and controlled by little instructions. This compensatory act eases one's conscience. After all, fundamental societal configurations cannot be changed by terse instructions like "Save energy, turn off appliances at night."

IT'S REALLY NOT ABOUT POLAR BEARS ANYMORE

To investigate the production of images by activists other than large NGOs such as Greenpeace, I would like to focus on two internationally known examples: Plane Stupid and Climate Camp,[9] two initiatives by activists, artists and graphic designers from London which play a particularly "style-setting" role. Plane Stupid, whose name alone communicates through wordplay the strategy behind their activism, use performative tactics to forcefully protest against air traffic—that is, against a specific cause of climate change. Through regional interventions and single-issue politics, they focus on one aspect of the extremely complex problem of climate change.

Colorful nylon camping tents are the "trademark" of the Climate Camp network. By choosing this form of protest, not only do they appropriate very real urban space, with their tents they also signal their solidarity with climate refugees. They bring together various social movements, such as environmental activists and leftist critics of globalization. The subject of climate change creates a meeting point for all sorts of groups—a new situation for those involved. Thus the climate movement was, or is, an umbrella movement.

So what do Climate Camp and Plane Stupid think of the simplistic calls for public action mentioned above? Rather than addressing consumers, through their actions they criticize those who are politically responsible (representatives of international government) and who profit financially (airlines). Both groups have succeeded in introducing radically new images and forms of protest into the established landscape of environmental movements. Their image politics

9 See http://climatecamp.org.uk and http://www.planestupid.com.

evinces a new way of thinking in images. Whether these new images will serve to inspire new actions, however, is an altogether different question. At the very least, these groups have found a new visual language for the subject matter.

Image production in their context is partly based on choreographed interventions in public space. For example, Plane Stupid has occupied landing strips and aviation companies' offices, and Climate Camp occupied London's city center during the G8 summit. Both groups employ artistic strategies to choreograph visual attractions that contrast with the mostly random visual diversity of demonstrations, at which it is usually the biggest or most artisanal costume or protest banner that catches the lens of a journalist's camera.

POSTPRODUCTION

A one-minute video clip commissioned by Plane Stupid[10] is also quite similar in approach to Rosler's photomontages (Figures 3 and 4). The clip was produced as a cinema ad and released on YouTube just before the 2009 climate summit in Copenhagen.

Plane Stupid has been protesting against the continued development of airports and airline companies since 2005. Unlike the initiators of Climate Camp, these are relatively young artists, designers and art educators, who are advancing new, creative forms of protest: skating on an ice rink in penguin costumes, occupying easyJet's London offices or a runway at Stansted Airport.

Joss Garman, one of the group's founders, has been an environmental activist since age 14 and spent years volunteering and campaigning for Greenpeace UK.[11] According to Garman, their biggest success was to stop the runway project at Heathrow airport. "The most powerful tool in the armory of the critics of the runway was the fact that a movement existed, comprising all sectors of society. From the local residents and their councils, to WWF and the RSPB, through to Greenpeace, Plane Stupid and the Climate Camp. Working together they took on the combined might of British Airways, the CBI, and the government, and won. It was the galvanizing of this coalition, which explains the success of the Heathrow campaign."[12]

---

10 Mother Advertising Limited, "Polar Bear," 2009.

11 Knight, "Profile: Joss Garman," 2008.

12 Garman, "The Crucial Role of Activism in Scrapping Heathrow's Third Runway," 2010.

This clip differs from the mass of activist videos in that it is not a documentary clip like most of the others, but rather a produced film, a brutal cinema ad. In the space of 55 seconds, the viewer watches as a contemporary business district becomes the stage for a rather surreal happening: polar bears fall from the sky, bouncing off building facades and landing dead on cars and asphalt. The clip ends with a pair of titles reading: "an average European flight produces over 400kg of greenhouse gases for every passenger [...] that's the weight of an adult polar bear." Thus, rendered images of polar bears are placed into a context of climate-change issues. In principle, the clip functions according

Figure 3 and 4: Stills from "Polar Bear," a YouTube video by Mother Advertising Limited, directed by Daniel Kleinman, 0:56, 2009.

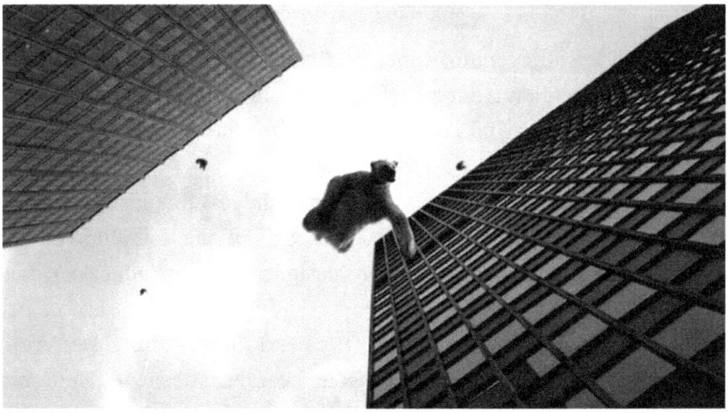

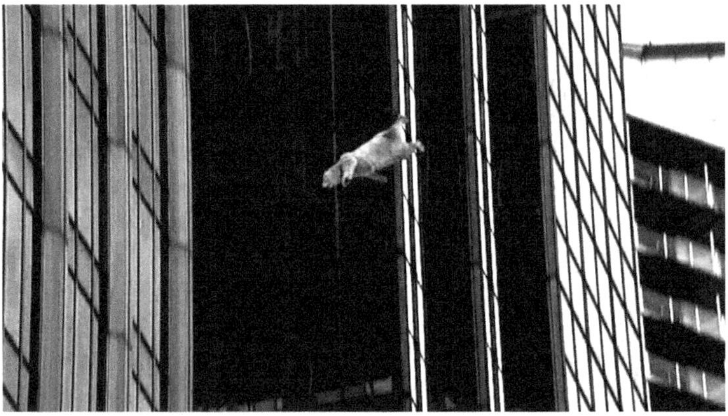

to the rules of advertising: A staged situation is followed by text that retroactively makes the meaning clear.

Here the distinction between "videohigh" and "videolow," which media scholar John Fiske drew in his essay "Videotech"[13]—that is, between low-tech and high-tech video recordings and their respective political implications—is deployed in a way that defies expectations: by deliberately opting for videohigh so as not to be immediately classed as agitprop. For "the credibility of video depends upon the social domain of its use. In the domain of the low (low capital, low technology, low power), video has an authenticity that results from its user's lack of resources to intervene in its technology. When capital, technology, and power are high, however, the ability to intervene, technologically and socially, is enhanced."[14]

As a rule, political activists[15] do not possess the resources to produce media that meet current standards for film and television. Activist videos can normally be identified by the fact that they are low-tech, which Fiske defines as follows: "The choices at the moment of shooting are reduced to those of framing, angle, and distance."[16] This certainly applies to the majority of videos shot with mobile phones. In this case, with the advantages of a good social network and the availability of digital technologies, it was quite possible to produce a clip that initially seems to have been made from a position of hegemony. Which, technologically speaking, has been possible for everyone for a few years now, thus largely invalidating the hegemonic position—with the exception of the expensive postproduction stage, which requires professional skills, the right software, processing power, etc. The Plane Stupid clip was created exclusively in postproduction.

---

13 Fiske, "Videotech," 2002, 383–92.

14 Ibid., 387.

15 Since the civil rights movement, organizational forms of activism have changed. Today, political activists—and more specifically in the context of this text, environmental activists—use the internet as a primary tool. With digitization, it has become easier for activists to organize, document and publicize direct action. Wiktionary defines "activist" as "one who is politically active in the role of a citizen; especially, one who campaigns for change." Cf. http://en.wiktionary.org/wiki/activist.

16 Fiske, "Videotech," 2002, 387.

RENDERING CLIMATE CHANGE

What is the relationship between rendered reality and analogue reality? With splatter action and a Biblical motif of an animal plague, the clip's director has appropriated the visual power of Hollywood to critique discount airlines and greenwashing. It is the "physicality" of the polar bears' collision with the cold, hard surface of capitalist architecture, the way nature and the built environment are slammed together that creates such strong discomfort for the recipient. The designation of individual "perpetrators" and "victims" gives way to a critique of capitalist structures. Up to now, the polar bear has been a popular advertising motif, but that was the polar bear in its habitat, the Arctic. That environment was seen as endangered, but for viewers it remained far, far away. Because of that distance, the polar bear on the ice remained a pure and beautiful image, never coming into contact with the world of humans. The use of a Surrealist method—the juxtaposition of two things that at first have nothing in common—renders visible the psychological state of ignorance.

In terms of moving images—a medium rarely employed by Surrealist artists—Jacques Rancière has analyzed Godard's montage practice as combining symbolic and dialectical montage. Among numerous examples from the fields of film and art, he also mentions Rosler's montage series, noting that "it is the clash of heterogeneous elements that provides a common measure."[17] For Rancière, "what is involved is revealing one world behind another: the far-off conflict behind home comforts."[18] Hinting in a similar direction, one YouTube user puts it more cynically in a comment on the Plane Stupid clip: "Lenin would be so proud of you for following his lead with the use of shock propaganda."

The clip relies entirely on the physical discomfort created by the situation. Bears and architecture are representatives of the two opposing sides in the problems of climate change: nature and human culture. The architecture stands for the relentless consumption of resources by industrialized nations, places where decisions are made and business is done according to the logic of growth. Another YouTube comment reveals the clip's location: "Vancouver, Canada, a common location for filming generic slick modern city in commercials and movies. You can look up 1052 W Hastings street there look east—it's where the bear falls on a car."

17 Rancière, *The Future of the Image*, 2007, 55.

18 Ibid., 56–57.

WHITEWASHING

It is an iconoclastic tale: an attack on the "generic icon" of climate change—the white polar bear, as white as the white in detergent ads—such an ideal of cleanliness, so much potential for projection. "A kind of theatrical excess in the rituals of smashing, burning, mutilating, whitewashing, egg and excrement throwing turns the punishment of images into a spectacular image in its own right."[19]

The polar bear, then, is used as an established icon in order to destroy it as such. "Against all 'real' scientific data, all that can be done is appeal to people's emotions," writes another polemical commenter on YouTube. The clip, however, is intentionally aimed at the connection between popular climate-change icons and consumer behavior. That political statement can stand without scientific data.

The Plane Stupid clip's producers are relying on recipients' collective memory of images, pictures of iconic polar bears in the media.[20] "Experience [...] is filtered through 'already seen' images," wrote Umberto Eco.[21] In this case, the aim is to make reference not to a particular film, but rather to the typical Hollywood aesthetic—which occurs in postproduction, via rendering.

Taking a different slant on familiar media images, Barbie Zelizer writes that images showing impending death, "their recycling over time [...] from the news into other modes of visual representation [...] art installations, fiction, and theater,"[22] are a typical recurring phenomenon. She adds that "about-to-die images often take on an afterlife through mnemonic practices that create a curious separation of the images of impending death from the events themselves. It is significant that this afterlife is not necessarily associated with the image's iconic status."[23] An undefined aesthetic reference is quite enough to make clear to viewers that a symbolic borrowing from the culture industry has taken place. What W.J.T. Mitchell describes as a "supply of magical images"[24] has been reassembled by Plane Stupid.

19 Mitchell, *What Do Pictures Want?*, 2005, 126–27.

20 "We used Polar Bears because they're a well understood symbol of the effect that climate change is having on the natural world." "Polar Bear," VisionOnTV (blog), 2009.

21 Umberto Eco, "A Photograph," 1986, 214.

22 Zelizer, *About to Die*, 2010, 64–66.

23 Ibid., 66.

24 Mitchell, *What Do Pictures Want?*, 2005, 128.

This explains the strong reactions, namely rejections, shown by YouTube viewers in comments like "it's not real" or "PROPAGANDA" even though the clip is an animation. These comments are aimed at the filmmakers' decision to tell the story in this way.

METAPICTURES: SELF-REFERENTIAL PICTURES

W.J.T. Mitchell characterizes the "living image" as a "secondary, reflexive image of images," also calling it a "metapicture."[25] It is a statement about other pictures—in the case of the Plane Stupid ad, about the polar bear as an established image in climate-change discourse. For Mitchell, these are "not merely epistemological models, but ethical, political, and aesthetic 'assemblages' that allow us to observe observers."[26]

Polar bears falling from the sky? What are spectators to make of this seemingly Biblical reference? Is it a parable for the exploitation of nature by humans? The clip illustrates this quite graphically, just as the plagues in the Bible, as a dramatizational device, served to make the Exodus of the Israelites from Egypt part of a larger history.

The image of falling polar bears also recalls a very different media picture that has been burned into our collective memory: the traumatic image of the Falling Man of 9/11.[27] Like the bears in the clip, people plummeted from the World Trade Center's towers. For ethical reasons, the only images shown were of bodies falling, never hitting the ground. But the mental image was gruesome enough. Now, the Plane Stupid clip has broken that taboo with polar bears as surrogates, thereby transforming the last ten years of public image production. At the same time, it steps outside the narrowly defined repertoire of climate-catastrophe images. For the clip refers to an event that has often been said to have changed the way we view media images and the representation of reality. The repertoire of 9/11 contains no images of a body hitting the ground. In the last shot of Plane Stupid's video, a polar bear crashes into a parked Cadillac. A viewer comments: "the only thing i find disturbing about this video is the polar bear landing on the cadillac."

Pop culture, too, contains examples of the citation of this ancient motif. In the 1999 film *Magnolia* by Paul Thomas Anderson, there is a scene in which

25 Ibid., 10.

26 Mitchell, *Picture Theory*, 1994, 49.

27 Zelizer, *About to Die*, 2010, 44–48.

hundreds of frogs rain from the sky. The animals have just as much psychoanalytical weight there as in the Plane Stupid clip. The repressed unpleasantly intrudes, inevitably returns. It falls heavily from the sky and crashes into solid urban space—in both cases onto automobiles, civilization's units of individual mobility. Considerable attention was paid to the sounds of the impacts in postproduction so that they would induce a feeling of unease. One can hear heavy bodies smashing into hard metal structures, so that the images of the fall become very physically present. In *Magnolia*, the Biblical scene remains unexplained, appearing more as an unpleasant dream that the protagonists experience individually. One drives around the city at night, another sits at home on the couch. Because each experiences the metaphorical events alone, these events remain somewhat unreal in the film. Above all, the scene provokes physical discomfort in the face of a higher power—nature's revenge?—just as Alfred Hitchcock's *The Birds* does. The Bible references, it seems, serve to create sources of aesthetic authority that can be linked to Sigmund Freud's concept of the uncanny.[28] For Freud, the uncanny symbolizes the return of the repressed.

In 2001, when the images of the terror attacks on New York's World Trade Center were broadcast around the world, those images looked like science fiction. Then came discussions of whether those media images resembled images from Hollywood directors, whether there were parallels with *Independence Day*, *Armageddon* and other science-fiction films.[29] Therein lay the shock of no longer being able to distinguish between fiction and reality. The next possibility would be to ask whether the images had perhaps anticipated Hollywood. After 9/11, the Bush administration met with leading Hollywood directors to discuss images of the future, a scenario that would fit perfectly into Baudrillard's simulation theory.[30]

Thus the Plane Stupid video not only comments on the iconology of the polar bear, but also treads on the problematic terrain associated with the media sensation that is 9/11.

28 Freud, *The Uncanny*, 2003.

29 See for example "Fernsehen I-III," three episodes of the ZDF talk show *nachtstudio* from September 2001 featuring Rainald Goetz, Moritz von Uslar and Volker Panzer. The second episode (accessible at http://www.youtube.com/watch?v=UoNcIZA5Jlw) was recorded and aired on September 12, 2001, the day after the World Trade Center attacks.

30 Baudrillard, *Simulacra and Simulation*, 1995.

In the tradition of iconoclasm, the icon gets smashed, while at the same time the feelings of shock and discomfort connected with the cultural memory of the attacks on the Word Trade Center are touched on. Through the combination of two "mental attacks," killing emotionalized polar bears and recapitulating and completing the Falling Man image, the clip reverberates in the viewer's awareness.

GUERRILLA COMMUNICATION

The clip would be unthinkable without the history of guerrilla communication, the strategy of a network of left-wing activists who have been turning the mechanisms of the mainstream into a subversive artistic practice since the 1990s.[31] They have made affirmation, over-identification and alienation their tools to intervene in the ongoing communication process. Armed with knowledge of the semiotic theories of Roland Barthes and Umberto Eco, they take a conceptual approach centered on devising a new architecture of the sign. However, their appropriations of media material have always remained visible. Low-tech alteration remains part of their activist aesthetic—think of *Adbusters*, for example. This do-it-yourself aesthetic creates a distance from the images being used or quoted, through which their normality and normativity can be called into question.

The reception of the clip is also determined by the location—YouTube—at which it may be accessed. The clip is available at any time and can be viewed more than once. For this type of campaign, however, activists must adapt to the functional mechanisms of the mainstream media. Professionalization, efficiency and media compatibility become essential criteria for political action. In this case, professionalism is represented by the clip's director, Daniel Kleinman, who had previously directed the trailers for several James Bond films (*Goldfinger*, among others). In a more recent advertising spot directed by Kleinman, the lead character is a polar bear who walks all the way from the Arctic through New York City to a suburban neighborhood to hug a man outside his house because he drives an electric car. That clip depicts the polar bear as an overly emotionalized human being.[32]

31 Thereby following in the footsteps of earlier movements such as Dada and Situationism.

32 "Polar Bear," 2010, advertising film for Nissan, 1:03, directed by Daniel Kleinman, TBWA\Chiat\Day.

POLAR BEARS NEVER HAVE TO CRY

In a brief description on their website, Plane Stupid makes the necessary paradigm shift clear: "It's really not about polar bears any more." The polar-bear motif has been used extremely frequently in the media in connection with portrayals of climate change. While the motif does have climate connotations, these go beyond its documentary import.

As the climate-policy consultant Anja Köhne says, "The polar bear presents an innocuous image, but the Arctic will melt and the polar bear can no longer be saved. This icon has devolved into kitsch. The image is insufficiently complex and no longer commensurate with the situation." Her suggested solution: "Basically, one would have to mount an assault on the icons."[33] Other, more graphic imagery which could be more effective, which enters "taboo areas," is not used; the preference, instead, is for aerial photographs as "opening shots" that show the catastrophe from a distance and "as a whole" in an attempt to objectify events, as if the viewer would thus be better able to understand and assess what had happened.

Yet the polar bear has already become a figure of cultural history. This is illustrated by the book *Polar Bear*, by the American artist Mark Dion, whose work deals with ecology and the cultural representation of nature. Dion photographed displays of stuffed polar bears in natural-history museums in Europe and the U.S., the majority of which portray the polar bear as a cute animal.[34]

The polar bear used to function as a metaphor for solitude and extreme conditions, but also for life untouched by civilization. In the Arctic it has no predators. The polar bear in the Arctic seems to fulfill a longing; it stands for the myth of survival in the wilderness, in the place where conditions are most extreme. The Neue Deutsche Welle song "Eisbär," by the Swiss band Grauzone, celebrated precisely this heroic aspect of the polar bear. What does this longing represent? The deeper meaning, not necessarily a given in songs of this genre,

33 Telephone interview by author, April 24, 2009. At the time, Köhne was climate policy consultant for Greenpeace and the World Wide Fund for Nature, as well as a research associate at the Freie Universität Berlin, working on a dissertation project on environmental integration in EU foreign policy.

34 Mark Dion, *Polar Bear,* 2003. Only at the Natural History Museum of Los Angeles County is it shown standing upright, a gigantic wild beast. Only then does danger enter the picture. Perhaps Californians have a different relationship to bears because bears are actually found in the wild there.

declares itself in the refrain: "Polar bears never have to cry." This is the impassioned coldness of the synthpop of the early 1980s, heard millions of times by teenagers seeking solace in the lyrics after their first disappointments in love. Be brave like a polar bear and endure your feelings: the song suggests identifying with the animal. Back then that double whiteness, the polar bear in its icy environment, signified innocence and regeneration. Today the polar bear also stands for a solitary fate, an individual, a personalized problem for which, conversely, there could be many personalized solutions. At any rate, the problem should not be solved as a society, the motif suggests. It is thus a living picture that, as W.J.T. Mitchell says, can be made to talk like a ventriloquist's dummy.[35] The polar bear images are made to "speak" to the viewer.

EXTENDED ACTIVISM

In an interview, Martha Rosler said of the original distribution of her Vietnam War montages, "I kept those works outside of the art world while the war was still raging. I published them in the alternative press at the time, or simply distributed them as photocopies, so the question of reception would have to be posed regarding the anti-war community, which seemed appreciative."[36] That was around 1970. Today, YouTube is to moving images what photocopies then were to photomontages: quick, inexpensive, with maximum reach—the distribution takes care of itself. What opportunities have emerged for activists as a result of new technological developments? What has changed as a result of YouTube? What conditions attend the reception of images there?

The climate movement documents its actions in YouTube videos and blog posts. The object is not to create new visual icons to succeed the polar bear as climate emissary. For Plane Stupid, the important thing is to call for more complexity to undermine the short-circuit arguments of a self-betraying society of the spectacle. Like the institutional networks, the activists are rapidly building networks as well, and because they are not centrally organized, there is no longer a need for hegemonic image politics in the internet age.

As of this writing, the clip has been on YouTube for two years and has been clicked 1,195,521 times—more than a million views. It is notable that the numbers of likes and dislikes are nearly the same: 1,612 likes, 1,509 dislikes. At least YouTube permits negativity, unlike Facebook. So what does it mean that so

35 Mitchell, *What Do Pictures Want?*, 2005, 140.

36 Lange, "Bringin' It All Back Home," 2005, 94-95.

many have bothered to click to express their displeasure with the clip, when social networks are not really made for unfavorable judgments? At any rate, the video seems to be polarizing, as the likes and dislikes show. One viewer writes: "Why does 1327 dislikes? The polar bears aren't real. And if it's because of the blood, it's not real. Everything is fiction, it's all done by computers." View counts like these are typically achieved only by Madonna and other pop stars. A Greenpeace video gets no more than 10,000 views on average.

SELF-CRITICISM

For Climate Camp, the issue is not primarily a matter of public attention on the web, but rather a discussion of how to translate the group's political demands into public interventions. After numerous camps, and in connection with the protests in the Arab world, Climate Camp's London activists decided to hold a retreat to reconsider their techniques. Their trademark, the colorful tents, no longer seemed appropriate and effective. Interestingly, Climate Camps, which have been taking place since 2008, look quite similar to Occupy camps. Like Climate Camp, the Occupy movement is also multisited. The difference lies in the temporal dimension. Climate Camps exist for a set time period, determined by the duration of a G8 summit, for example. The Occupy tents were set up for an unspecified length of time, taking over public space in the vicinity of public buildings representing the financial market. Their media-ready icon was Guy Fawkes, or rather the friendly-faced mask that represented him in the 2005 movie *V for Vendetta*, by James McTeigue. Some Occupy activists used the mask because their movement, too, is nonhierarchical. Therein lies the only point of commonality between the two groups.

The Climate Camp network is committed not to a specific object of protest, but to a style-setting format. It began with the idea of creating a space for discussion that would be open to various groups. An ecosystem is made up of many different populations, and Climate Camp's initiators sought to build theirs in mimetic fashion. Observations of nature became a prerequisite for design and activism. The first Climate Camp took place in 2006, with 600 people camping outside Britain's largest coal-fired power plant in Yorkshire for ten days. It was meant to be the start of a social movement. Since then, Climate Camps have come to be thought of as an alternative to summit meetings and have taken place all over Europe—for example, in connection with the 2009 G20 summit in London. The issue there was emissions trading, capitalism's obdurate commodification of nature. "Stopping carbon markets" was the campers' appeal to delegates. Like papal indulgences, emissions permits are acquired to build a new runway at Heathrow Airport, or to get approval for a new coal-fired power station. This systemic ambivalence is what the campers were criticizing. Emissions trading only swells the accounts of rich countries such as Norway with oil money.

CRISIS?

The discussions around climate change can no longer be wished away; the problem has become an established theme in society. No one today ignorantly asks, as the band Supertramp did in 1975 with their thoroughly ironic album title, *Crisis? What Crisis?* On the cover of the record we see a man sitting on a rooftop deck chair with sunglasses, transistor radio and tall mixed drink—a "perfect" vacation scenario. All around him, though, the entire city has degenerated into a garbage dump, with smokestacks on the urban horizon pouring forth smoke. The hedonist is unconcerned by his catastrophic surroundings. This discrepancy between the state of the environment and individual behavior bespeaks an ignorance that is still acted out today with regard to climate change. Climate change, however, cannot be conveyed solely with images of urban decay, since it must serve as a catchall term for all kinds of environmental problems and catastrophes. That makes the process potentially too abstract. The social psychologist Harald Welzer and the political scientist Claus Leggewie have acknowledged "the difficulty of translating scientific evidence into a narrative"[37] that is generally comprehensible. The best equivalent to the Supertramp cover are amateur videos on YouTube, which are quickly found via the search terms "glacier" and "cruise ship." Tourists, whose connection to the environment seems to have gone missing after days on the deck of an Arctic cruise ship, call out "Oh my goodness, wow, fantastic" as the ice breaks apart in movie-worthy fashion. Not a word about global warming; instead they record themselves and their enthusiasm for the collapse of nature: "nature snuff."

There cannot be one single icon for climate change for the climate movement. That would represent something hierarchical, whereas the climate movement consists of nonhierarchical networks comprising the environmental movement, opponents of globalization, and others. Occupy is different from

37 "Die Schwierigkeit der Übersetzung naturwissenschaftlicher Evidenz in ein Narrativ." Leggewie, Welzer and Heidbrink, "KlimaKultur—ein transdiziplinärer Projektverbund," 2008, 181.

climate protests, because Occupy has no spokesperson position. Recently it has become even clearer that previously stable signs can quickly drift. The future should no longer be thought of in terms of old categories such as hierarchical structures and iconic images. In the complexity of the global sphere, activist interventions, courses in activism and protests in the footsteps of performance art cannot develop long-term prospects and strategies, but they can return in ever-changing constellations and reframe or destroy ineffective, conventionalized icons.

AFTER THE POLAR BEAR

An activist group that briefly appeared in the summer of 2011 in Berlin performed symbolic protests under the name "Black Locust Movement." Perhaps their video[38] merely documents a one-off series of actions, but the idea of staying power forms the core of their protests. It is a protest that is aware of its impermanence. Above all, the protest fiction is transmitted to viewers' minds—and perhaps it is only there that it plays out and takes hold. But this trashy video makes an attempt to transfer to the viewer the idea of protesting, as opposed to the guilty conscience to which Greenpeace alludes. Species is manifesto[39] here, for the locust is a typical Berlin tree that grows wild, can survive in nearly any terrain and is known for its emission resistance. These trees, acting here on behalf of nature—not as victims, as in the Greenpeace video mentioned above—were to "break up the foundations of climate criminals, their technical facilities, their decision-making centers and their myths of clean coal." The locust's revenge is coming. With the aid of its "resilient and invasive qualities," the group aimed to achieve "an irreversible organic undermining of the following facilities: the Berlin headquarters of Vattenfall Europe, the Nordic embassies in Berlin, the lignite-fired power station of Jänschwalde and the roads around the lignite mine."[40] Several hundred trees were planted at night, by wild characters in bulky tree costumes. They came by night and planted: oppositional ideas.

38 "Black Locust," YouTube video, 1:45, 2011.

39 Cf. Donna Haraway, *The Companion Species Manifesto,* 2003.

40 Black Locust communiqué posted August 24, 2011 at Nerdcore, http://www.crackajack.de/2011/08/24/planting-trees-into-climate-crisis.

This ephemeral example shows how activists can experiment with paying tribute to the popularity of animals while avoiding the polar bear-human being metaphor. In this case, the wild animal costumes function in reverse, introducing nature as an active agent.

Translation: Patrick Hubenthal

BIBLIOGRAPHY

Barchfield, Jenny. "Celebs Speak out at Rio Earth Summit." June 21, 2012. http://bigstory.ap.org/article/celebs-urge-drilling-free-sanctuary-arctic-0.

Baudrillard, Jean. *Simulacra and Simulation.* Translated by Sheila Faria Glaser. Ann Arbor: University of Michigan Press, 1995.

"Berlin's Cute Knut: Polar Bear Cub Explores the Outside World." Spiegel Online International. Last modified on March 5, 2007. http://www.spiegel.de/international/berlin-s-cute-knut-polar-bear-cub-explores-the-outside-world-a-469860.html.

Black Locust communiqué. Posted August 24, 2011 at Nerdcore. http://www.crackajack.de/2011/08/24/planting-trees-into-climate-crisis.

"Black Locust." YouTube video, 1:45. Uploaded by "BlacLocust" on August 19, 2011, http://www.youtube.com/watch?v=edVqWPBZKwg.

Climate Camp Website. Accessed July 4, 2013, http://climatecamp.org.uk

"Das ZDF nachtstudio vom 12.09.2001 (1/4)." YouTube video, 14:51. Originally aired on the ZDF television network on September 12, 2001. Uploaded by "alexomat2" on September 27, 2001.

Dion, Mark. *Polar Bear.* Cologne: Walther König, 2003.

Doyle, Julie, and David Harradine. Manifesto for Producing Artwork on Climate Change. Accessed July 4, 2013. http://arts.brighton.ac.uk/projects/here-today--moving-images-of-climate-change/manifesto.

Eco, Umberto. "A Photograph." Translated by William Weaver. In *Travels in Hyperreality.* San Diego: Harcourt, 1986.

Fiske, John. "Videotech." In *The Visual Culture Reader,* edited by Nicholas Mirzoeff. London: Routledge, 2002.

Freud, Sigmund. *The Uncanny.* Translated by David McLintock. London: Penguin Classics, 2003.

Garman, Joss. "The Crucial Role of Activism in Scrapping Heathrow's Third Runway." *The Guardian,* May 17, 2010. http://www.guardian.co.uk/environment/cif-green/2010/may/17/heathrow-third-runway-activism.

Haraway, Donna. *The Companion Species Manifesto: Dogs, People, and Significant Otherness.* Chicago: Prickly Paradigm Press, 2003.

Harraldine, David, and Sam Butler. "Fevered Sleep. It's the Skin You're Living In." Web video, 2012, 8:31 min. Accessed December 2013. http://fevered sleep.co.uk/films/its-the-skin-youre-living-in.

Knight, Lucy. "Profile: Joss Garman." *New Statesman.* October 14, 2008. http://www.newstatesman.com/environment/2008/10/garman-greenpeace-movement.

Lange, Christy. "Bringin' It All Back Home." *frieze 95,* November–December 2005.

Leggewie, Claus, Harald Welzer and Ludger Heidbrink. "KlimaKultur—ein transdiziplinärer Projektverbund." In *Zwei Grad. Das Wetter, der Mensch und sein Klima,* edited by Petra Lutz and Thomas Macho. Göttingen: Wallstein, 2008.

Mitchell, W.J.T. *What Do Pictures Want? The Lives and Loves of Images.* Chicago: University of Chicago Press, 2005.

—. *Picture Theory: Essays on Verbal and Visual Representation.* Chicago: University of Chicago Press, 1994.

—. *Cloning Terror: The War of Images, 9/11 to the Present.* Chicago: University of Chicago Press, 2011.

Moore, Tristana. "Baby Bear Becomes Media Star." March 23, 2007, http://news.bbc.co.uk/2/hi/europe/6486993.stm.

Mother Advertising Limited. "Polar Bear." YouTube video, 0:56. Directed by Daniel Kleinman. Uploaded November 20, 2009, http://www.youtube.com/watch?v=fxis7Y1ikIQ.

Partizan. "A Homeless Polar Bear in London." YouTube video, 1:36, directed by Michael Geoghegan, uploaded July 2, 2012. http://www.youtube.com/watch?v=4XpF04nximI.

Perlmutter, David D. "Hypericons: Famous News Images in the Internet-Digital-Satellite Age." In *Digital Media: Transformations in Human Communication,* edited by Paul Messaris and Lee Humphreys. New York: Peter Lang, 2006.

—. "Photojournalism and Foreign Affairs." *Foreign Policy Research Institute E-Notes,* January 27, 2005. http://www.fpri.org/enotes/20050127.media.perlmutter.photojournalism.html.

Planestupid website. Accessed July 4, 2013. http://www.planestupid.com.

"Polar Bear." YouTube video, 1:03. Directed by Daniel Kleinman. Uploaded October 10, 2012. http://www.youtube.com/watch?v=c7HxLD88_Nc.

"Polar Bear." VisionOnTV blog entry. November 20, 2009 http://climate camptv.blogspot.de/2009/11/polar-bear.html.
Rancière, Jacques. *The Future of the Image*. London: Verso, 2007.
"Image Politics: Pictur(e)ing Climate." Conference report. Uploaded March 21, 2012. http://www.climatepictures.net/Conference_Home_files/ConferenceReport.pdf.
Zelizer, Barbie. *About to Die: How News Images Move the Public*. Oxford: Oxford University Press, 2010.

IMAGE SOURCES

Figure 1: http://www.youtube.com/watch?v=4XpF04nximI.
Figure 2: http://www.youtube.com/watch?v=4XpF04nximI.
Figure 3: http://www.youtube.com/watch?v=fxis7Y1ikIQ.
Figure 4: http://www.youtube.com/watch?v=fxis7Y1ikIQ.

How Photography Matters: On Producing Meaning in Photobooks on Climate Change

ULRIKE HEINE

Photographic images play a crucial role in the global debate on anthropogenic climate change. Different actors, I argue, emphasize photography's status of authenticity and immediacy, and hence revitalize the photograph as a "message without a code."[1] This paper reflects on the use of photographic images in support of the global climate change debate by analyzing a set of photobooks on the topic. These photobooks have been produced and published by art photographers, photojournalists and editors from 2006 to 2010, but they stand in line with a much older tradition of using books illustrated with photograhps to promote ecological awareness. While early photobooks that were published in the US Sierra Club's Exhibit format series (1960-1968) promoted the beauty of nature in order to influence political decision-making, the recent photobooks on climate change are educational tools, which aim at raising public awareness and subsequently change individual behavioral patterns.

This paper is an explorative study within the field of visual studies, combining a synchronic and a diachronic approach. Deploying a historical perspective, I will define the global climate change debate as a visual discourse. Adopting the methodologies of visual and semiotic analysis, I will compare recent photobooks published on the topic of climate change. As a result of this analysis, I will explain how different modes of relations between images as well as between text and images can be detected and explained as "connotation procedures."[2] These procedures, in my opinion, are strategies that are applied to serve the didactic purpose of mediating climate change.

1 Barthes, "The Photographic Message," 1977, 17.

2 Ibid., 20.

PHOTOGRAPHY AND CLIMATE CHANGE

The extraordinary significance of photographic images within the climate change discourse became obvious to a global audience during the 15th Conference of Parties (COP 15) held in Copenhagen in December 2009. Prior to its official start, the conference was defined not only as a key event in the transnational discussion on anthropogenic climate change.[3]

As it took place, COP 15 itself became a visual event. Numerous exhibitions displaying photographs were organized in order to draw public attention to different aspects of climate change. For example, the Dutch photo agency NOOR presented its comprehensive body of photographic images "Consequences" in an exhibition organized by Greenpeace International in its "Climate Rescue Station", a temporary globe construction close to the main conference venue at the Bella Center. In addition, the NOOR photographers presented their works as live slideshows in a street poster exhibition as well as in the special issue of the free Danish newspaper "Information."[4] At the same time, Collectif Argos, a French group of photojournalists and writers presented their project "Climate Refugees" on the streets of Copenhagen.[5]

Another street exhibition, "100 Places to Remember before they Disappear,"[6] operated as the visual figurehead of "Hopenhagen," a festival organized by the City of Copenhagen. Furthermore, a mobile version of the multimedia project "Conversations with the Earth" was presented in the city's Nationalmuseet. The exhibition comprised photographic portraits of "eye witnesses of climate change" and plates displaying their oral accounts of the

3 In discussing an international agreement on carbon emissions, the summit was even seen as a "preliminary decision for the future of mankind," see Dowideit, "Kopenhagen," 2009.

4 For a short summary of all public events on the occasion of the launch of „Consequences by NOOR“ see the video documentation by Abreu, "Consequences by NOOR. Project Launch in Copenhagen," 2010.

5 Giant canvases juxtaposing images, texts and witness accounts were displayed on representative buildings throughout the city, for example the French Embassy building and the Royal Danish Academy of Fine Arts, see "Refugies Climatiques—Exhibition in Copenhagen," 2009.

6 For installation shots from the street exhibition in Copenhagen see the project's homepage "100 Places to Remember before they Disappear."

impact of climate change on their daily lives.[7] Additionally on the occasion of COP 15, many photographers, editors and curators used the attention of the global media and presented photographic projects dealing with various aspects of global warming.[8]

Photographic images depicting climate change have been produced by photojournalists as well as by photographic artists. All of the photographers working on the topic of climate change are more invested in creating rich bodies of photographic images than in producing a single compelling image. These convolutes are distributed through various channels: as photo stories in magazines and newspapers, as exhibition displays or as multimedia essays and slideshows on the Internet. Most of the images that are used in the discourse have been specially produced for the topic of climate change. Fewer images are re-framed according to the topic. Here, editors compile photobooks and exhibition displays selecting from a variety of databases.

In essence, photographs are used to trigger an overall interest in climate change and to support the debate with visual arguments. They are utilized not just with respect to their epistemic, but also their aesthetic values. As epistemic means, photographs are given a direct heuristic function in the understanding of climate change as a physical phenomenon. On the other hand, the photographs' aesthetic value is called upon when climate change is defined as a social and cultural phenomenon. Both epistemic and aesthetic values are used in interplay in order to bridge the "climate gap," as a "gap between knowledge and action."[9] This concept describes a paradox crucial to the perception of climate change. Despite the fact that there exists a complex body of scientific knowledge on

---

7 For more information on the project see "Conversations with the Earth." A small set of photographs and captions was also presented on the "Klimaforum09—The People's Climate Summit" in Copenhagen. Here, the exhibition "Vulnerable. The Face of Climate Change in the Maldives," which comprised photographs by "talented Maldivian photographers" was also shown, see "Vulnerable. The Face of Climate Change in the Maldives."

8 Thus, for example, Hutchins (ed.), *Climate Change*, 2009, was published on the occasion of COP 15. Also, Gary Braasch openend the exhibition "Climate Change in Our World" on November 10, 2009 and presented the revised edition of his book *Earth Under Fire*, 2009.

9 Krznaric, "Empathy and Climate Change," 2010, 155.

climate change that facilitates a pro-ecological attitude, necessary actions regarding profound lifestyle changes remain rare.[10]

In this paper, it is argued that photographic projects on climate change take the "knowledge-action gap" as a starting point and strive to overcome it by either stressing the photograph's documentary purpose, and thus its conceived evidential status,[11] or by referring to its assumed emotional immediacy.[12]

However, defining the visual cosmos of climate change is a somewhat difficult task, as it embraces a wide range of photographic subjects as well as a variety of photographic genres. Climate change does not easily translate into linear visual tropes. As it will be discussed below, the concept of climate change enters the photographs through textual connotations. However, the visual representation of aspects related to climate change is deeply rooted in the genre of "environmental photography,"[13] which defines the inherent link between photography and environmentalism. The genre developed with the emergence of the Conservationist Movement in the US during the first half of the twentieth century. Early conservationist photography was used politically, and also triggered a popular cultural movement. It sought to define the natural environment as a protectable asset that also conveys an aesthetic experience.

---

10 For a general discussion see Kollmuss and Agyeman, "Mind the Gap," 2002, 239–260.

11 Thus, the aim of the photojournalistic project "Consequences" by Noor is "a showcase for creativity in photography and an eyewitness record of the causes and the humanitarian effects of world-wide temperature changes. […] These reportages show not what might happen in the future but what is happening today, emphasizing the urgency of addressing the issues at stake.", see "Climate Change by NOOR," 2010.

12 As Nadine Barth, editor of the photobook *Vanishing Landscapes*, writes in the preface of her photobook: "The photographers and artist whose work is represented in these pages offer us something else: they present us with images that familiarize us with the topic [of climate change] on an associative level. Our subconscious notes these images, comparing them with what we are familiar with and adding this unknown element to our existing repertory of images. Certain images can trigger a flash of recognition on a more metaphysical, even emotional level." Barth, "Preface," 2008, 6.

13 Stuart Franklin, "Environmental Photography," blog entry posted October 2, 2008, http://www.blog.magnumphotos.com/2008/10/environmental_photography.html (accessed July 25, 2012). Even though the genre has been practiced for more than four decades now, the term "environmental photography" still lacks an adequate definition as well as a historical reflection. Franklin's blog has been the only attempt so far to bridge this gap.

Photographers such as Ansel Adams (1902–1984) followed the modes of depiction established in romantic landscape art, which tried to align with the sublime in introducing untamed nature as a subject matter. Largely depicting the "absence of man," these photographic landscapes negated any form of human intervention and opposed the world shaped by industrial development. The nature/culture dichotomy manifested in early American landscape photography slowly began to dissolve in the 1960s, with the dawn of the second environmental movement that focused on a profound redefinition of the relationship between man and nature. This ideological shift was reflected in photographic practices that moved from idealistic representations of landscape on to more conceptual approaches that reflected on the presence of man.

Appropriately, the photographer Stuart Franklin defines two main approaches of environmental photography. First, there is an "emphatic eco-centric approach" bearing a "misanthropic" streak. Second, and more contradictory, is the anthropocentric approach that centers on the vulnerability of the human existence. Both approaches coexist in photographic projects of climate change. Criticizing these two approaches as "didactic evangelism", Franklin emphasizes a third approach that he has observed in recent photographic practices. He describes it as a "cultural turn" in environmental photography. Here, the photographer plays a significant role as both witness and author, subjectively questioning the interface between nature and culture.[14] This approach has been adopted by many photographers working on climate change as they combine the eco- with the anthropocentric perspective in order to create rather complex visual narratives. Following the format of the photo story, these narratives unfold in a sequence of images, rather than in one single image.[15]

The photobook as a medium has played a crucial role in the evolution of environmental photography. By launching the famous Exhibit Format series, environmentalist David Brower who served as the Sierra Club's first Executive Director from 1952 to 1969, established the ecological coffee table book as a distinct medium.[16] The Exhibit Format series' large format photobooks contained high quality productions of photographs by renowned US-American photographers such as Ansel Adams, Eliot Porter (1901–1990), and Minor

14 Ibd.

15 The projects "Consequences" by Noor and "Climate Refugees" by Collectif Argos are examples for this integrative approach.

16 Altogether, between 1960 and 1968, twenty Volumes were published in the series. Its name goes back to Brower's idea to develop a book format, which assumes the character of an exhibition, see Dunaway, *Natural Visions*, 2008, 118.

White (1908-1976). The books were intended to promote the diversity of American landscape and were explicitly used to support the club's political campaigns. Hence, *This is the American Earth* (1960), the first book in the series, was published on the occasion of the Club's campaign supporting the "Wilderness Bill." As a preliminary step to the enactment of the "Wilderness Act" in 1964, the bill became an important landmark for the legal establishment of environmental protection in the United States.

Although photography was clearly the focus of the Exhibit Format series publications, one can find a diverse range of accompanying texts giving the images additional meaning. For *This is the American Earth*, the curator Nancy Newhall wrote texts in verses, which accompanied the convolute of images by prominent American photographers. In her text, she retraces the history of mankind from its beginning to the profound and devastating impacts of industrialization, thereby composing an early manifesto of anthropogenic environmental pollution. The volume *In Wildness Is the Preservation of the World* (1962) combines color photographs by Eliot Porter with quotations from Henry David Thoreau's writings.

Contributing to the popular discourse on climate change, the ecological photobook has been re-launched by different groups such as governmental boards, non-governmental organizations, and independent editors. These players utilize the inexhaustible range of book formats connecting text and images. The books chosen for analysis in this paper represent the varied spectrum as they adapt different concepts of photographically illustrated books. While some resemble artistic photo albums, where the text is clearly subordinate to the images,[17] others are textbooks that prominently integrate photographic images.[18] An important landmark for the visual mediation of climate change was the release of Al Gore's documentary *An Inconvenient Truth* in 2006. The release was accompanied by the launch of a book by the same title. Presenting a combination of photographic images, informational graphics visualizing climate data, and explanatory text, the book became a prototype for the production of photographically illustrated books on climate change and a landmark in educational entertainment dealing with ecological issues.

---

17 As applied in Barth, *Vanishing Landscapes*, 2008, and Braschler and Fischer, *The Human Face of Climate Change*, 2010.

18 Examples for this approach are Schmidt and Wolfe, *Climate Change*, 2009 and Braasch, *Earth Under Fire*, 2009.

CONNOTATION PROCEDURES

The aim of this paper is to analyze how meaning is produced through different configurations of text and images in the aforementioned books. Defining these configurations as structural relations is a crucial aspect of understanding the strategies used by photographers and editors to tackle the task of representing climate change.

By combining text and images a photobook can offer a wide range of representational modes. In different ways, the book's layout visually frames the images as well as the succession of images within the book. The layout also prestructures image-text and image-image relations, creating different layers of context. Thus, different forms of text, such as headlines and captions, are used to indicate the images' context of origin, to explain their content, or to provide a connotation on the level of association. Often, forewords and additional essays create an overall narrative frame.[19]

In his essays "The Photographic Message" (1961) and "Rhetoric of the Image" (1964), Roland Barthes analyzes press and promotion photographs and develops "connotation procedures." [20] These procedures unfold within a photograph or in a sequence of photographs ('syntax'). Barthes' reflections on text-image relations can be seen as a critical account of the subordination of text to image, as unfolded in the post-war picture press: "[...] the image no longer illustrates the words; it is now the words, which structurally are parasitic on the image [...] in the relationship that now holds, it is not the image which comes to elucidate or 'realize' the text, but the latter which comes to sublimate, patheticize or rationalize the image."[21] In his second essay the "Rhetoric of the Image" Barthes conducts a thorough analysis of a poster advertisement of French food manufacturer "Panzani," elaborating all of the pronounced connotation procedures. Adopting Roland Barthes' idea of different "procedures of connotation" and originating from a thorough analysis of a sample of photo-

19 Research on photobooks is still marginal. In her dissertation Andrea Jeannette Nelson tackles the latter by applying narratology to analyze structural characteristics and the medium's use within pedagogical practices, see Nelson, *Reading Photobooks*, 2007. The two volumes entitled *The Photobook*, 2004 and 2006, edited by Gerry Badger and Martin Parr serve as comprehensive compendiums on the history of photobooks.

20 Barthes, "The Photographic Message," 1977, 20.

21 Ibid., 25.

books on climate change,[22] I propose a categorization of text-image relations that comprehends four different modes. I will refer to these modes as "illustration," "inscription," "comparison," and "contemplation" and elaborate them separately in the following paragraphs. With regard to photobooks, the analysis is conducted on the open page as a distinct compound of image, text and layout elements.

ILLUSTRATION

Illustration is the most common strategy used in the books analyzed here. By definition, the term "illustration" suggests the subordination of image to text.[23] In most books, the procedure of illustration is applied as a juxtaposition of text and image: the text is printed on the left pages while the photographs are printed on the books' right pages. Given the fact that in the Western world the direction of reading is from left to right, a subordination of text to image is already suggested by the book's layout. Due to the proximity of text and image, the viewer is encouraged to relate the written word to the visual information.

The photobook *Climate Change. Our Warming World,* compiled by editor Christina Hutchins adapts this mode of text-image relation as its core principle. The book (6.3 x 9.7 inches) was published by the British non-governmental organization "Our World Foundation" as a free handout for politicians and activists on the occasion of COP 15.[24] In the book, facts, quotations and rhetorical questions related to climate change are printed on the left page and are accompanied by frameless full-page images on the right page. Arranged in six subsequent chapters, Hutchins compiled information about visible symptoms of climate change, their consequences for the environment and possible solutions. She also outlined detailed instructions for possible actions.

22 These photobooks are: Braasch, *Earth Under Fire*, 2009; Braschler and Fischer, *The Human Face of Climate Change*, 2010; Collectif Argos, *Climate Refugees*, 2010; Gore, *An Inconvenient Truth*, 2006; Schmidt and Wolfe, *Climate Change*, 2009.

23 Illustration is defined as "every graphic depiction, including photography, that works as an descriptive addition to a linguistic text." Jahn and Lieb, *Wörterbuch der Kunst*, 2008, 388 (translated by the author).

24 The book was also sponsored by the British energy company Ecotricity and actively promotes switching to the company's green energy tariff.

Hutchins almost exclusively uses photographic works that are available on various image databases rather than having been produced only for this photobook.[25] Among the images used are editorial images and numerous stock images. These images are characterized by their "polysemy," which, according to Roland Barthes means that they have an extensive 'floating chain' of signifieds, the reader [being] able to choose some and ignore others".[26] The latter enable manifold but vague connotations, of which we can get an inkling when considering the key word indexing of the databases. However, key to the concept of illustration is the idea that the text firmly establishes the image's interpretation.

Figure 1: Pages 16 and 17, in Hutchins, Climate Change. Our Warming World, text by Christina Hutchins. Photograph by Yanik Chauvin.

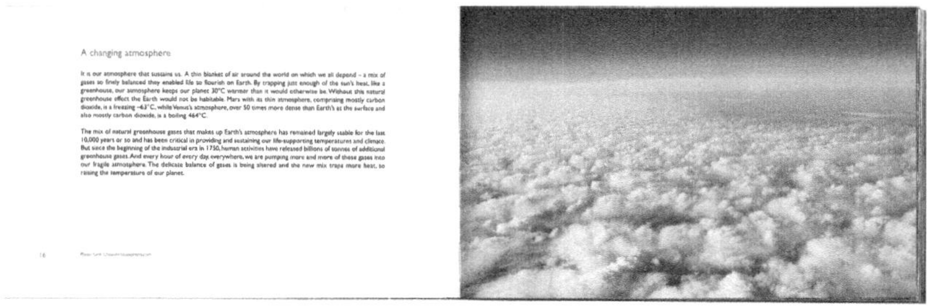

A changing atmosphere

It is our atmosphere that sustains us. A thin blanket of air around the world on which we all depend – a mix of gases so finely balanced they enabled life to flourish on Earth. By trapping just enough of the sun's heat, like a greenhouse, our atmosphere keeps our planet 30°C warmer than it would otherwise be. Without this natural greenhouse effect the Earth would not be habitable. Mars with its thin atmosphere, comprising mostly carbon dioxide, is a freezing –63°C, while Venus's atmosphere, over 50 times more dense than Earth's at the surface and also mostly carbon dioxide, is a boiling 464°C.

The mix of natural greenhouse gases that makes up Earth's atmosphere has remained largely stable for the last 10,000 years or so and has been critical in providing and sustaining our life-supporting temperatures and climate. But since the beginning of the industrial era in 1750, human activities have released billions of tonnes of additional greenhouse gases. And every hour of every day, everywhere, we are pumping more and more of these gases into our fragile atmosphere. The delicate balance of gases is being altered and the new mix traps more heat, so raising the temperature of our planet.

On page 16, following the headline "A changing atmosphere" (Figure 1), the first paragraph elaborates on the condition and physical composition of the outer atmosphere, explaining the greenhouse effect in colloquial terms. The second paragraph refers to the changes that have happened to the atmosphere throughout the past 150 years as part of an overall anthropogenic change in the world's climate. The photograph shows an aerial view taken from above the clouds, probably from an airplane. The fluffy clouds make up a hilly landscape that stretches into the distance. The layer of clouds is interrupted by holes through which one can see the earth's surface. Geometric structures of dwellings can be distinguished in the lower left corner of the image. Also on the left side of the image, just below the line of horizon a lake or river comes into sight. The horizon itself divides into a light blue glow of atmospheric gases and the dark blue of the outer atmosphere, which extends, one can guess, much further beyond the frame of the image. Strikingly, the reduced tonal range of the

25 Such as Corbis, Getty Images, Associated Press Images, Aurora and Reuters.

26 Barthes, "Rhetoric of the Image," 1977, 39.

photograph is limited to white and blue. In the accompanying text, the phrase "blanket of air around the world" is used to describe the atmospheric layers. This metaphor fundamentally connects the text to the image, as indeed, the depiction of the layer of clouds gives the impression of a soft blanket or a blanket of snow that smoothly covers a landscape underneath. The mode of illustration here relies on the indexical reference, as the photo shows at least a part of the physical phenomenon of the atmosphere. By naming and explaining the depiction as such, other possible connotations are revoked.[27]

In their photobook *The Human Face of Climate Change* (2010), photographers Monika Fischer and Mathias Braschler also adopt the juxtaposition of text on the left and images on the right. The photographs, taken by Braschler and Fischer in 2009, were captured with a middle-format camera and show full-length portraits of people that live in areas affected by climate change (Figure 2). The text begins with a short introduction of the depicted person (including country of origin, name and occupation) and concludes with a transcript of her/his verbal records. The book's main idea is to emphasize the personal assessment of symptoms and impacts of climate change. Placing the person's portrait next to their quotations and basic information on their place of origin, visually establishes the concept of the eyewitness.

Figure 2: Pages 92 and 93, in Braschler and Fischer, The Human Face of Climate Change. Interview transcript and photograph by Mathias Braschler and Monika Fischer. (→ color plate 13)

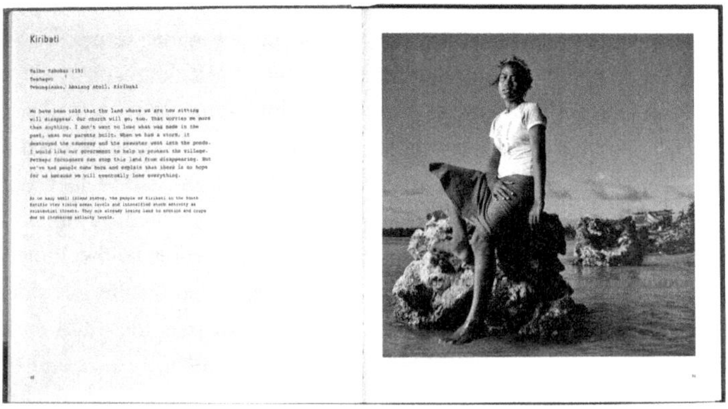

27 The keywords under which the image is listed with "shutterstock" give an idea of alternative possible connotations: "clouds, plane, sky, fly, areoplane (sic!), white, travel, business, peace, horizon, moisture, heaven, transport, high, blue, weather, above, trip, nature," see Chauvin, "Clouds" (Image ID: 125312).

Despite the comparison of text and image in this order, this mode of illustration works by a principle different to Hutchins' approach. Braschler and Fischer's images appear more stylized and coherent in their composition and design, which makes them more memorable than the stock images chosen by Hutchins. Despite their chronological order, the striking portrait photographs overcome their subordination to the text, which itself appears in very contained typewriter letterpress. The images' pithiness originates from their careful composition and their detailed resolution—both properties of the middle-format camera—as well as from effects of light composition. The portraits' formal characteristics dominate their informational content. The white frame around the images emphasizes the physical distance between image and text. With *The Human Face of Climate Change,* Braschler and Fischer have created a classic photobook, where text and layout complement each other without weakening the images' own procedure of connotation.

The photobook *Climate Refugees* (2007/2010) by the French Collectif Argos serves a third mode of illustration. Here, verbal journalistic accounts are followed by the photo report that emerged in the same context. Captions above and next to the images establish the connection between text and image. However, the text and the sequence of images stand side by side, each having the possibility of having a meaning of its own.

INSCRIPTION

There is a second practice of text-image-relation, which I call "inscription." It immediately integrates text and image with each other so that the written word becomes a label, "being inserted into the natural disposition of the scene."[28] The term "inscription" is used here to emphasize the acquisition of the image by language. The term "inscription" encompasses the process as well as the result, which is the inseparable union of visual and textual information in a joint space. Inscription as a strategy of visualization has a long tradition in visual history. Medieval block books, popular prints in early modern times as well contemporary advertisement posters are considered different forms of *Text-Bilder* (text-images).[29]

---

28 Barthes, "Rhetoric of the Image," 1977, 33.

29 Sabine Griese coined the term Text-Bilder to stress the subordination of text to image and therefore distinguish it from the mode of illustration, see Griese, *Text-Bilder und ihre Kontexte*, 2011, 20.

Figure 3: Pages 16 and 17, Climate Change. Picturing the Science. Text Gavin Schmidt and Joshua Wolfe. Photo Joshua Wolfe. (→ color plate 14)

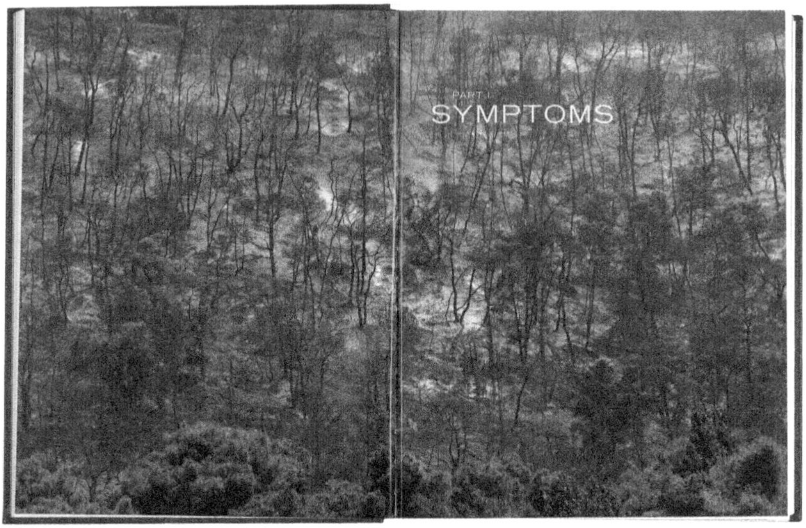

In the books analyzed here, inscribed images are mostly used to structure the overall narrative in those publications that favor text before image. In their edited book *Climate Change. Picturing the Science* (2009), photographer Joshua Wolfe and scientist Gavin Schmidt use textually connoted images as title pages for the chapters. In these chapter the depictions of a burnt forest (pages 16–17), a scientist taking measurements at a glacier (pages 132–133), and a hydroelectric power plant (pages 210–211) are connoted with metaphorical expressions describing steps of the physical process of healing, i.e. "symptoms," (Figure 3) "diagnosis" and "possible cures." Anthropogenic climate change as a disease that can be cured is a metaphor reaching back to the 1970s.[30] In an anthropocentric outlook on the world, it supports the notion of the "fragility" of planet earth, which is put at par with that of man. While the frameless image stretches across the page spread, the text is printed in white sans serif letters within the upper third on the left page.[31] The upper line states the chapter's number while the lower part, in bold type, contains the chapter's headline. The

30 See article by Fleming in this volume.

31 The book combines a sans serif typeface for headlines, epigraphs and captions with serif letters for the text.

writing is not integrated into the image's composition but seems rather to breach it intentionally.

The medical terminology supports the interpretation of climate change as a temporary problem that can be solved. The inscriptions connect key words of this healing process with three photographic representations of landscape that are characteristic of the spectrum for the visual climate change discourse: the destroyed natural landscape ("symptoms"), the uninhabited landscape being meticulously observed by the scientist ("diagnosis"), and the industrially shaped landscape ("possible cures"). By using singular key words instead of complex phrases that signify the three essential stages of solving the climatic crisis, the book becomes a well-structured educational tool.

Through the clear-cut layout, Schmidt and Wolfe's inscriptions also work as convincing attempts to coin icons for each of the steps. The "symptoms" are clearly exemplified through the striking juxtaposition of green-leaved—healthy—trees in the lower foreground of the image and the burnt—sick—trunks that dominate the upper two thirds. The very letters of the word "symptoms" arise from the smoke that signifies how recent the destructive fire was, and therefore emphasizes the "here and now"[32] of the depicted scene. According to the image introducing the book's second chapter, the "diagnosis" lies in the hands of a female scientist, leaning over a steep slope and fully concentrating on her measurements. Fieldwork is shown as a life-threatening and courageous act. The photograph thrives on dramatic lighting that recalls the romantic painting tradition. The scientist is set against an overclouded mountainous valley backdrop, which also shows a main site of climate research: alpine glaciers. The word "diagnosis" is purposefully set between the top of the mountain peaks and the impending clouds. The depiction of a hydropower plant to signify "possible cures" visualizes the authors' belief in technological solutions to the climate crisis. However, the word "possible" hints to a critical attitude implied in the solutions discussed in this chapter. This ambivalence is already situated within the chosen image: vast, industrially shaped landscapes photographed from a bird's eye perspective have a rather negative connotation within the climate change discourse as they symbolize the developed world's hunger for energy that has fueled industrialization.[33]

32 Doyle, "Seeing the climate?" 2009, 294.

33 With his project "Oil", the Canadian photographer Edward Burtynsky deployed this mode of visual representation to coin the idea of "carbon landscapes", i.e. landscapes that have been shaped by the drilling, refinery and use of oil as a major fossil fuel, see Burtynsky, Oil, 2009.

The lucidity of the inscribed photographs used by Schmidt and Wolfe is obvious when comparing *Picturing the Science* to photographer Gary Braasch's publication *Earth Under Fire* (2007/2009).[34] Braasch uses inscribed photographs as front pages of his chapters. As titles for the chapters he chose phrases that linguistically embody the idea of global warming: "Fire on the Ice" (pages 4-5), "Polar Thaw" (pages 46-47), or "Choosing a Safer, Cleaner, and Cooler World" (pages 158-159). Other headlines, such as "Breaking the Boundaries of Life" (pages 76-77) and "Tomorrow's Climate Today" (pages 112-113) refer to the loss of biodiversity and the increase in extreme weather events as anticipated effects of climate change. Although Braasch's publication also follows the three-step line of argumentation, the rationale does not become visible through the verbal imagery deployed in each chapter's title. The interplay of verbal and visual codes doesn't work as well as in *Picturing the Science,* a good example being the inscription used to mark the first chapter (Figure 4).

Figure 4: Earth on Fire, pages 4 and 5. Photo and text by Gary Braasch.

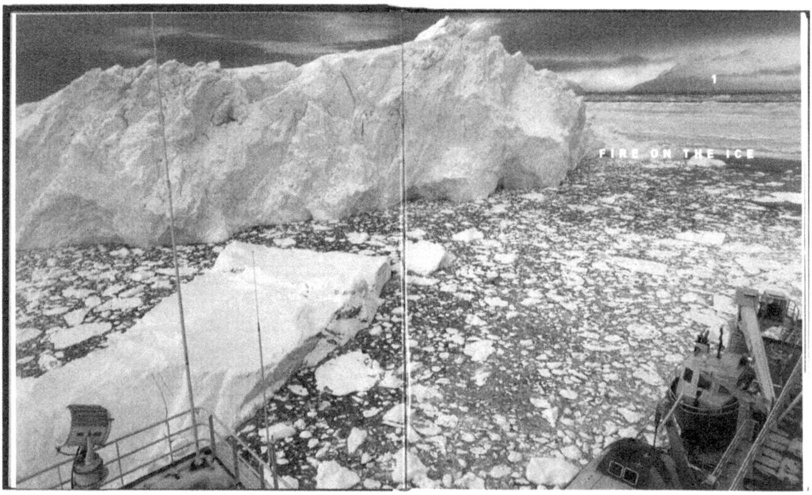

The image, taken from a high vantage point, shows a field of shattered sea ice. In the foreground, parts of a ship are depicted. The left side of the photograph is dominated by bluish-white compact icebergs that probably recently broke off a glacier located just outside the photograph's frame. Together with the smaller,

34 In a conversation, Joshua Wolfe pointed out that Braasch's book served as a role model for his own project.

shattered ice, the larger ice blocks seem to flow into the ocean, melting into water. On the point where the fjord opens up, the first chapter's title "Fire on the Ice" is printed in white, discreet font. While the ice is visible in the photograph in its diverse forms, one finds it difficult to associate the verbal code of "fire" with any part of the image. In his book, Braasch does not apply the term "fire" in a literal sense to describe a visible symptom of climate change,[35] but rather as a metaphor for the potentially dangerous rise of global average temperature. Braasch establishes this connotation through the inscription of the very cover of his book: The book's title, "Earth under Fire" is accompanied by the explanatory subtitle "How global warming is changing the world." The picture shows an immense gas flare released from a chimney in an oil refinery. The flare's structure corresponds strangely with the cumulus clouds in the background, contradicting yet harmonizing with it. Following this strong visual coding of "fire," the "Fire on the Ice" caption is delusive—even more so as fire in its actual aggressive and uncontrollable physicality has become a consequence of drought-related events due to climate change in other parts of the world.

Inscriptions relate to discourses beyond the inscribed image and are thus powerful tools to coin and reinforce iconic representations of climate change. It is important, however, that the inscription is done carefully, with regard to the photograph's compositional structure and being mindful to the different connotations already in the image. With this in mind, Roland Barthes states that "[it] is true that there is never a real incorporation since the substances of the two structures (graphic and iconic) are irreducible, but there are most likely degrees of amalgamation."[36]

COMPARISON

While the terms "illustration" and "inscription" describe relations between text and image, "comparison" refers to the juxtaposition of two images. Here, the text is reduced to its informational value and emphasizes the image's denotative

35 Gavin Schmidt and Joshua Wolfe use a photograph which one could indeed think as a literal visualization of the phrase "fire on ice": it shows a scientist lightning highly combustible methane, which emanates from holes in a frozen lake in Alaska, see Schmidt and Wolfe, *Climate Change*, 2009, 63.

36 Barthes, "The Photographic Message," 1977, 26.

message. The comparative display evolved early in the history of photography and constituted in the medium's use as an auxiliary means within the sciences.[37]

Applied within the mediation of global climate change, visual comparisons tackle an inherent problem of the physical aspects of the phenomenon. Climate change evolves only over a certain period of time and the concept therefore suffers a time lag regarding perception of the problem. Documenting two moments of the ongoing process visually, comparisons emphasize the changes that have taken place in between and thus virtually seek to bridge the time lag. Comparative displays also address a second gap, which unfolds between scientific research and public perception: they create an immediate visibility and thus establish evidence that can be read by both, experts and laypeople. The indication of place and date of image capture contributes to the strategy's aspirations to also work as a scientifically valid method of visualization.

In his slide presentation on climate change, which is documented in his book *An Inconvenient Truth* (2006) the former politician and environmental activist, Al Gore extensively uses comparative displays. For instance, he introduces a series of image comparisons (pages 42–57) showing glaciers from all over the planet, stating: "It is evident in the world around us that very dramatic changes are taking place"[38] and thus empowering the photograph as a means to make these changes visible and evident. Within the series, a sequence shows the decreasing glaciers of Mount Kilimanjaro (Figure 5). A color photograph from the 1970s showing the snow-capped mountain ridge is placed alongside a photograph taken from the same position in the year 2000. What is clearly visible in the blue shimmering background: the snow has receded and has only left a small cap of ice and snow on the very top of the mountain. The identical vegetation in the foreground is an important marker for the image's evidence. A bush with green leaves bears witness to the fact that the two pictures must have been taken during the same season. The clear composition and geometry of the older photograph finds its analogy in the more recent photograph on the right side of the page. The images also build upon the color contrasts between the blue sky and mountain, the white snow and the red and green leaves on the bushes to highlight the changes.

Several comparative displays in Gore's book stem from the project "World View of Global Warming," which was initiated by American photographer Gary Braasch in 1999 as a long-term project. Inspired by the German project

37 On visual comparison as a method, see Dünkel, "Comparing Images," 2014, 24–27.

38 Gore, *An Inconvenient Truth*, 2006, 42.

Gletscherarchiv ("glacier archive"),[39] Braasch collects comparative displays of various symptoms of climate change worldwide. While Gore charges his comparisons emphatically (the basis of his photobook is the transcript of a power point presentation), Braasch restricts the verbal connotation of his comparisons to short explanatory captions.

Figure 5: Page 42 and 43, in Gore, An Inconvenient Truth. Photographs by Bruno Messerli. Text by Al Gore. (→ color plate 15)

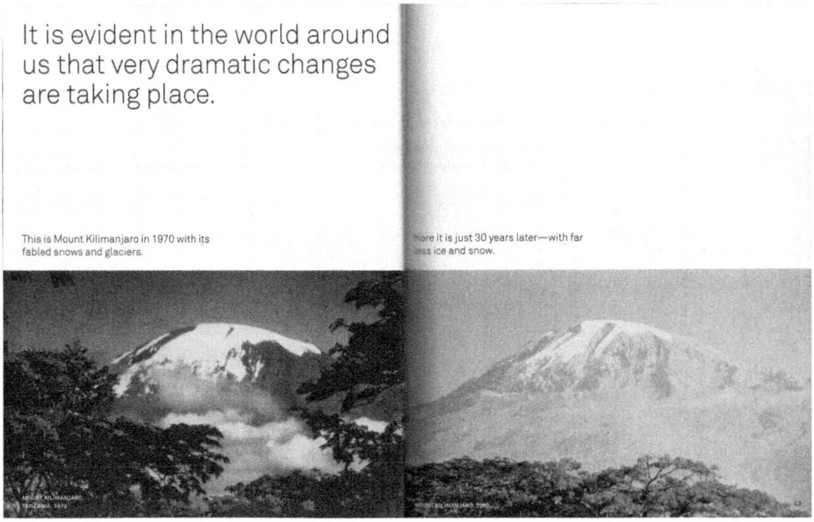

The strategy of comparison draws upon its immediate reference to the denotative status of photographic images. In the juxtaposition of two almost identical photographs, the connotation unfolds in the differences emerging from the shift in its denotative messages. Visual codes such as composition, tonality, and lighting replace the need for textual connotation. Mediation and explanation of the visual impression becomes irrelevant. The comparative display enacts the very idea of "seeing is believing." It is the aesthetic manifestation of the

39 The image collection "Gletscherarchiv" (engl.: glacier archive) was initiated in the early 1990s by Sylvia Hamberger and Wolfgang Zängl. It comprehends a great number of historic photographs of alpine glaciers that were produced for different purposes. Hamberger and Zängl have facilitated the photographs' exact re-production. Old and new images are comparatively displayed within exhibitions and in photobooks, see Zängl and Hamberger, *Gletscher im Treibhaus*, 2004.

photograph as an indexical document and a time capsule at the same time.[40] Through a standardized practice of displaying, it creates the illusion of visual evidence. Three main elements of this practice can be detected in *An Inconvenient Truth*: firstly, the change in time is emphasized through the use of photographs of equal size and composition. Al Gore consistently translates the chronological display of his images into juxtapositions in the book and thus stresses the "before and after" notion. Secondly, the composition of the images chosen for comparisons has to aim at displaying a subtle, yet notable change by erasing all other possible modifications. This staging is the result of a meticulous reenactment of the original photographic momentum.[41] Thirdly, the minimal

Figure 6: Pages 36 and 37, in Braasch, Earth on Fire. Text by Gary Braasch. Photographs by Lonnie Thompson and Gary Braasch.

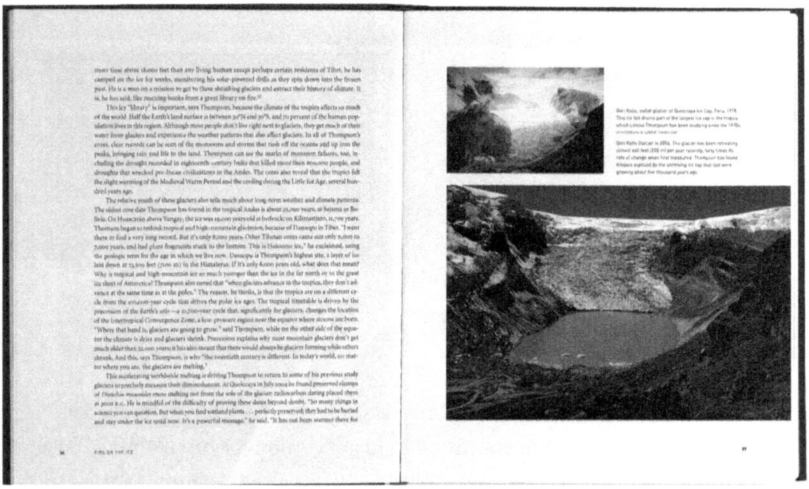

captions have to provide local consistency compared to the volatile nature of time. Avoiding these predispositional settings impedes the immediate, non-mediated visibility. Such impeded comparisons can be observed in some of Gary Braasch book *Earth on Fire*. On pages 36 and 37 (Figure 6), Braasch arranges a photograph taken by scientist Lonnie Thompson above his own. The photographs are reproduced in two different sizes. The captions, printed in the

40 Barthes, *Camera Lucida*, 1981.

41 Which involves taking the same vantage point, choosing similar light conditions, and even taking the same technical devices to capture the photograph, see Zängl and Hamberger, *Gletscher im Treibhaus*, 2004, 8f.

blank space between the two images, extend the factual insistence on the "this was there then."[42] The information provided shifts the focus from the shrunken glacier bed to a detail in the foreground of the re-enacted image, the green mosses that according to Lonnie Thompson have emerged due to the temperature rise. As the textual information is essential to understand the connotation attached to the depiction of the mosses, the comparative turns into an illustrative display.

CONTEMPLATION

A fourth procedure of connotation favors the single image while avoiding textual commentary and thus embodying the much-quoted power of images. In the analyzed photobooks this procedure is implemented as frameless images, which are printed over double pages devoid of any immediate textual connotation. This display triggers a different mode of perception as the reciprocal, relational consideration of image and text information is replaced by a thorough observation of the visual information. To stress the emphasis put here on the process of individual and subjective perception, I refer to this procedure of connotation as "contemplation."[43] What distinguishes the contemplative image from the "pensive" image that constitutes the core of most art photography practice is that applying contemplative displays is a purposefully chosen strategy. The analysis of two examples shows that these displays are applied to contrast and complement other connotation procedures, such as illustration and inscription.

The photobook *Climate Refugees* (2007/2010) comprehends up to two full-page images per photo story. The photo stories evolve as a combination of image and text, which is largely pre-structured by the book's layout. In spite of different authorship and content, each story in the book contributes to the book's overall message: to promote the idea of people who are displaced because of the changing climate. Every chapter begins with a text essay that does not contain any images. Each text essay is followed by a photo essay in which the captions are printed at a sufficient distance from the images. These captions link the images to the essay by repeating names or using immediate illustrations of narrated scenes. Frameless full-page images interrupt the text-image parts. These

42 See the article by Morris and Sayler in this volume.

43 The term is detached here from the connotation it has earned within religious and philosophical reflections.

full-page images are very diverse: close-ups create proximity to the scenery depicted while snap-shots that are aesthetically enhanced by motion blur support the photographs' connotation of immediacy (for instance, pages 30–31). Many images are shot from a low-level perspective, creating the sensation of the photographer's proximity to the subject (Figure 7). The different visual means employed support the essential idea of the photojournalistic image by invoking its indexical authenticity and with it, the photographer's role as an eyewitness of climate change.

Figure 7: Pages 30 and 31, in Collectif Argos, Climate Refugees. Photograph by Hélène David.

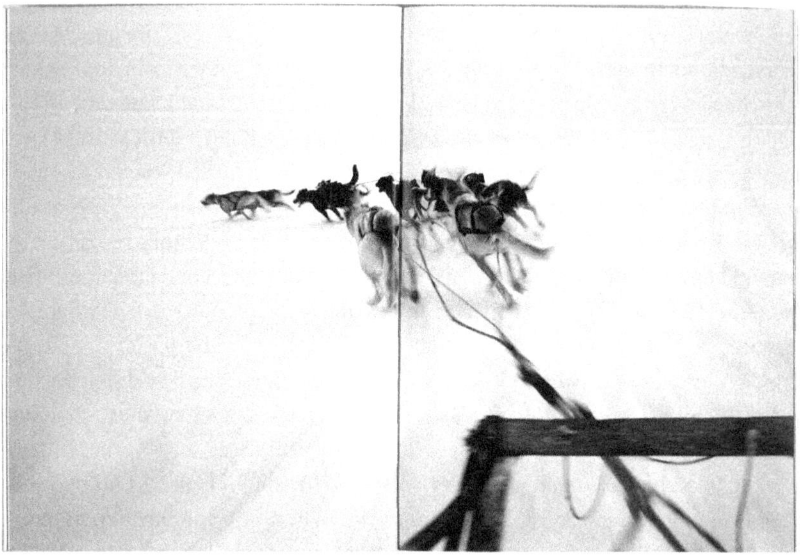

Al Gore incorporates only one contemplative image in *An Inconvenient Truth*: a snapshot of the earth taken from space by a member of the Apollo 17 in 1972 (pages 30–31, Figure 8). Shortly after it was taken and brought back to earth, the image became a powerful symbol for the global environmental movement. On the occasion of the first "Earth Day" in 1972, the photograph of the "Blue Marble" was coined as a symbol for the systemic idea of earth as a self-contained, living organism ("Gaia"), that is threatened by human activity.

Figure 8: Page 302 and 303, in Gore's, An Inconvenient Truth. Photograph by NASA, 1972.

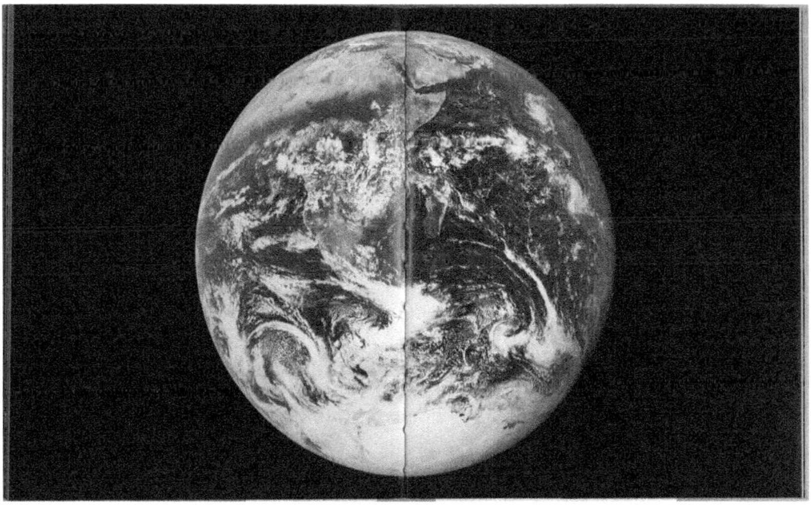

However, the very image of the earth bears conflicting codes, which evolved throughout the extensive distribution of the photograph. Thus, the ecological "whole earth" reading is challenged by the interpretation of the NASA image as "one earth." According to Denis Cosgrove, the latter is a "geopolitical concepttion coeval with the European and Christian sense of imperium."[44] Thus, the image bears both, notions of ecology, equity and humanity, as well as of technocracy, superiority and hegemony. Throughout the book, the image is repeatedly reproduced and textually connoted. Gore recalls the image's history of iconization (pages 14–15) by calling the photograph the "most commonly published image in all of our history." Through a small circular cut-out on the book's sleeve, a detail of the "Blue Marble" is seen. The fold-out inner side of the sleeve shows a photograph of Al Gore presenting his slide show. He is raising his right hand in an explanatory gesture and carries a presenter device in his left hand. His facial expression suggests that he is talking. In the background, a large image of the "Blue Marble" is projected against the wall. A small caption in the lower left corner is marked as a quote by Al Gore. It reads: "We have everything we need to begin solving this crisis, with the possible exception of the will to act. But in America, our will to take action is itself a renewable resource." The reference to the idea of the "One Earth" reads as an expression of

44 Cosgrove, "Contested Global Visions," 1994, 289–90.

cultural hegemony that awkwardly thwarts the idea of human and ecological equity inherent to the whole-earth discourse.

CONCLUSIONS

In spite of the explicit proclamations of photography's power to compel and prove, confidence in photographic images seems to be limited within the global climate change discourse. The ambiguity of meanings, which is inherent to the photographic image, interferes with the use of photographs for didactic purposes. It is this use, however, which aims high, at providing both comprehensive information and incentives for changes in mentality and behavior.

Many authors, who use photography to mediate aspects related to climate change, refer to its apparent "natural code," which is believed to constitute the medium's immediacy and veracity. The two notions are crucial to the theoretical framing of photography as an adequate means to document reality. These notions are also core values to the genre of documentary photography, to which we can allude all off the photographic projects analyzed in this paper. The emphasis on the documentary status of photography within the climate change discourse is linked to the epistemic and perceptive difficulties inherent to the concept of climate change. Photographers and editors have decided to create extensive bodies instead of single iconic images in order to address its complexity.

In their search for adequate and effective visual representations of the phenomenon of climate change, photographers also discuss the status of photography as a means to promote social change. This meta-discourse should be examined and addressed more openly by the photographers themselves.

Furthermore, the ambiguity of the "photographic message" should be acknowledged rather than suppressed by the use of textual connotations. In omitting layers of text or reducing it to denotative information, the viewer is urged to actively "read" the photograph. It is hoped that the process of analyzing the ambiguity of the image's messages will trigger a deeper understanding of the topic of climate change. With their project "A History of Future," artists Susannah Sayler and Edward Morris (The Canary Project) aim at establishing the visual literacy that is crucial to the much sought-after deeper understanding. Creating varying exhibition displays such as "Double Blind"[45] where large-scale

45 "Double Blind" was shown at the Illges Gallery at Columbus State University in April 2010, see "A History of the Future."

photographs were covered with a black cloth and the captions left to the viewer's visual imagination, Sayler and Morris fundamentally interrupt the procedures of revealing the connotation through text and image. The interruption has a profound ideological implication as it brings to fore that "climate change is a proposition and not a fact."[46]

In contrast, all the projects analyzed in this paper utilize photography to present climate change as a reality that can be seen and therefore, be depicted. The photobooks were designed to persuade the viewer of the veracity of the climatic changes and their possible grave implications for the course of human history. Authors look for procedures of connotation such as inscription, or comparison as well as for the use of symbolic images and therefore apply strategies all mastered by commercial advertising. While adopting these modes of representation and display, manifested codes attached to these modes are also brought to fore. These connotations have to be taken into account when combining text and images in order to make climate change not only visible but also credible as an impending source of concern for humanity.

BIBLIOGRAPHY

"100 Places to Remember Before they Disappear." Accessed October 20, 2012. http://www.100places.com/.

"A History of the Future." Accessed 30 October 2012. http://www.history-of-the-future.com/?p=311

Badger, Gerry, and Martin Parr, editors. *The Photobook. A History*—Vol. I. London: Phaidon Press, 2004.

—. *The Photobook. A History*—Vol. II. London: Phaidon Press, 2006.

Barth, Nadine. "Preface." In *Vanishing Landscapes*, edited by Nadine Barth. London: Frances Lincoln, 2008, 6–7.

—. *Vanishing Landscapes*. London: Frances Lincoln, 2008. (also published as: Barth, Nadine: *Verschwindende Landschaften*. Cologne: Dumont, 2008).

Barthes, Roland. "The Photographic Message." In *Image Music Text*, trans. Stephen Heath. London: Fontana Press, 1977.

—. "Rhetoric of the Image." In *Image Music Text*. Translated by Stephen Heath. London: Fontana Press, 1977.

—. *Camera Lucida. Reflections on Photography*. New York; Hill and Wang, 1981.

46 See article by Morris and Sayler in this volume.

Braasch, Garry. *Earth Under Fire: How Global Warming Is Changing the World.* Berkeley / Los Angeles: University of California Press, 2009 (first edition 2007).

Braschler, Mathias, and Fischer, Monika. *The Human Face of Climate Change.* Ostfildern: Hatje Cantz Verlag, 2010. (Also published as: *Schicksale des Klimawandels.* Ostfildern: Hatje Cantz Verlag, 2010.)

Bredekamp, Horst, Birgit Schneider, and Vera Dünkel, editors, *The Technical Image. A History of Styles in Scientific Imagery.* Michigan: Michigan Press, 2014.

Burtynsky, Edward. *Oil.* Essays by Paul Roth, Michael Mitchell, and William E. Rees. Göttingen: Steidl/Corcoran, 2009.

"Climate Change by NOOR." (was: "Consequences by NOOR" and "Solutions by NOOR"), Accessed October 25, 2012, http://noorimages.com/project /climate-change-by-noor/.

"Clouds." Photographed by Yanik Chauvin. Accessed March 10, 2013. http:// www.shutterstock.com/pic-125312/stock-photo-clouds.html.

Collectif Argos. *Climate Refugees.* Cambridge: Massachusetts Institute of Technology, 2010. (First published as: Collectif Argos. *Réfugiés climatiques.* Gollion: Infolio editions, 2007).

"Consequences by NOOR. Project Launch in Copenhagen." Filmed by Ivan Abreu, December 2009. Accessed October 24, 2012. http://vimeo.com/ 24721215.

"Conversations with the Earth." Accessed October 20, 2012. http://www.conver sationsearth.org/.

Cosgrove, Denis E. "Contested Global Visions: One-World, Whole-Earth, and the Apollo Space Photographs." *Annals of the Association of American Geographers* 84, (1994).

Dowideit, Anette. "Kopenhagen: Eine Vorentscheidung für die Zukunft der Menschen?" Welt am Sonntag, November 29, 2009. Accessed July 5, 2012. http://www.welt.de/welt_print/wirtschaft/article5364760/Kopenhagen-Eine-Vorentscheidung-ueber-die-Zukunft-der-Menschen.html.

Doyle, Julie. "Seeing the climate? The Problematic Status of Visual Evidence in Climate Change Campaigning." In *Ecosee: Image, Rhetoric, and Nature,* edited by Sidney Dobrin and Sean Morey, 279–98. New York: State University of New York Press, 2009.

Dunaway, Finis. *Natural Visions. The Power of Images in American Environmental Reform.* Chicago: The University of Chicago Press, 2008.

Dünkel, Vera. "Comparing Images," In *The Technical Image. A History of Styles in Scientific Imagery*, edited by Horst Bredekamp, Birgit Schneider and Vera Dünkel, 24–27. Michigan: Michigan Press, 2014.

Franklin, Stuart. "Environmental Photography." Blog post, October 2, 2008. Accessed 25 July 2012. http://www.blog.magnumphotos.com/2008/10/-environmental_photography.html/.

Gore, Al. *An Inconvenient Truth: The Planetary Emergency of Global Warming and What We Can Do About It*. Emmaus: Rodale, 2006.

Griese, Sabine. *Text-Bilder und ihre Kontexte: Medialität und Materialität von Einblatt-Holz- und Metallschnitte des 15. Jahrhunderts*. Zürich: Chronos-Verlag, 2011.

Hutchins, Christina, editor. *Climate Change. Our Warming Planet*. Bristol: Alastair Sawday Publishing Co. Ltd., 2009.

Jahn, Johannes, and Stefanie Lieb. *Wörterbuch der Kunst*. Stuttgart: Kröner, 2008.

Kollmuss, Anja, and Julian Agyeman. "Mind the Gap: Why Do People Act Environmentally and What Are the Barriers to Pro-environmental Behavior?" *Environmental Education Research* 8, no. 3 (2002): 239–260.

Krznaric, Roman. "Empathy and Climate Change. Proposals for a Revolution of Human Relationships", in *Future Ethics. Climate Change and Apocalyptic Imagination,* edited by Stefan Skrimshire, 153–172. London: Continuum International Publishing Group, 2010.

"London Futures." Accessed October 20, 2012. http://www.postcardsfrom thefuture.co.uk/.

Nelson, Andrea Jeannette. *Reading Photobooks. Narrative Montage and the Construction of Modern Visual Literacy*. PhD diss., University of Minnesota, 2007.

"Refugies Climatiques—Exhibition in Copenhagen." Accessed October 20, 2012. http://www.climaterefugees.eu/nw,News, 239.

Schmidt, Gavin, and Joshua Wolfe, eds. *Climate Change: Picturing the Science*. New York: W.W. Norton & Company, 2009.

"Vulnerable. The Face of Climate Change in the Maldives." Accessed October 15, 2012. http://www.bluepeacemaldives.org/vulnerable/.

Zängl, Wolfgang, and Sylvia Hamberger. *Gletscher im Treibhaus. Eine fotografische Zeitreise in die alpine Eiswelt*. Steinfurt: Tecklenborg Verlag, 2004.

IMAGE COPYRIGHTS & SOURCES

Copyrights and sources are given in the image captions.

The Pensive Photograph as Agent: What Can Non-Illustrative Images Do to Galvanize Public Support for Climate Change Action?

EDWARD MORRIS AND SUSANNAH SAYLER

This essay examines the value of pensive, non-illustrative photographs in generating public sentiment for addressing climate change. The authors use their own work with the arts collective The Canary Project as a case study, with a particular focus on a photography project titled A History of the Future. They argue that a fundamental disbelief in the potential trauma of climate change is a greater impediment to political action than ignorance of the science. With reference to critical thinking on the nature of photography and art history, they describe the process by which the pensive photograph (and, by extension, other types of cultural production) can open a space for belief in such trauma within receptive viewers. They go on to describe how such cultural production can contribute to societal conceptions of what is normal, intolerable, possible, urgent, etc., and how this building of public sentiment, in turn, becomes the prerequisite to concrete political action. Their analysis bears both directly: on the question of visualizing climate change and indirectly: on the nature of art and activism in general.

1.

Our project, *A History of the Future,* consists of a set of photographic images, all landscapes, showing places around the world where scientists are observing the impacts of climate change. The basic method of scientific observation is measurement. However, these measurements are not visible in the photographs. The images are still images and singular (i.e. we took no time-lapse or before-and-after images).

In launching this project, we wanted to find out what could be seen of climate change in still landscape images but, more importantly, what could be conveyed through these images that might bridge the gap between scientific data and public understanding of the issue and also between public understanding and political will. We wanted to find out if photographs could make us or anybody else more seriously invest in the proposition that climate change is real, urgent and a serious risk to life.

The project consists of images, but is not the images themselves. The project is the display and distribution of those images. The images live many lives. They have been art objects, framed in a gallery with an edition number and the aura of secrecy and value. They have been pedagogical tools in science museums and classrooms, openly disclosing (delimiting) their content. They have been billboards and bus ads. They have been editorial content. In each case, we struck a different bargain between illustrating the facts of climate change (which is always contingent on captioning and context) and laying bare a state of disorientation and collapse of scale that seemed the only reliable opening into the trauma of climate change.

Figure 1: Exhibition installation view, Double Blind, Illges Gallery, Columbus State University, Columbus, GA, 2010; Figure 2: Billboards from A History of the Future series, Columbus, GA, 2010.

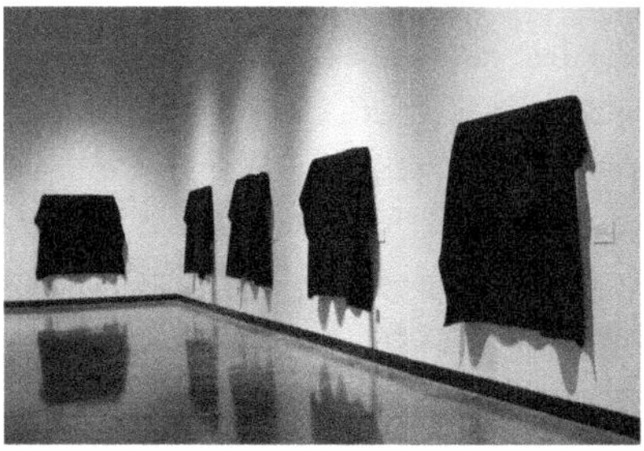

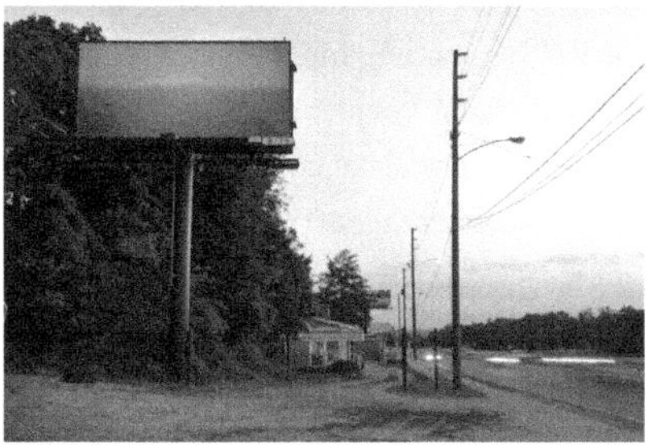

Our sense of what our images do best changed substantially in the course of the project that, in turn, produced a transformation in our understanding of what activist art or research-based art could be. In one of our more recent exhibitions of this work, we simply hung black fabric over the photographs; didactic text next to the cloaked images gave their context. Meanwhile, outside the gallery, we put the same images on billboards throughout the metropolitan area without any text or explanation whatsoever. Where there is normally seeing (the gallery space) there would be only believing; and where there is normally believing (the advertising space) there would be only seeing. Why did we do that? What good was it?

2.

"Seeing is believing" is the general form of most justifications for the efficacy of photographs concerning climate change, the tacit logic being that photographs present evidence that you cannot deny. However, this logic is actually a muddle and based on false premises.

First of all, a photograph is not merely seeing. A photograph is captured sight that is also seen—an indeterminate status that accounts for its pensiveness. This pensiveness is a quality intrinsic to all photographs that can either be rendered subordinate to some textual content or given free reign. Jacques Rancière describes the state of being pensive as being full of thoughts but not necessarily the one thinking them. It is a state between active and passive. When applied to an image he writes: "It contains unthought thought, a thought that cannot be attributed to the intention of the person who produces it and which has an effect on the person who views it without her linking it to a determinate object."[1] This occurs precisely because photography is not simply seeing, it is seeing another's sight. A memory that feels like yours but is not. However, this intrinsic pensiveness of the photograph is defeated when the specificity of the thing seen is subverted or when the thought is too readily dispensed with and fails to permeate the viewer like smoke filling a room. Often, dispensing with the photograph's pensiveness is a crucial tactic, notably anytime where the efficiency of information delivery is paramount. Generally speaking, this is the case with the images of photojournalism, which tend to be illustrative rather than pensive for good reason: they need to communicate to the reader quickly and clearly. What we are at pains to describe here, however, is another type of image in which the intrinsic pensiveness overwhelms or outruns its immediate context.[2]

---

1 Rancière, *The Emancipated Spectator*, 2009, 107.

2 We have chosen the term "pensive image" because it has more substantial critical precedent (notably Rancière, who is referenced throughout this essay, and Barthes). Nearly equivalent terms would have been contemplative image or indeterminate image. The pensive image has the tendency to be viewed as art. See section 7 of this essay for a more detailed discussion of the illustrative image as distinct from the pensive image.

It is important to note that there is something deceptive about the illustrative image when it comes to climate change. While all photographs are, in a sense, evidentiary, what they verify is fundamentally limited only to what can be seen in the image without textual or con-textual supplement. (The context simply tells you where to place the thing verified, into what proposition). The expansive formulation of this thought is: the photograph is proof of the past and of something specific existing in the past (no more than this, but also no less). This truth claim resides solely in the image, but does not rest in the printed image's relationship to the thing photographed (i.e. its indexicality). Each photograph also asserts absolute presence (space) and absolute absence (time). The pensive image stands in an uncertain relationship between this absolute specificity and the necessary context of each image within the larger of system of images. As Rancière says, "an image never stands alone."[3] We will return to this thought below in our discussion of the semiotics and phenomenology of the photograph, as it bears directly on the key questions of this publication.

Moreover, the language of evidence suggests knowledge, but knowledge and belief are not the same—even if we frequently mistake one for the other. We commonly say "I know this" or "I know that," and yet as Duchamp pointed out, "In general, when one says 'I know' one doesn't know; one believes."[4] Knowledge implies the possibility of verification, whereas belief does not. Belief can only be tested internally and is a function of the relative presence or absence of doubt. Belief is involved with a system of assumptions and determines how we see the world and how we function within it. Duchamp was right about the frequency of our using "know" when we really mean "believe" because it is much harder than it seems (if not impossible) to provide a clear criteria for verification. The difficulty of objective verification obviates the sort of knowledge assumed by the statement "I know," but we don't acknowledge that in common speech. "I know" often means, "I am certain (without a doubt)," which has everything to do with belief and nothing to do with the possibility of objective verification. So, "seeing," which is empirical, cannot be "believing," which is not empirical. Seeing is actually closer to knowing, in part because both can be construed as a process rather than a state. Wittgenstein on this point: "'I know' has a primitive meaning similar to and related to 'I see' ('wissen', 'videre') [...] 'I know' is supposed to express a relation, not between me and a proposition (like 'I believe') but between me and a fact. So that the fact is taken

3 Rancière, *The Emancipated Spectator*, 2009, 99.

4 de Certau, *The Practice of Everyday Life*, 1984, 177, citing Duchamp, Duchamp du signe, 185.

into my consciousness [...] This would give us a picture of knowing as the perception of an outer event through visual rays which project it as it is into the eye and consciousness."[5]

Lastly, and most fundamentally to the topic of this essay, climate change cannot be seen in order to believed. At the end of the day, it must simply be believed. That is because climate change is a proposition and not a fact (no matter how empirically grounded that proposition might be). In the words of our editors, climate change is an "abstract, statistically created, long-term research object." Needless to say, you can't exactly photograph such a thing.

3.

So what *did* we photograph for our project? A mountain, a stick with numbers on it, an aqueduct, a church in snow, a fallen water tower, etc. If these things were not "climate change" (because what is an image of a proposition?), then were they at least impacts of climate change? Not exactly that either.

Our method was as follows: First, we conducted research into what changes are occurring on the planet that: 1) had been attributed, at least in part, to climate change on the basis of two or more peer-reviewed articles in relevant journals; 2) were visible to the human eye and thus ostensibly photographable with a non-specialized camera and lens.

We then interviewed experts about these changes and where they were most evident or likely to be evident in the near future. We selected 13 locations on five continents (we didn't make it to Australia); and five categories of impacts (glacial, permafrost and icecap melting; fires and droughts; sea-level rise; extreme weather events; and disrupted ecosystems). We then developed partnerships with local experts who could guide us in the field. These experts showed us around on location, pointing to this and that. We photographed what they showed us in solitude, either forcing a break in the tour or returning later in better light. Initially, we were completely focused on impacts—on what could be seen of climate change—and we expected our project to be more or less documentary in this sense, even if we often gave in to the time-honored trick of shrouding the image in the golden light of early morning and evening.

5 Wittgenstein, *On Certainty*, 1969, 13e-14e.

Figure 3: Glacial, Icecap and Permafrost Melting XLVII: Cordillera Blanca, Peru, 2008, archival pigment print, 40 x 50 inches.

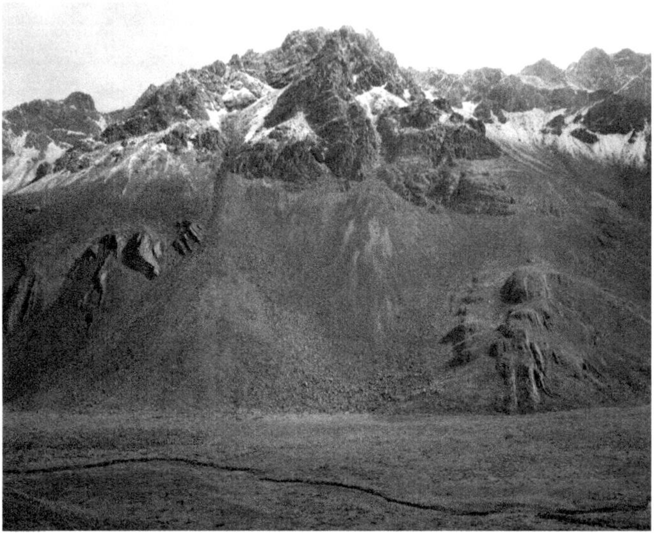

Figure 4: Glacial, Icecap and Permafrost Melting XXXVI: Bellingshause Base, King George Island, Antarctica, 2008, archival pigment print, 40 x 50 inches.

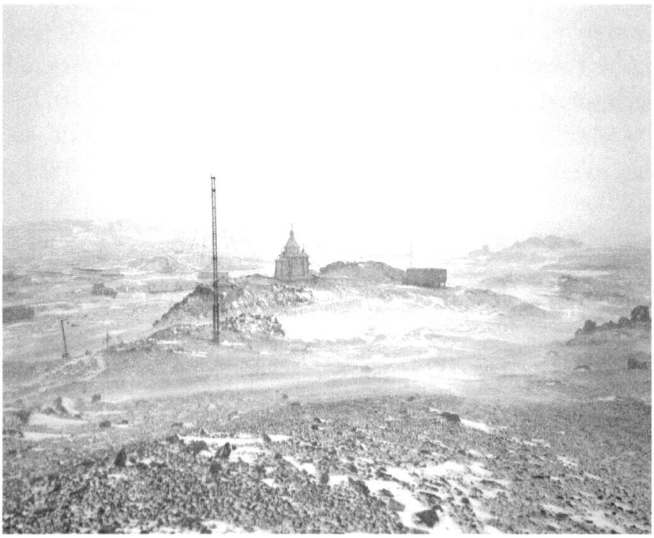

Figure 5: Extreme Weather Events XV: Plaquemines Parish, Louisiana, 2005, archival pigment print, 40 x 50 inches. (→ color plate 16)

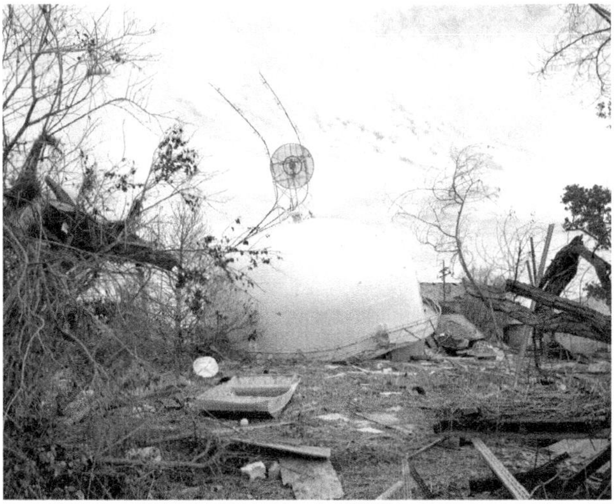

However, once in the field, we received a series of unexpected shocks. The shock was not some apparent violence being done to the land. On the contrary, the shock was the indifference of nature to such putative violence. Our shock was the silence in these places, not the noise. Our shock was the invisibility of the impacts to the human eye and not their glaring visibility. A glacier is melting? But actually, this is too slow to see in process without the aid of some technology. A tree is burning? Well, what made it burn? A coral reef is "bleached?" But it looks so beautiful. A water tower has been crushed? But, where is that storm now and did climate change actually cause it? We traveled to all these places that scientists had identified as harbingers of catastrophe and found them devoid of any clear sign of danger. The land was just there, imperceptibly changing. Paradoxically, this shock strengthened, rather than diminished, our belief in the proposition we set out to investigate: *that anthropogenic climate change, as benign and boring is it may sound, is real, urgent and of grave threat to life on the planet.* We could see only the footprints and not the beast walking. This hardly made the beast seem less threatening.

The various images we settled on to articulate this experience have one central underlying quality. Within the areas demarcated by our expert guides, we shot whatever made us feel, rather than understand (believe, rather than know), that a change was truly beginning—i.e.: that the history of the earth is long and the history of humans infinitesimal; that time is irreversible; that unimaginable forces are constantly erasing and writing the land; that there is a life outside of ourselves; that this thing referred to as a "climate change impact" could translate into suffering; that the real exists beyond our symbolic order; that seeing the past is seeing the future, etc. In short, we photographed what, to us (in the field, in real time) seemed to manifest not the facts of climate change per se, but its trauma.

4.

Treating trauma of any sort photographically is a vexing issue, in part because of the oversaturation of images in contemporary culture, in part because the seeing of a photograph is never commensurate with direct experience, and in part because, as Errol Morris and Susan Sontag observe in different ways, believing precedes the seeing of a photographic image and always mediates it.[6] With respect to climate change, these problems are compounded by the fact that its impacts, in addition to being putative, are not directly experienced, and therefore have only an uncertain claim to trauma. We ought, therefore, to define more closely what we mean by trauma in this instance and in what sense this became the governing principle of our project, both in terms of the images we decided upon and their various modes of exhibition. To do so, we must first return to the subject of belief vs. knowledge.

Belief refers to a relationship between myself and a proposition and carries with it no possibility of absolute verification, whereas knowledge refers to a relationship between myself and a fact and does hold forth the possibility of verification. Yet, curiously, we speak of certainty in regard to both knowledge and belief. With respect to knowledge, we say we are certain in order to describe a knowledge is total, with no chance of error (if this is indeed possible, as some dispute), i.e. we reference the exterior. With respect to belief, we say we are certain to describe a mental state without doubt, i.e. we reference the interior.

Certainty is not absolute. There are degrees of certainty, as evidenced by how we use the word. We say of a belief that it is strong or weak depending on the degree of certainty. Thus, we say "I am fairly certain," "I am totally certain," etc. Accordingly, it is important to note that belief in something (for example climate change) can be strong in one respect (for example confidence in

6 See: Morris, *Believing is Seeing*, 2011, and Sontag, *Regarding the Pain of Others*, 2003.

the science) but weak in other respects (for example that "the flow of everyday reality can be perturbed" to the extent predicted by the science[7]).

Michel de Certeau usefully defines belief as an investment in a proposition. For de Certeau that investment is exhibited in action.[8] Whatever our strength of belief concerning the science of climate change, we exhibit a very weak belief overall, as evidenced by our lack of action. The Lacanian Slavoj Zizek diagnoses this as follows: "Our attitude here is that of the fetishistic split. 'I know very well (that global warming is a threat to the entire humanity), but nonetheless [...] (I cannot really believe it). It is enough to see the natural world to which my mind is connected: green grass and trees, the sighing of the breeze, the rising of the sun [...] can one really imagine that all this will be disturbed?'"[9]

Zizek's evocation of fetishism here is telling. A fetish develops as a response to trauma, specifically when that trauma is too great to be assimilated into life as we have come to expect it, i.e. our symbolic order. A break occurs, a rupture. The experience of trauma is one of profound dislocation and disorientation.

With respect to climate change, this fetishistic split is particularly vexing, because the only thing that can cause a rupture in our sense of normally proceeding life (sun rising and falling, snow falling and melting) is belief itself and not anything that can be directly experienced. This is the case in the first instance because even if a traumatic event can be attributed to something called climate change, we require a belief in that very attribution in order to link the direct experience of the event to this larger proposition. Furthermore, we must link this disturbance to the more unimaginable proposition that such events will cease to become disturbances at all, but the normal stuff of life. This, too, requires belief and everything mitigates our holding onto that belief in any real way for very long. Our bodies and minds will not permit it. In other words, we return to Zizek's conclusion: we can very well know climate change, but we cannot believe it. To truly believe it (and act accordingly) requires a rupture, a trauma, and this trauma must override our direct perception of the world.

Such a belief extends beyond a mere acceptance of the soundness of the science we are told about but cannot independently verify. More significantly, this belief requires certainty that such a rupture is even possible, that there is a precedent for it: some firm ground to think it could happen and what that would

7 Zizek, *End Times*, 2011, 445.

8 De Certau, *The Practice of Everyday Life*, 1984, 177–189.

9 Zizek, *End Times*, 2011, 445–446.

mean.[10] It is the difficulty in appreciating the latter that best accounts for the gap between scientific understanding and public sentiment or political will. Plenty of people know about climate change, even know that it is a valid proposition, but few really believe it. There is a circularity here: you have to believe your way into the trauma that is necessary to truly believe. In this sense, the subject of the photographs in *A History of the Future* is not real trauma, or even vicarious trauma.[11] It is more the intimation of trauma and can never be wholly grasped. To become receptive to this intimation of trauma (literally "wound" in the original Greek), we need an opening, a piercing. What we want to say is that this piercing certainly can come from a photograph (and, by extension, other media that we may want to call art), yet no photograph can guarantee it, for it also depends on the viewer.

5.

At some point, we began referring to our photographs for this project as blank stares. That is how they came across to us when viewed individually. Was the landscape looking at us, or were we looking at the landscape? This effect of an eerily vacant but relentless gaze was produced by one of two things: 1) the discrepancy between the neutrality and apparent indifference of the landscapes in contrast to their presumed meaning as harbingers of catastrophe, and/or 2) the simultaneous intimacy and distance (presence and absence) we felt when looking at something particular, which we were only doing in the formalized manner of the project because some expert had told us to be where we were.

Another way to describe this is that the actuality of the landscapes themselves was so much stronger (so much more present) than the discourse surrounding

10 Climate change activists often suggest that catastrophic weather events are needed to rupture the collective sense of normalcy. While events like Hurricane Sandy bring communities together and rally civic pride, researchers suggest that people usually form opinions in keeping with their cultural group, so as not to drive a wedge between them and their peers. Kahan, "Why We are Poles Apart on Climate Change," 2012, 255.

11 The notion of vicarious trauma, however, is pertinent to the argument of this essay, because if accepted it shows that trauma does not have to be experienced directly to be felt. We have not made a survey of the professional literature on the topic, but are aware of its extent. McCann and Pearlman, "Vicarious Traumatization," 149.

climate change that had brought us there in the first place. We then faced an essential choice, which we have articulated elsewhere as follows[12]:

1) Dramatize the landscape in some way, such as indulging in a myth about damaged earth, in an effort to activate feelings of indignation, disgust, wonder or dismay (whether or not this motivation is frankly admitted or not)

or

2) Confess to a feeling of incomprehension, disorientation, loss of scale, seduction and panic.

Figure 6: Extreme Weather Events XV: XXIII: Coney Island, NY, 2012, archival pigment print, 40 x 50 inches. (→ color plate 17)

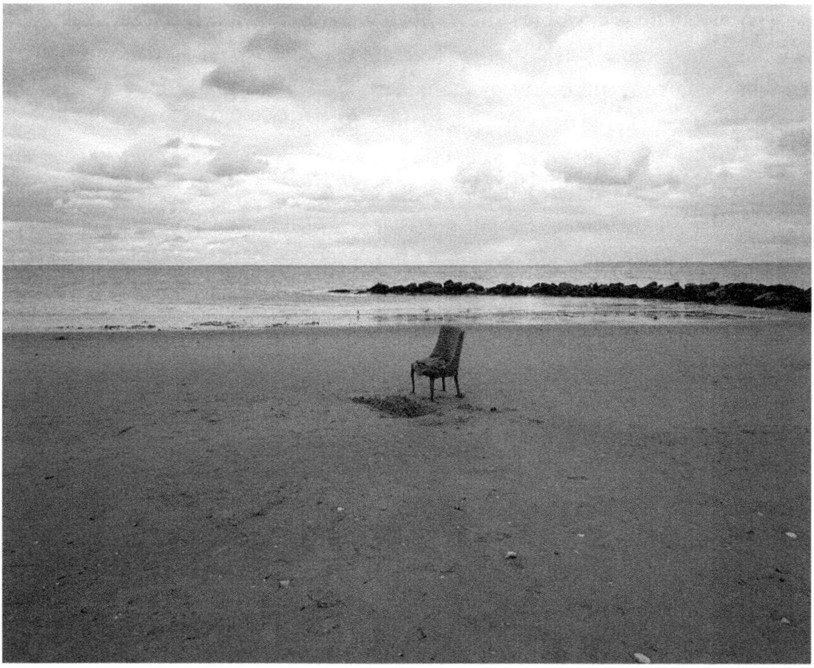

One is a gesture of mastery and the other of submission. Most photography dealing with environmental themes employs some version of the former. You can see this technically in the color saturation, the single-minded themes, the

12 Morris and Sayler, "Blank Stare," 2012.

bold compositional choices which preference symmetry and an elevated perspective that is dehumanizing and gives the impression of omniscience. You can also see it in the way the work is framed and discussed, usually in the vein of heroism. In other words, most landscape photography has still not escaped the long shadow of landscape photographer Ansel Adams.

Adherence to the lofty formal techniques of Adams is not incidental; it is part and parcel of a philosophical outlook that regards nature as transcendental, "a religious idea," to quote Adams himself.[13] Most contemporary landscape photographs tend to be more self-consciously ambiguous in their statements of purpose than Adams, yet still operative is the underlying dogma that nature and man are opposed and that, as such, we are damaging nature. This leads to a simplistic symbolic order, which is often evident in the compositional structure of their photographs. More significantly, it also leads to adamant disavowals of any political intent or responsibility. Such work creates a (visual) sanctuary and a locus of mourning that dispenses with any need to act; in part because of the certainty and conventionality of the underlying beliefs and in part because the mere act of viewing is held to be sufficient. Such work assumes no future or, at best, longs for a future that is conservative—a future where wild places are preserved with only privileged access.

Against this tendency, we felt a perverse attraction to the following idea of nature that is even older and more out-of-vogue (from Rilke):

> "To see landscape thus, as something distant and foreign, something remote and unloving, something entirely self-contained, was necessary […] For we began to understand Nature only when we no longer understood it; when we felt that it was the Other, indifferent towards men, which has no wish to let us enter […]"[14]

Whatever this romantic perspective lacks in terms of sensitivity to humanity's position as part of an ecosystem or foreknowledge that we would come to call this era the Anthropocene, it is perspicacious and intellectually rigorous on one point: the supreme indifference of the inorganic world.

13 Adams, *On Our National Parks*, 1992, 113–117.

14 Hass, Introduction to The Selected Poems of Rainer Maria Rilke, citing Rilke, "Concerning Landscape," 1989, XXV.

6.

So if the photographs from *A History of the Future* on their own were blank stares born of disorientation, expressing sublime indifference to the implications of climate change, what application did they have to the original, essentially activist, premise of the project? We need to open up our understanding of "activism" to answer this.

Paul Chan, a self-described activist and artist, is often quoted for his firm distinction between his own artistic and activist impulses. He speaks to it in many places. Among the more succinct is this quote from an interview in Bomb magazine:

> "Collective social power needs the language of politics, which means, among other things, that people need to consolidate identities, to provide answers [...] to make things happen. Whereas my art is nothing if not the dispersion of power [...] And so, in a way, the political project and the art project are sometimes in opposition."[15]

The implication here (fleshed out elsewhere by Chan) is that:

- Activist — Singular, constructive, affirmative, un-nuanced, instrumental, common, contractual and with demands
- Art — Manifold, deconstructive, interrogative, nuanced, purposeless, unique, free and unending

Chan's oppositional dichotomy between art and activism is still a broadly shared bias in the art world and beyond. If one wants to attack Chan's dichotomy, one usually chooses to joust with the definition of art (cf. Nato Thompson and others[16]). That is a fair game, but not the battle we want to fight here. In fact, we think this definition of art is a very useful description of the way it (often) functions. There ought to be a word reserved in our language for a cultural product able to hold apparently contradictory ideas in balance. There should be a special semantic place for things we read and view that have claims to negative capability or that are able to speak to our moral intuitions rather than our reason, that blossom questions faster than they shed answers. We believe the old standby, art, serves for this purpose. Another word we might use when speaking

15 McClister, "Interview with Paul Chan," 2005, 26.

16 See Thompson, *Living as Form*, 2012 or *Seeing Power*, 2013. As one of many examples, see Enwezor, "The Artist as Producer in Times of Crisis," 2005.

of a photograph in this context is the word we have used throughout this essay: pensive.

On the contrary, what is objectionable to us about Chan's oppositional duality is his far too restrictive definition of activism.[17] Chan's definition of activism holds true only if we restrict the term to those actions that are exclusively in the service of immediate political goals and appreciable, near-term change, which, as Chan indicates, does in fact require consolidation of identities around a central reductive argument. However, the current use of the word activism does not necessarily imply any such restriction. Interestingly, dictionaries, which always lag behind actual usage, are divided on this point. From a survey of online and offline definitions, we find:

1) Many, including the Oxford English Dictionary (OED), do not even have a definition of activism outside of a reference to an obscure early 20th Century philosophical theory that "assumes the objective reality and active existence of everything." Further, there is no entry in the OED for the word "activist" at all!
2) Those dictionaries that do have a definition of activism in the more contemporary sense often assume the necessity of vigor, militancy and/or direct action. For example, this definition from freedictionary.com: "a policy of taking direct and often militant action to achieve an end, esp. a political or social one."
3) In our very quick and anecdotal survey, Wikipedia alone offered a definition that conforms more closely to the way the word is actually used today: "Activism consists of efforts to promote, impede, or direct social, political, economic, or environmental change. Activism can take a wide range of forms [...]"

Activism is concerned with effecting change in a society, full stop. Militancy is not required. For doesn't change happen in many different strata of society and at many different tempos? Isn't the formula for change partly direct political action, meaning specific advocacy for specific policies, and partly cultural,

17 It is especially aggravating that this narrow view of activism continues to hold such sway over the art world—a sort of modernist hangover that makes fools of many artists who talk out of both sides of their mouth, espousing Guatarri to burnish their intellectual bona fides, and at the same time promoting a version of art that is merely the "portrayal of bourgeois self-understanding." (That final delicious phrase is from Burger, *The Negation of the Autonomy of Art*, 1984).

meaning the establishment of what people think is moral, important, normal, intolerable, possible, etc.? Don't artists make culture?

In reference to enacting legislation, the much-mythologized Abraham Lincoln said: "With public sentiment nothing can fail; without it nothing can succeed." In other words, there has to be sufficient popular acceptance of an idea for legislators to act (or to even be in office in the first place). Take the gay rights movement, for example. We have recently had Don't Ask Don't Tell and Gay Marriage on the legislator's docket in the U.S., but how long was that in the making, and what happened to get to that point? People needed to start regarding gay people as just people. Well, how did that happen? The notion of being gay entered the culture, first radically, and then gradually became normalized. And how did it enter the culture and how did it become normalized or gain wider acceptance? Are we to believe that Robert Mapplethorpe (or Gran Fury) had absolutely nothing to do with that? Or that Ansel Adams, Elliot Porter, Joseph Beuys or Agnes Denes had nothing to do with environmental legislation or reforms? Or that Martha Rosler, Judy Chicago and Barbara Kruger had nothing to do with progress on equal rights for women? Etc., etc. etc.: the list is virtually endless.

The point is not that hordes of people came into contact with, or were directly influenced by, these works. The point is that these works manifestly changed the cultural landscape and opened up cultural spaces. At minimum, this opening up was seized upon by taste making youth in urban centers, then popularized in the media, and then taken up more broadly by consumers of that media across the country and globe. This is how culture works. Once more people became sympathetic to a given viewpoint (because such a viewpoint had become part of their social identity), then, and only then could the issue be taken up politically in terms of changing the law of the land, etc.

Video artist and critic Tom Sherman has a way of explaining this process by employing the rubric of "memes." As Sherman defines them: "memes are cultural building blocks, the message forms and contents at the base of our cultures."[18] Artists, in Sherman's opinion, make "a-typical memes" that provide a vision or aspiration for an entirely new culture.[19] If memes make up the cellular structure of our culture and a new, a-typical meme is introduced into that structure, then the entire composition of the culture has changed, no matter how minutely. What happens next is anybody's guess. It depends on the life force of

18 Sherman, "Memory and Hope," 2006, 190.

19 Sherman, "Prosthetics and Memetics," 2006, 207.

these new cultural cells. Sometimes they get sick and die; sometimes they spread and mutate.

Or we might turn Chan against himself. Chan beautifully defines the art he is interested in as "critical thinking in sensuous form."[20] But what critical thinking is innocent of persuasion? Even if concerned with deconstruction, the thinking piece walks a reader/viewer into and through a point of view that productively alters her understanding of the world and thus her belief system.

We can also say it this way: art makes a space for belief and belief makes a space for change. To map this formula explicitly on the argument we made above: art produces ways of thinking and seeing that open a path toward the accumulation of public sentiment (through the creation of subjectives sensitive to this or that issue), which in turn can precipitate political change. Is this activism? If we want to reserve the word activism for a description of those activities that seek some direct, leveraged outcome and depend on the "vigorous" mobilization of masses of people, so be it. In that case, we propose to re-purpose the old term "constructivist" to describe any art that is able to strengthen belief (in whatever manner and no matter how diffusely). Either way, the duality proposed by Chan disappears and art can be seen for what it is: a (potentially) productive force in society or else a luxury commodity.

If we must go to such lengths sketching how a non-rhetorical, non-instrumentalist art might still function as activism (or at least constructivism) that is because it is not very easy to see. When you get right down to it, not many people see or know about art, and those that do tend to be pretty progressive anyway. So of course, it is much easier to see how an African-American sitting down at a segregated lunch counter in 1960 has an impact. Our argument is not that art is better at changing the world than direct action. What we are saying is that art contributes to the necessary work of laying the cultural foundation (public sentiment) for this more apparent direct action to take hold. Art is good at this because it opens rather than closes thought. Yet, even making that slight claim invites a pretty decisive reconsideration of the nature of artistic practice.

7.

With respect to climate change and photography in particular, we make an additional and more specific claim: that certain photographs of climate change —those that we might term art because they have the indeterminate nature

20 Chan, "What Art is and Where it Belongs," 2009, 6.

requisite to art as defined by Chan and others—can strengthen belief in the potential trauma of climate change. This is important because, as already discussed, a lack of conviction in the possibility of such trauma is a far more serious impediment to political will than ignorance of the science.[21]

In giving special treatment to the pensive image (what we are calling art only on account of convention, as a sort of shorthand) we are distinguishing it from another image type: the illustrative.[22] There are many ways that we could understand the differences between these two modes, most of which are fairly self-evident or easily deducible. The illustrative image is speedy and delivers information. It depends for its efficacy on the unique truth claim of the photograph (THIS was THERE THEN), but makes quick work of it—employing it to gain our trust but then steering that trust towards a statement (journalistic, scientific or rhetorical). The image, in this instance, is subordinate to text, and therefore, the crushing specificity of the captured image is not allowed full unfolding (lest it distract us). The illustrative photograph is useful in strengthening a belief in the scientific justification of climate change, but less so the trauma. Particularly potent examples of the illustrative image are the before-and-after comparison photographs of melting glaciers (Balog, Arnold, Braasch and others), which are able to depict, and thus in a sense to verify, an actual measurement (even if the general proposition drawn from the measurement must still be supplied). Climate change certainly becomes more real to the public through the dissemination of this sort of image. The science becomes easier to believe because we must travel less distance to accept its premises, i.e. we are presented with a verification of the empirical propositions upon which the more general proposition is based. If this is not immediately self-evident, consider what the state of public acceptance of climate change would be without such illustrative photographs.

However, illustrative photographs do not get us very far with the more significant challenge of imagining the potential trauma of climate change with a

---

21 What will obviate the need to lay this ground is the clear, unquestionable presence of the trauma, for example, multiple Hurricane Sandies, but at that point it will be too late. In the meantime, art can help us understand the implications of events like Sandy, as per the HighWaterLine project, which our organization The Canary Project helped produce. This project was predominantly illustrative in nature, but had some contemplative elements, notably in the performance of a lone person marking 80 miles of coast with a warning.

22 Here, we are following the lead of Ulrike Heine and her presentation at the conference that led to this publication, see article by Heine in this volume.

force sufficient to infuse belief. For example, it is telling that the illustrative image in no way curtails the skeptic's point of view that glaciers are melting as part of a natural cycle. Such an image doesn't forebode anything, etc. For some intimation of the trauma and a statement of its plausibility, we must turn to the pensive image.

8.

What makes the pensive image function in this regard, and how is it distinct from the illustrative image? Clearly, the pensive image is not utterly void of context and thus some form of textual inscription. No image is innocent in this way (as Barthes disingenuously maintains in *Camera Lucida*[23]). However, the point is not its purity, but its indeterminacy, or un-fixedness. Paradoxically, the indeterminacy of the pensive image depends for its efficacy on the unique value of the truth claim (THIS was THERE THEN), just as much as the determinacy of the illustrative image. In the case of the pensive image as opposed to the illustrative, the truth claim is not choked off and made subordinate, but instead allowed a full unfolding. This opening is essential to a depth (rather than an accuracy) of feeling.

All photographs are tied to propositions (i.e. have a context), but the truth claim of the photograph (i.e. what can be verified by it) is independent of any such relationship. The truth claim resides in the image itself and as stated before is simply the claim THIS was THERE THEN. This is often referred to as the photograph's inherent indexicality, i.e. it points to (index finger) something actual in the world. However, as art historian, Tom Gunning has argued, this language of semiotics is insufficient to an understanding of the photograph's truth claim, because a photograph does not merely point to (re-present) the actual, it physically *presents* it.

As Gunning points out, referencing Bazin and Barthes, we must move past a semiotic analysis of the photograph to a phenomenological description to appreciate this, because semiotics only sheds light on the mechanism for recording the real and not the presentation of it in the printed or digitally manifested photograph. As Gunning writes, "whereas signs reduce their reference to a signification, I would claim the photograph opens up a passageway to its subject, not as a signification but as a world, multiple and

23 Barthes, *Camera Lucida*, 1981.

complex."[24] Barthes refers to this opening up as an emanation. It has to do with the materiality of the photograph, the fact that we are not merely signifying, we are *actually seeing* what was once seen by somebody else and is now being seen by us *at a different time*. We are looking at something that absolutely was but will absolutely never be again. We are looking at both presence: the literal shining forth of a person or object in the world (the quality of space), and absence: the disappearance of that thing (the quality of time). This hovering between presence and absence, familiarity and unfamiliarity, death and life, present and past, the self and other is completely contingent on a full activation of the specificity of an image tearing away from its ever-present context. Our way to this opening is blocked (partially or wholly) if the specificity of the image is subsumed by a generality, as in the illustrative photograph. Simply put: there is no space in such a photograph, because its content is (over)determined.

A coming-to-terms with the potential trauma of climate change depends on sensitivity to a number of truths that emanate from this sort of forced contemplation of specificity, particularly in reference to a landscape as opposed to a portrait. A partial catalog might include:

- The discrepancy between geologic and human time
- The irreversibility of time
- The insignificance of human history against the scale of geologic history
- The indifference and independence of the inorganic world
- The existence of the real beyond our symbolic order

These abstractions would remain remote if not for the discomfort produced by the fissure in our spatio-temporal understanding. It hardly seems incidental that the language of photography is the language of violence ("shoot," "take," "dodge," "burn," "crop"), as Matthew Thompson and others have noted.[25] We rip the object from the flow of time—a blade across the eyeball, "an instantaneous abduction of the object out of the world into another world, into another kind of time."[26] Whatever was in the image is no more and can never be again. It is in this sense that every photograph is a death mask (for Bazin),[27] a

24 Gunning, "What's the Point of an Index?" 2004, 39–49.

25 Thompson, "The Object Lost and Found," 2011, 71. This entire paragraph borrows ideas from this essay.

26 Ibid, citing Metz, "Photography and Fetish," 83.

27 Ibid, citing Bazin, "The Ontology of the Photographic Image," 1960, 4.

catastrophe (for Barthes).[28] In other words, photographs can help us get closer to the trauma of climate change when their own inherent traumatic properties are given fuller realization, as in the pensive image.

Figure 7: Exhibition installation view, The Canary Project Works on Climate Change 2006 - 2009, Grunwald Gallery, Indiana University, Indiana. In this exhibition, the photographs from A History of the Future series were paired with objects from various of the university's collections to suggest a contemplation of human versus geologic time.

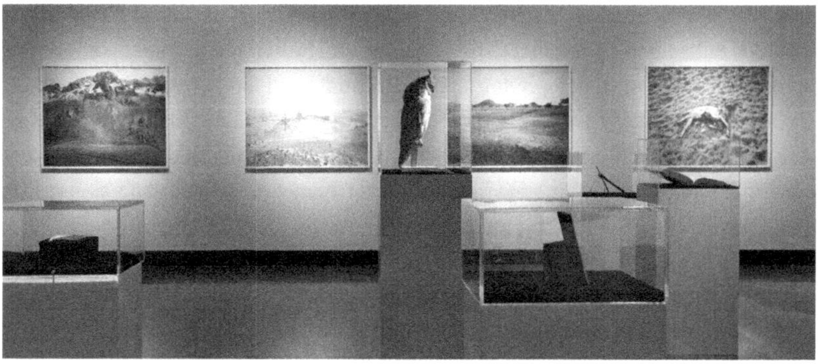

Since all these effects derive from a full unfolding of the truth claim which resides solely in the image, and since the illustrative image chokes off the truth claim by making the image subordinate to text, the illustrative image does not have the same opportunity for piercing the viewer and opening a space for a belief in trauma as the pensive image.

9.

Of course, there is no question of success here. One always fails. That also sets the pensive image apart from the illustrative—there is no mark to hit. What the viewer does with the experience of contemplation, or even whether she accepts the invitation, is, of course, up to her (and her responsibility). As Rancière has pointed out there is no "straight line from the intolerable spectacle to awareness of the reality it was expressing; and from that to the desire to act in order to change it." This is the central fallacy of much "activist" art and presumably the

28 Ibid, citing Barthes, *Camera Lucida*, 1981, 96.

reason why Paul Chan needed to make the distinction he made between art and activism. It is also why the making of art (in this case pensive images), however necessary and however efficacious in the production of spaces for new(found) belief, is not enough when it comes to issues of justice and survival such as climate change.

In our own practice, we produce work as photographers under the name Sayler / Morris, but simultaneously direct a collective, The Canary Project, to which Sayler / Morris is but one of more than 100 contributors. The Canary Project produces artworks and media that foster a contemporary ecological consciousness with a particular focus on deepening public understanding of climate change. Projects range from a poster and public messaging campaign called Green Patriot Posters to educational initiatives such as workshops, to the type of art-making discussed in this essay. But even this is not enough, we also engage, as citizens, in protests, collaborative work with the organization 350.org, social media advocacy, and last but not least, we vote.

Unwilling to disown our activism or be plainly disingenuous about it (as many artists are), we began mentally dividing our projects per Chan's definition: this one is direct activism (consolidation of voice and power); this one is art (diffusion of voice and power). Or to put it in a way that avoids Chan's problematic dichotomy: this one is about direct advocacy and collapses complexity in exchange for broad impact and specific cultural leverage; this one is about something we cannot adequately describe but which nonetheless feels necessary to express and which has a more speculative cultural impact. If our exhibition strategies for works like *A History of the Future* became more and more abstruse to the point of blacking out the photographs all together, it was not in order to create a self-aggrandizing aura around the work. It was to create the type of opening up we have discussed at length here. Such work stands beside and not in front of our more directly activist projects.

Other professionals involved in the struggle to address climate change (if not create a new, more sustainable society) have it even tougher than artists. For artists, advocacy creates a barrier to the work, but this barrier can be transcended. For scientists, on the other hand, advocacy threatens to undermine the very foundation of their credibility. Lynda Walsh has shown how in making probabilistic statements about the future impacts of climate change, scientists are put in the position of prophets and are interpreted as making value judgments.[29] This situation is obviously exacerbated when a scientist like Jim Hansen

29 Walsh, "Ethos and Stasis," 2012.

likewise takes a very public position as an advocate for a specific policy (such as vetoing the Keystone Pipeline). Yet, what is the alternative? Are we meant to hope that sentences like the following get us where need to be: "Most of the observed increase in global average temperatures since the mid-20th century is very likely due to the observed increase in anthropogenic GHG concentrations."?[30]

A profound sadness frequently washes over us during conferences or professional discussions of climate change. A dispassionate tone of analysis (even bordering on professional delight) seems to preside. All professionals working on this issue must find ways to overcome the decorum and limitations of their field and to articulate the catastrophe lying in wait amid the numbers.

BIBLIOGRAPHY

Adams, Ansel. *Our National Parks.* Boston: Little Brown and Company, 1992.

Barthes, Roland. *Camera Lucida.* Translated by Richard Howard. New York: Noonday Press, 1981.

Bazin, André. "The Ontology of the Photographic Image." *Film Quarterly* 13, no. 4 (1960): 4–9.

Burger, Peter. *The Negation of the Autonomy of Art.* Minneapolis: University of Minnesota Press, 1984.

Chan, Paul. "What Art is and Where it Belongs." *e-flux*, no. 10 (2009), http://www.e-flux.com/journal/what-art-is-and-where-it-belongs/.

de Certau, Michel. *The Practice of Everyday Life.* Berkeley: University of California Press, 1984.

Enwezor, Okwui. "The Artist as Producer in Times of Crisis." In *Empires, Ruins + Networks: The Transcultural Agenda in Art*, edited by Scott McQuire and Nikos Papastergiadis. Melbourne: Melbourne University Press, 2005.

Gunning, Tom. "What's the Point of an Index? Or, Faking Photographs." *NORDICOM Review* 5, no. 1/2 (September 2004): 39–49.

Hass, Robert. Introduction, *The Selected Poems of Rainer Maria Rilke.* Translated by Stephen Mitchell. New York: Vintage, 1989; citing Rainer Maria Rilke, "Concerning Landscape".

IPCC, 2007: Climate Change 2007: Synthesis Report. Contribution of Working Groups I, II and III to the Fourth Assessment Report of the Intergovernmental Panel on Climate Change [Core Writing Team, Pachauri, Rajendrak

30 IPCC, "Attribution of Climate Change," 2007.

and Andy Reisinger, editors]. IPCC, Geneva, Switzerland, 2.4 "Attribution of Climate Change."

Kahan, Dan. "Why we are poles apart on climate change." *Nature* 488, no. 7411 (2012): 255.

McCann, I. Lisa and Laurie Anne Pearlman. "Vicarious Traumatization: A Framework for Understanding the Psychological Effects of Working with Victims." *Journal of Traumatic Stress* 3, no. 1 (1990): 131–149.

McClister, Nell. "Interview with Paul Chan." In *BOMB Magazin,* no. 92 (Summer 2005).

Morris, Edward and Susannah Sayler. "Blank Stare: 10 Images and Corresponding Notes on the Pensive in Photography and its Utility in the Face of Catastrophe." In *Making the Geologic Now.* Brooklyn: punctum books, 2012.

Morris, Errol. *Believing is Seeing. Observations on the Mysteries of Photography.* New York: Penguin Press, 2011.

Rancière, Jacques. *The Emancipated Spectator.* Translated by Gregory Elliott. New York: Verso, 2009.

Sherman, Tom. "Memory and Hope." In *Before and After the I-Bomb.* Banff: Banff Centre Press, 2006.

Sherman, Tom. "Prosthetics and Memetics." In *Before and After the I-Bomb.* Banff: Banff Centre Press, 2006.

Sontag, Susan. *Regarding the Pain of Others.* New York: Picador, 2003.

Thompson, Matthew. "The Object Lost and Found." In *The Anxiety of Photography*, edited by Matthew Thompson. Aspen: Aspen Art Museum, 2011.

Thompson, Nato. *Living as Form, Socially Engaged Art from 1991 to 2011.* Cambridge, MA: MIT Press, 2012.

Thompson, Nato. *Seeing Power: Art and Activism in the Age of Cultural Production.* Brooklyn, NY: Melville House, 2013.

Walsh, Lynda. "Ethos and Stasis: What Classical Rhetoric can Teach us about the Graphics in the IPCC Climate Change 2007 Report." Image Politics: Pictur(e)ing Climate—Visualizations, Imaginations, Documentations [Conference]. Postdam January 19, 2012.

"Why Photography Matters—Role of Photographic Images within the Global Climate Change Discourse." Image Politics: Pictur(e)ing Climate—Visualizations, Imaginations, Documentations [Conference]. Postdam January 20, 2012.

Wittgenstein, Ludwig. *On Certainty.* Translated by Denis and G.E.M. Anscombe. New York: Harper & Row, 1969.

Zizek, Slavoj. *Living in the End Times.* New York: Verso, 2011.

IMAGE COPYRIGHTS

Chapter 5
Images of Climate Control

Picturing the State of the Nation's Environment: Early Aerial Photography in the United States from the 1930s to the late 1960s

GISELA PARAK

This essay argues the importance of aerial photography in the United States—in the period from 1930 to the late 1960s—as a means of exploring the nation's topographic, climatic, and environmental conditions. Aerial photographs have long been integrated into federal investigations of the land, and disseminated by popular magazines and specialist books discussing man's impact on nature. The essay sheds light on a crucial shift in how the aerial view was used: whereas in the first half of the century it apparently confirmed modernity's technological utopia, in the later half of the century, it was employed to emphasize environmental devastation.

"Could there be a prototype here for the broad forms of our cities and open spaces of the twentieth-first century? The magnificence of this valley held something unique, offering a basis for a fresh American aesthetic, which in turn could inspire greater efforts in ways yet untried to develop cities that could support general great populations and yet thrive in harmony with the surrounding land. Untried because unperceived, or because the idea was too enormous to comprehend. But whether or not we have the imagination to translate these lessons into guides for creating great cities, the fact remains that there lies the seed of a truly American Aesthetic."[1]

In architect Nathaniel Alexander Owings' (1903–1984) enthusiastic interpretation of a photograph of Monument Valley, shot by William Garnett in 1956 (Figure 1), nature is celebrated as a source of inspiration for the American

1 Owings, *The American Aesthetic*, 1969, 14.

people. He declares that gigantic rock formations should be taken as models and imitated by man in his construction projects, since their agglomeration anticipates that of modern cities. Owings envisages the creation of a canon of shapes that could be reproduced in iron, steel, and concrete. Thus nature would imbue culture with its forms and patterns, serving as a blueprint for human settlements and living conditions.

Figure 1: William Garnett: Monument Valley. William Garnett: Monument Valley, Navajo Tribal Park, in: Nathaniel Alexander Owings, An American Aesthetic, (New York / London: Harper & Row 1969): 14.

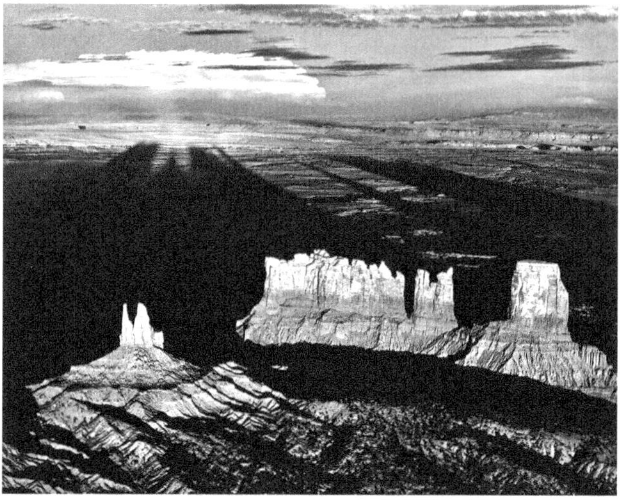

However, Owings' description blends out the manmade transformation of the land, something that is deeply entangled in this *American Aesthetic*. The plotting of smaller units arranged within a quadrangle grid was applied to the greater part of the country and this, rather than its famous wilderness, constituted the norm. But what is even more important here is that the aerial view and, more precisely, a whole book of such views captured by photographer William Garnett (1916–2006), served both to illustrate and provide a means of investigating the human-nature dichotomy. In short, the aerial view was conducive to detailed observation and to firing the imagination when it came to creating a vision of the human relationship to nature.

In this regard, Denis Cosgrove asserted that "the American landscape makes sense from the air," thus illuminating the significance of aviation both as a twentieth-century revolutionary technology and as a catalyst for reassessing the human capacity for cognition.[2] Indeed, the merits of aerial photography in providing a means of understanding the landscape below and its significant impact on methods of land management and surveillance speak clearly for the medium as a technological accomplishment. However, it also provided a contested form of representation insofar as it determined the ways in which nature was perceived."

Historians of technology such as Daniel J. Boorstin (1914–2004), Lewis Mumford (1895–1990), and Thomas P. Hughes (*1923) have illuminated the importance of technological inventions to America in giving birth to the notion of an "American genius" and declaring the United States a "nation of inventors."[3] Technology played a crucial part in the conquest of the West and the enthusiastic reception of technology has fueled the story of continuous progress and enabled explorations of topographic and climatic conditions aimed at making the land productive. Aerial photography, in its dual function as a scientific tool and an enhancer of the imagination, has therefore helped shape the history of both technological determinism and the exploitation of natural resources, as well as the relationship between humanity and nature in the United States. Furthermore, promoting a pageant of technological improvement also affected the way that the aerial view was received: contemporaries exhibited a partial "blindness" or incapacity when it came to interpreting the evidence it provided free of ideological presumptions.

This paper argues that aerial photography became an indispensable tool of agricultural management in the period from the 1930s to the late 1960s and, at the same time, helped establish an aesthetic that was impossible to disengage from its practical application—an aesthetic that could never be viewed in isolation, within a purely artistic sphere. By representing cultural beliefs and, at the same time, creating objective, factual records, the ways in which photographic images "pictured" topographic conditions were ambivalent. On the one hand, its aerial, even divine, perspective gave credence to both accusations of human hubris and the claim to be able to "control" not only the land, but also its climatic settings.[4] On the other hand, the view provided of the surface of the earth and the structures of human settlements inevitably lead to reflection upon

2 Cosgrove, *Geography & Vision,* 2010, 87.

3 Boorstin, *The Democratic Experience,* 1974, 525; Mumford, *Technics and Civilization,* 2010; Hughes, *American Genesis*, 2004, 13.

4 See the article by Fleming in this volume.

the nature of these manmade landscapes and, beyond that, to a consciousness of civilization and a critique of the human relationship to the natural environment.

In reflecting on and, moreover, contributing to the generation of these divergent associations, the aerial view exemplifies a paradigmatic shift in the reception and understanding of "nature" in the twentieth century. Thus, changing notions mirror how Americans utilized, examined, and understood the environmental conditions of their country. As a means of surveillance, the aerial view was embedded in the development of modern science and the introduction of optical instruments to study, measure, and evaluate natural phenomenon in order to understand them.[5] And as Cosgrove wrote, it reveals the subliminal pattern of American settlements consisting of rectangular grids that became the norm among the Founding Fathers, who divided open lands into equal portions, thus expressing the Jeffersonian idea of space.[6]

Besides these associations, in its scientific applications, the aerial view is bound up with buzzwords such as land management, soil conservation, urban planning, the prevention of natural disasters, and—for the later part of the twentieth century—pollution. Thus, it is able to represent all the objectionable side effects of manmade landscapes.

The aerial view as technological imagery exemplifies the persuasive function of images in debates that take place at the intersection of scientific applications and the environmental and public spheres. Moreover, its use in the early representation of climatic conditions preempts today's debates about climate change and global warming.[7] Above all, the aerial photograph presents an image of the outer world that merges with cultural perceptions and causes friction between what one may want to see and what is out there.

MONITORING AGRICULTURAL LAND MANAGEMENT IN THE 1930S

Photographs such as those taken by William Garnett, one of America's air flight pioneers, embody the visual excitement that aerial views typically trigger. Landscapes take on the appearance of geometric abstractions. This abstraction of

5 Bredekamp et al., *The Technical Image,* 2014.

6 Cosgrove, *Geography & Vision*, 2010, 87.

7 For a discussion of the all-encompassing connotations and broad definition of the term climate as a statistical average of "individual types of weather in their geographical distribution," see Fleming and Jankovic, *Klima*, 2011, 1.

the natural rhythm of the landscape captured by the camera emphasizes the notion of the aerial view as revealing a mystery that eludes those on the ground.

Garnett's artistic landscape panoramas date back to the beginning of his career in the 1930s, when he started exploring the American landscape in his Cessna airplane and shooting the world in black and white. In 1969, a selection of his images was published under the aforementioned title, as promising as it was proud, *The American Aesthetic*. The book sought to establish the aerial view as a pertinent American experience, which enabled the viewer to perceive and grasp the vast dimensions of the nation in a way previously unavailable to contemporaries.

According to the book's accompanying interpretative texts, the aerial view fitted well with the narrative of the United States as a nation of great inventors and engineers.[8] Its picturing of vast open lands implied a feeling of technological superiority, since it was technology that was responsible for both the aerial view itself and the manmade landscapes below.

Given its application in the United States in monitoring and improving agricultural production (Figure 2), it is tempting to assume that the aerial view provided the perfect technological tool with which to drive history—to paraphrase Leo Marx's famous thesis.[9] Clearly, the aerial view became a significant tool for gathering data during the twentieth century, thus further reflecting the symbiosis of agricultural and technological progress in American history—wherein the myth of the frontier continued to serve as a cultural leitmotif.[10]

Since the seventeenth century, the map had been the visual tool for gathering and preserving knowledge of unknown territories, not to mention an indispensable tool for governing the colonies. By the time geological surveys of the 1860s and 1870s were undertaken, additional visual sources of information such as graphs, stats, and photographs began to be used. The U.S. government invested heavily in unlocking new territories for settlement, such that expeditions became vast logistical exercises with incredible manpower at their disposal. Agricultural and economic interests drove the expeditions, and beyond

8 Owings, *The American Aesthetic*, 1969, 16. For more on the discussion of the U.S. as a nation of inventors and technological engineers, see Perry, *The Great American Hall of Wonders*.

9 Marx, *The Machine in the Garden,* 2000; Smith and Marx, *Does Technology Drive History?,* 1994.

10 Turner, *The Frontier in American History,* 1921.

Figure 2: USA—Idaho, Our World from the Air. In: E.A. Gutkind, Our World from the Air. An International Survey of Man and His Environment.

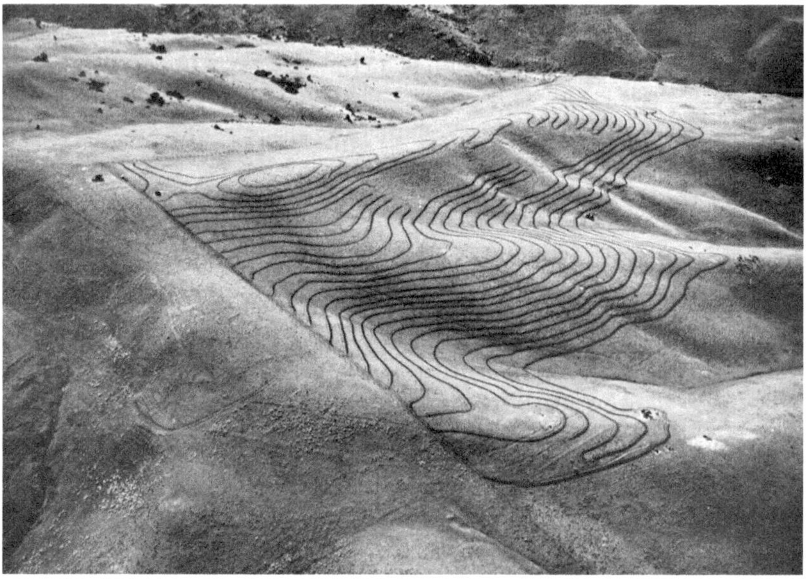

gathering related data, explorers such as John Wesley Powell (1834–1902), George Montague Wheeler (1842–1905), Clarence King (1842–1901) and Ferdinand Vandeveer Hayden (1829–1887) were advised to look out for open land. In 1878, Powell published his *Report on the Lands of the Arid Regions*, in which he warned his contemporaries of the limited scope for settlement in the West due to its climatic conditions. He insisted that certain regions would always be reliant on sophisticated irrigation systems because of aridity and lack of access to water.[11] Ever since, the fixed notion of unlimited resources has lurched between a belief in the nation's abundance and recognition of scarcity.[12] After the surveys, federal programs employed scientific tools to improve cultivation methods and visual tools, and became entangled in the fight for political power and economic strength.

When the manmade calamity of the Dust Bowls struck in the 1930s, politicians and intellectual leaders remained largely optimistic that technology would overcome natural limitations. Advances in irrigation systems, fertilizers, pesticides, and measures to prevent soil erosion such as terracing, wind shields

11 Powell, *Report on the Lands of the Arid Region of the United States,* 1879.

12 Worster, "Facing Limits," 2012.

or contour plowing were expected to help safeguard agricultural production. New Dealers drew on the empirical social sciences to ascertain the optimal allocation of inhabitants across the regions and resettled farmers in less hostile regions that promised to make the fulfillment of high production quotas easier (Figure 3).

Building on the experience of the 1860s and 70s, a huge resettlement and agricultural adjustment program was documented photographically under the auspices of the Department of Agriculture. Yet, though the Historical Section of the Farm Security Administration played a significant role in shaping New Deal ideology and the technological future of agriculture, the program only put photographers in the air in its later years and, even then, only rarely.[13]

Figure 3: USA—Washington. In: E.A. Gutkind, Our World from the Air. An International Survey of Man and His Environment.

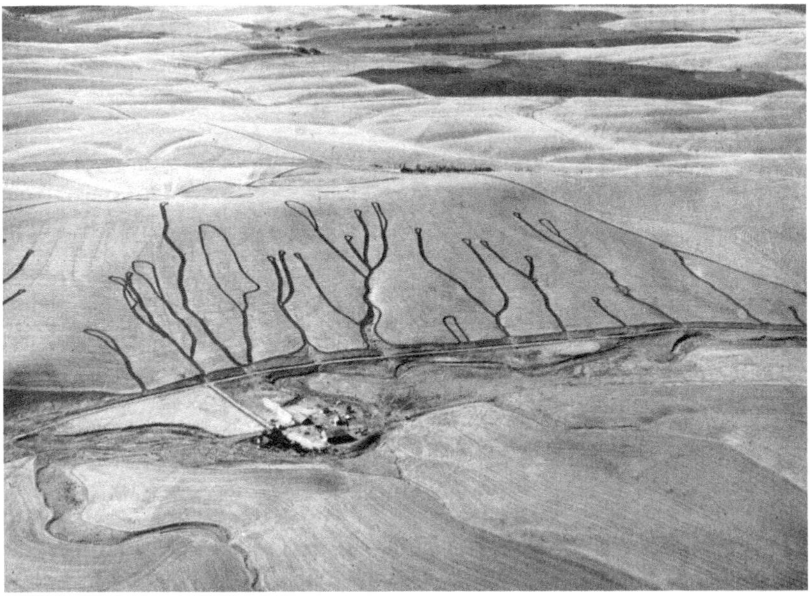

It was another subdivision of the U.S. Department of Agriculture, the Soil Conservation Service, that hired photographer Sherman M. Fairchild (1896–1971) in 1930 to document the state of topsoil on farming land in the Great

13 Weems, "Aerial Views," 2004.

Plains.[14] Fairchild had trained at the U.S. Army's School of Aerial Photography in Rochester and began photographing the land in the early 1920s as an independent entrepreneur. He was employed by cities such as Kansas City and Los Angeles to explore options for urban development.[15] The Soil Conservation Service also commissioned Fairchild to document the effects of new farming methods, like strip cropping, in hindering wind erosion. Thus, his photographs served as tools to evaluate the implementation of these new methods. Preventing further damage to the landscape became central to the New Deal's program of economic adjustment and the rehabilitation and modernization of agriculture.

In addition to the Soil Conservation Service, the Agriculture Adjustment Administration (AAA) employed aerial photographers in order to monitor farmers' compliance with production quotas and soil conservation efforts through the use of visual surveillance.[16] Cash subsidies were directly tied to strict compliance with governmental guidelines. In 1937, the AAA had an impressive 36 crews in the air. By 1941, it had documented more than 90 per cent of American fields.[17] After WWII, the opportunities for monitoring production through the use of aerial photography were in fact enhanced, and the blossoming American wheat fields became an expression of an era of prosperity and economic growth.

Until the first satellite images from the NASA program in the 1970s, aerial view reconnaissance steadily improved. In 1970 and 1971, a special program called "Corn Blight Watch" used remote sensing to detect corn blight. The damage to vegetation could be detected and documented since the crops with corn blight reflected less infrared light than healthy plants.[18] Infrared recording was also used to detect the infestation of crops by vermin and to adjust quantities of pesticides used. The Landsat program captured farmland from high altitudes, such that different shades of gray in the grid of the photograph indicated the level of damage to crops. The original images were processed using an innovative computer technology that summarized shades of gray in larger grids.

In the 1950s, scientists praised the value of the aerial view as a new source of visual information. Signs of their enthusiastic reception can be seen in numerous handbooks of the 1950s and 60s, which attest to the accuracy of photography as

14 Cosgrove and Fox, *Photography and Flight,* 2010, 49.

15 Ibid., 48.

16 Ibid., 50.

17 Ibid., 50.

18 Mack, *Viewing the Earth*, 1990, 147.

well as the possibilities it offered.[19] A publication entitled *Landscape from the Air* revealed what geographers and geologists could hope to learn from studying the views.[20] For example, a view of the Arkansas River demonstrated how the riverbed had altered with time. Another book, *The Uses of Air Photography*, drew attention to opportunities for various disciplines, ranging from geography to geology, cartography, soil conservation, plant disease management, zoology, game management, and archaeology. Its introduction confirmed that there was no more effective medium "for the recording of annual changes." It predicted that air surveys would "assist the exploration of man's natural environment."[21]

These scholarly applications point to an understanding of the aerial view as a synoptic technological vision capable of supporting a modernist utopia. In the late 1950s, a radically different interpretation of aerial views emerged. A new layer of meaning was discovered in the images that Garnett and Fairchild had produced in the 1930s and not one that inspired trust in permanent technological progress. Aerial views started to provide a form of visual representation that sparked the critique of American economic growth and of the way in which natural resources were exploited. Contemporary debates and politics, and particularly the nascent environmental movement and the Cold War, informed this shift in reception.

HIGH MODERNISM, COLD WAR SCIENCE AND THE BIRTH OF AN ENVIRONMENTAL CONSCIENCE IN THE LATE 1950S

The New Deal's intention of controlling nature through large-scale landscape planning, contoured fields and engineering projects, such as its massive dam-construction program, carried the modernist vision into the 1950s. Cold War rhetoric intensified the persuasive use of technological imagery and major

---

19 Bagley, *Aerophotography and Aerosurveying,* 1941; Hart, *Air Photography Applied to Surveying,* 1943; Smith, *Aerial Photographs and Their Applications,* 1943; Abrams, *Essentials of Aerial Surveying and Photo Interpretation,* 1944; Lobeck, *Military Maps and Air Photograph,* 1944; American Geological Institute / Committee on Education, *Outstanding Aerial Photographs in North America,* 1951; Walker, *Geography From The Air,* 1953; Miller, *Photogeology,* 1961; National Research Council / Highway Research Board, *Photogrammetry and Aerial Surveys,* 1966; Allum, *Photogeology and Regional Mapping,* 1966.

20 Monkhouse, *Landscape from the Air,* 1960.

21 St Joseph, "The Scope of Air Photography," 1977, 26.

illustrated papers visually celebrated accomplishments made as a result of the arms race. By the 1960s, however, a shift in public opinion had occurred, such that scientists' control over nature came to be considered a menace to humanity, particularly with regard to the atomic bomb and knowledge of the harmful impact of nuclear fallout.

The value of imagery relating to atomic energy as propaganda was complemented by photographs celebrating the mechanization of American agriculture from the same period—with both contributing to Cold War rhetoric. The knowledge that America's prosperity and economic strength depended on highly effective methods of wheat and corn production, among other features of American farm life, countered the Soviet claim to have genetically manipulated wheat so that it grew more efficiently than regular crops.[22] *Life* magazine started a media campaign to celebrate the mechanization of American agriculture and published features on agricultural improvements, among which an article entitled "Wheels are Its Basic Tools," published 15 December 1952, is exemplary.

Whereas popular imagery as disseminated by *Life* magazine embraced scientific achievements and mirrored a general euphoria in terms of technology, John Brinckerhoff Jackson's (1909–1996) *Landscape* magazine established a critical platform from which to address problems of economic growth and the world's rising population in the early 1950s.[23] Jackson, a writer and publisher, introduced a whole new perspective on the issue of land administration. The magazine focused on the topic of urban development and the relation between nature and civilization, reflecting upon the impact of man on the natural environment and the creation of manmade landscapes. This resonated with pressing contemporary debates concerning structural change in American cities and so-called urban sprawl as a major contributor of environmental pollution. Furthermore, the magazine extensively used aerial photographs to illustrate its concerns.

In his article in the first issue of *Landscape* in spring 1951, Jackson used an aerial view by the Soil Conservation Service to explain how to use this kind of image for the purposes of archaeological study, just as it was described in the handbooks of the time. An image of Rio Grande Valley, for example, enabled him to study the landscape and observe its changing structures.[24] His article in

---

22 Jones, *Science, Politics and the Cold War*, 1988, 3.

23 Osborn, *The Limits of the Earth*, 1953; Brown, *The Challenge of Man's Future*, 1954; Thomas, *Man's Role in Changing the Face of the Earth*, 1956.

24 Jackson, "The Need of Being Versed In Country Things," 1951, 4.

the second issue pointed out the special function of the aerial view as a tool for studying landscapes and, in particular, the influence of human settlements on nature. To this end, he juxtaposed two different genres of landscape. Whereas the reproduction of a nineteenth-century painting reflected the concerns of an era of expansion, and the common faith in the unlimited, inexhaustible resources of the land, for Jackson, the aerial view offered new opportunities. It presented the condition of the land roughly a century later and offered a far better aid to perceiving, describing, and analyzing a given site. Moreover, the production, study, and interpretation of visual sources became an important tool in Jackson's analysis of altered landscapes and changing settlement structures. Thus, his method of studying landscapes revolutionized architectural debates, making visual analysis an essential part of urban studies. Besides its unique function in initiating modern landscape studies, *Landscape* provided a multifunctional platform for discussing related topics and, as such, contributed to the emerging environmental debate of the 1960s by establishing one of the first environmental scholarly networks.

As previously mentioned, in the post-WWII era, photographic images celebrated an unprecedented period of prosperity and economic growth. At the same time, specialists started to voice their suspicions regarding the "sheer abundance enjoyed by the average American," which seemed to them to be "disproportionate," and published their concerns over Americans taking more than their fair share of the world's resources.[25] *Landscape* magazine's spring issue of 1960 not only sported a cover that referred to conservation topics, it also contained an apprehensive article on pollution by an author named Clyde Orr (1921–2010), a professor of chemical engineering.[26] The article's author no longer promoted technology and science as the solution to environmental damage, but instead slammed the ideology of progress for its grave contribution to environmental degradation.[27] Orr's article was illustrated by a photograph showing air pollution in Los Angeles, very likely taken by William Garnett, although the editors provide no details of the photographer (Figure 4).[28] The

25 Tietge, *Flash Effect*, 2002, 97. Tietge refers here to an article published in the April 1953 issue of *Scientific American*, entitled "Science and the Citizen."

26 It presented a photograph by Eliot Porter, who, at the same time, became one of the famous photographers for the Sierra Club publications. Images by Porter and Ansel Adams started to accompany the association's preservationist claims, and were part of a visual campaign to nurture the environmental consciousness of ordinary Americans.

27 Orr Jr., "Our Changing Environment and How to Survive it," 1960, 29.

28 This figure shows a similar photograph taken by William Garnett of the same period.

author pointed out that in a global atmosphere, the disposal of toxins, or diseases would also affect people globally, and that mankind was now exposed to many more such influences than ever before. With this assertion, he joined influential writers of the 1960s like Rachel Carson (1907–1964), René Dubos (1901–1982) and Barry Commoner (1917–2012), all of whom contributed to the era's new environmental awareness by enlightening the public as to the sources of pollution.

Figure 4: William Garnett: Smog, Los Angeles 1949. In: Ansel Adams and Nancy Newhall, This is The American Earth (San Francisco: Sierra Club, 1958): 58.

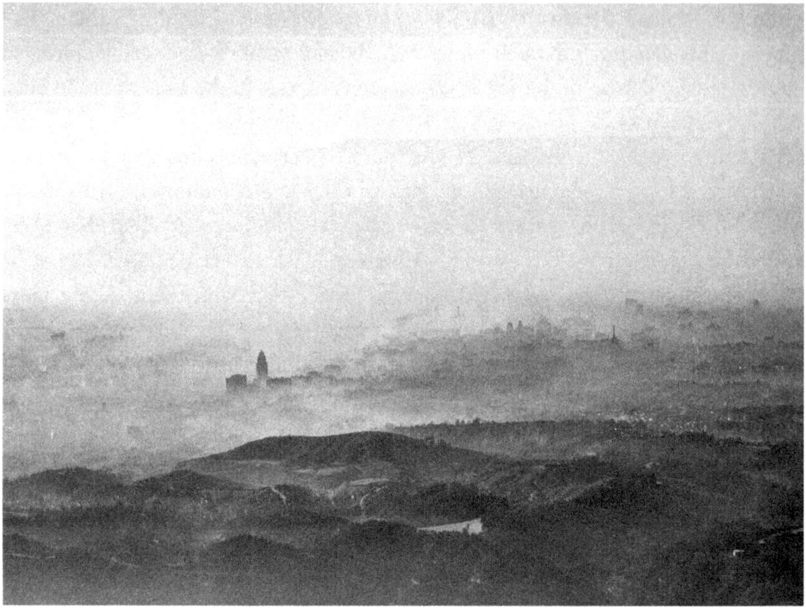

The entanglement of architectural, ecological, scientific and political debates exemplified by the discussion in *Landscape* represent a larger, critical shift in the use made and the perception of the environment that took place in the late 1950s. This growing common awareness led to a critical turn in the perception of the aerial view, as demonstrated by its use in popular books of the period. Prior to the recognition of the atomic threat, the aerial view was considered an expression of technological progress but, by the late 1950s, it became an anchor for the discussion of problems of urban sprawl and environmental degradation.

THE AERIAL VIEW AS A MEANS OF ENVIRONMENTAL CRITIQUE IN THE LATE 1950S

In 1955, one of the first international environmental symposiums ever took place at Princeton, entitled "Man's Role in Changing the Face of the Earth." A keynote address was held by architect Erwin Anton Gutkind (1886-1968), who had been involved in the physical reconstruction of Europe after World War II. Three years prior to the conference, Gutkind had published a selection of aerial views entitled *Our World from the Air*,[29] which provided the basis for his Princeton lecture. Gutkind summarized the umbrella theme of the symposium, which was dedicated to the energies of man and nature, as the "twin agents of the perennial revolution which shapes and reshapes the face of the earth and the character of man's activities."[30]

For him, the aerial view "enabled mankind for the first time in its history to experience this interaction in all its innumerable ramifications."[31] Ultimately, he argued for the synoptic perspective incorporated in the aerial view to be substituted for an organic world view that provided a "purely analytical approach to numerous problems."[32] But he also wished to move beyond analytical applications of the aerial view, such as those used in land management, in order to harmonize its dual features. For him, the aerial view represented a double challenge in that it "forces us to see the transformation which man has wrought from the earth as a perennial revolution and to experience their dynamic unity as a whole. Without the synoptic view from the air this would be impossible, and without the trend toward unitary ideas we would fail to interpret correctly what we now can see."[33]

Gutkind's sense of a turning point in history was confirmed by George Peabody Gooch's (1873–1968) foreword to the 1952 edition of the book *Our World from the Air*. Therein, Gooch claimed that the aerial view had prompted an epistemic turn.[34] In the accompanying collection of images, this understanding of aerial photography was demonstrated by the juxtaposition of very different images, including those depicting soil erosion in the United States during the 1930s and a view of atomic annihilation after the bombing of

29 Gutkind, *Our World from the Air,* 1952.

30 Gutkind, "Our World from the Air," 1956, 1.

31 Ibid, 1.

32 Ibid., 1.

33 Ibid., 2.

34 Gooch, "Foreword," 1952.

Hiroshima. In Gutkind's own words, collages of this kind "lead the way, at least for the first few steps, towards a world-wide comprehension of the dependence of man on the successful adaptation of himself to his environment, and of his environment to his material and spiritual needs."[35] The same thought was echoed in Nathaniel Owings' 1969 book *The American Aesthetic*, in combination with aerial photographs picturing air pollution. Publications such as those by Gutkind and Owings confirm a shift in the use and interpretation of aerial views. Moreover, they reflect the rise of a wider, critical environmental discussion, which simultaneously affected aesthetic reflections.

GARNETT'S LAKEWOOD SERIES AS ENVIRONMENTAL ICON IN THE 1960S

In the early 1960s, William Garnett's Lakewood series was used to draw attention to a nascent environmentalism. Garnett took the photographs in 1949 while flying over Lakewood, a suburb of Los Angeles. One photograph shows standardized tract houses, like little boxes in the landscape, arranged in accordance with a strict pattern, each alike. By the early 1960s, the picture was frequently reproduced, most notably in two pivotal publications: Ansel Adams' *This Is the American Earth*, and Peter Blake's *God's Own Junkyard.*

This Is the American Earth was the first campaign with significant visual content, an exquisitely illustrated photography book in the Sierra Club exhibit format series. It contained not only Adams' own photographs, but numerous pictures commissioned by the Farm Security Administration. Its introduction offered a carefully written survey of the nation's relationship to the land since the era of exploration. The fourth chapter, "The Mathematics of Survival," contains four photographs by Garnett, presented in combination with an alarming text that pointed out the negative effects of civilization: "Hell we are building here on earth. Headlong, headless, we rush—to pour into air and water poisons and pollutions until dense choking palls of smog lie over cities and rivers run black and foul,—to blast down the hills, bulldoze the trees, scrape bare the fields to burn predestined slums; until city encroaches on suburbs, suburb on country, industry on all, and city joins city, jamming the shores, filling the valleys, stretching across the plains."[36] Three photographs from Garnett's Lakewood series (Figure 5) represented the degradation of the earth.

---

35 Gutkind, "Author's Preface," 1952.

36 Adams and Newhall, *This is The American Earth,* 1960.

Figure 5: William Garnett: Lakewood series, California 1950. In: Ansel Adams and Nancy Newhall, This is the American Earth (San Francisco: Sierra Club, 1958): 59.

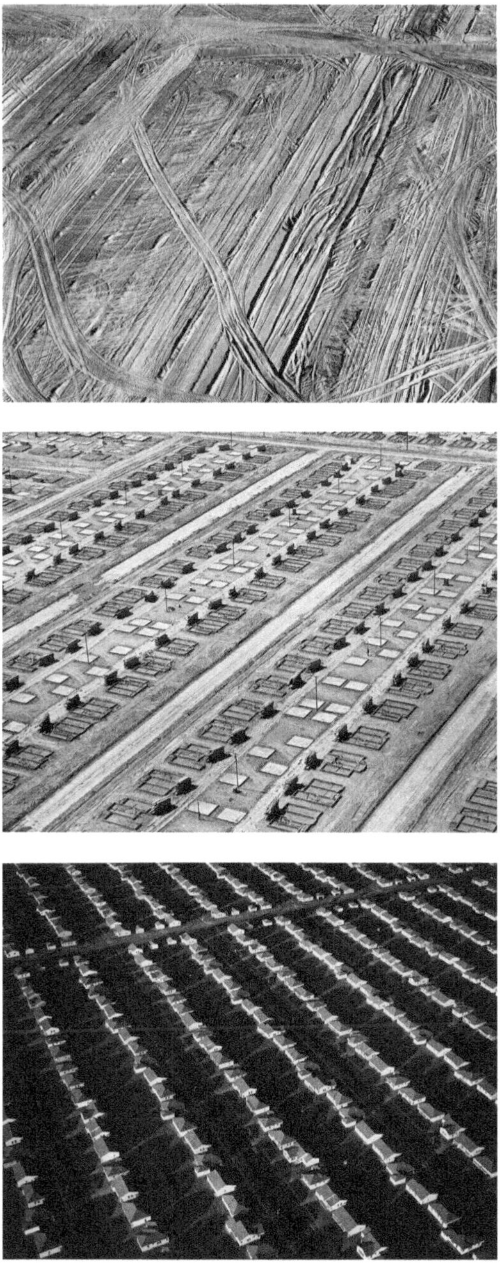

They showed leveled and bulldozed territory on which identical squares were created. The bigger images on the left presented a view of air pollution in Los Angeles.

Architect Peter Blake's *God's Own Junkyard*, with its damning subtitle, "The planned deterioration of America's landscape," also aimed at drawing attention to the dimensions of environmental degradation in the United States. The Lakewood series again served as the appropriate illustration. Blake presented all four of Garnett's aerial views and, in his conclusion, he branded the mutilation of landscapes wrought by cutting trees, planting billboards and telephone poles and bulldozing hills, or turning a place into a dump, as the destruction of the "future of civilization in America."[37] Additional aerial photographs enabled Blake to demonstrate the degradation of the land, and another photographic combination pointed to the direct correlation of population growth, suburban sprawl, and the increasing number of dumps produced by society, by presenting a view of a suburb next to another showing a car dump site.

Blake painted a dark picture of the environmental conditions in the United States and denounced the manner in which the once beautiful land had been turned "into the biggest slum of the face of the earth."[38] He considered contemporary manmade America a disgrace "of [...] vast proportions" and thought that only a "concerted national effort [could] now hope to return physical America to the community of civilized nations."[39] Interpretations such as these helped establish the Lakewood photographs as a symbol of landscape degradation and shifted the perception of the aerial view from an expression of utilitarian and progressive thinking towards an allegory of destruction.

Both publications used Garnett's Lakewood photographs in support of their thesis, but where *This Is the American Earth* used the image to set up a contrast with a generally less radical notion of the human relation to nature (the book also included many picturesque photographs), *God's Own Junkyard* was dedicated solely to showing the threat that pollution posed. However, both publications connected an emerging environmental conscience with a pictorial mode of representation. This was a trend that would be further developed, but not until some years later, with the photographic representation of environmental disasters and news coverage of incidents like the Santa Monica oil spill and the Cuyahoga River fires.

---

37 Blake, *God's Own Junkyard,* 1964, 69.

38 Ibid., 8.

39 Ibid., 8.

CONCLUSION

This analysis of the aerial views taken by the pioneers of aerial photography Sherman M. Fairchild and William Garnett illuminates the changing cultural use of photographs, which were not uncommonly the subject of opposing interpretations during the period from 1930 to the late 1960s. In the 1930s, the aerial view represented high modernity's euphoric, imperial worldview.[40] It was used as a tool to monitor land management, and furthered the belief in the abundance and manageability of American agriculture, as well as serving as a symbol of modernist visions. During the forty years that followed, the Cold War, the debate concerning American suburbia, and the nascent environmental movement altered the use of aerial views. This entailed a paradigmatic shift that extended into the 1960s; William Garnett's Lakewood series became a symbol of environmental degradation, helping reveal humanity's depletion of the earth's resources. By then, for contemporary interpreters, aerial surveillance of the earth had started to point to the negative environmental impacts of civilization. Not that this new awareness put an end to the expectation that technological advances would boost fertility; indeed, the aerial view was displaced by satellite images, offering an even better means of surveillance.

In this history of cultural vicissitudes, the same photographs have been used with multiple, and even contradictory intentions. Their visual expressiveness is closely bound up with framing narratives and a plurality of contexts, as has been alluded to by Vilém Flusser.[41] On the one hand, the aerial photograph offers visual evidence as a trace of the reality, which is then integrated into the discussion of environmental and climatic topics. On the other, its expressive aesthetic runs the risk of supporting a naïve technological determinism. In the American example, however, the aerial view prompted intellectuals to reflect on its subliminal meaning. This led, over the course of the twentieth century, to what was formerly considered an expression of modernist utopia being interpreted as quintessential evidence that raised awareness of destructive dystopias. Understood as a basis for critical investigation, the aerial view may help to mediate between human desires and nature.

40 For the definition of the term "High Modernism" see Scott, *Seeing Like A State,* 1998.

41 Flusser, *Into the Universe of Technical Images,* 2011.

BIBLIOGRAPHY

Abrams, Talbert. *Essentials of Aerial Surveying and Photo Interpretation.* New York: McGraw-Hill, 1944.

Adams, Ansel and Nancy Newhall. *This Is the American Earth.* San Francisco: Sierra Club, 1960.

Allum, J. A. E. *Photogeology and Regional Mapping.* New York: Pergamon Press, 1966.

American Geological Institute / Committee on Education. *Outstanding Aerial Photographs in North America.* Washington D.C.: American Geological Institute, 1951.

Bagley, James W. *Aerophotography and Aerosurveying.* New York: McGraw-Hill Book Company, 1941.

Blake, Peter. *God's Own Junkyard: The Planned Deterioration of America's Landscape.* New York: Holt, Rinehart, and Winston, 1964.

Boorstin, Daniel J. *The Democratic Experience.* London: Vintage, 1974.

Bredekamp, Horst, Birgit Schneider, and Vera Dünkel, editors, *The Technical Image. A History of Styles in Scientific Imagery.* Michigan: Michigan Press, 2014.

Brown, Harrison. *The Challenge of Man's Future: An Inquiry Concerning the Condition of Man During The Years That Lie Ahead.* New York: Viking Press, 1954.

Cosgrove, Denis. *Geography & Vision. Seeing, Imagining and Representing the World.* London: I. B. Tauris, 2010.

Cosgrove, Denis and William L. Fox. *Photography and Flight.* London: Reaktion Books, 2010.

Fleming, James Rodger and Vladimir Jankovic. *Klima.* Chicago: University of Chicago Press, 2011.

Flusser, Vilém. *Towards a Philosophy of Photography.* London: Reaktion Books, 2000.

—. *Into the Universe of Technical Images.* Minneapolis: University of Minnesota Press, 2011.

Gooch, G.P. "Foreword." In *Our World from the Air: An International Survey of Man and His Environment,* edited by Erwin Anton Gutkind. Garden City, N. Y. / Doubleday: British Institute of Sociology, 1952.

Gutkind, Erwin Anton. *Our World from the Air. An International Survey of Man and His Environment.* Garden City, N. Y. / Doubleday: British Institute of Sociology, 1952.

—. "Author's Preface." In *Our World From The Air. An International Survey of Man and His Environment.* Garden City, N. Y. / Doubleday: British Institute of Sociology, 1952.

—. "Our World from the Air: Conflict and Adaptation." In *Man's Role in Changing the Face of the Earth. An International Symposium,* edited by William L. Thomas. Chicago: University of Chicago Press, 1956.

Hart, Augustus Cecil. *Air Photography Applied to Surveying.* New York: Longmans / Green, 1943.

Hughes, Thomas P. *American Genesis.* Chicago: University of Chicago Press, 2004.

Jackson, J. B. "The Need of Being Versed In Country Things." *Landscape* 1, no. 1 (Spring 1951): 4.

Jones, Greta. *Science, Politics and the Cold War.* London: Routledge, 1988.

Lobeck, Armin Kohl. *Military Maps and Air Photographs; Their Use and Interpretation.* London: McGraw-Hill, 1944.

Mack, Pamela. *Viewing the Earth. The Social Construction of the Landsat Satellite System.* Cambridge: MIT, 1990.

Marx, Leo. *The Machine in the Garden: Technology and the Pastoral Ideal in America.* Oxford: Oxford University Press, 2000.

Miller, Victor C. *Photogeology.* New York: McGraw-Hill, 1961.

Monkhouse, F.J. *Landscape from the Air: A Physical Geography in Oblique Air Photographs.* Cambridge: University Press, 1960.

Mumford, Lewis. *Technics and Civilization.* Chicago: University of Chicago Press, 2010.

National Research Council / Highway Research Board. *Photogrammetry and Aerial Surveys.* Washington, 1966.

Orr Jr., Clyde. "Our Changing Environment and How to Survive it." In *Landscape: Human Geography of the Southwest* 9, edited by John Brinkerhoff Jackson, no. 3 (Spring 1960): 29.

Osborn, Fairfield. *The Limits of the Earth.* Boston: Little, Brown, 1953.

Owings, Nathanael Alexander. *The American Aesthetic.* New York: Harper & Row, 1969.

Perry, Claire. *The Great American Hall of Wonders: Art, Science, and Invention in the Nineteenth Century.* Washington D.C.: Smithsonian American Art Museum, 2012.

Powell, John Wesley. *Report on the Lands of the Arid Region of the United States, with a more detailed Account of the Lands of Utah.* Washington: Govt. Print. Off., 1879.

Scott, James C. *Seeing like a State: How Certain Schemes to Improve the Human Condition Have Failed.* New Haven: Yale University Press, 1998.

Smith, Harold Theodore Uhr. *Aerial Photographs and Their Applications.* New York: D. Appleton-Century Company, 1943.

Smith, Merritt Roe and Leo Marx. *Does Technology Drive History? The Dilemma of Technological Determinism.* Cambridge: MIT Press, 1994.

St Joseph, John Kenneth Sinclair. "The Scope of Air Photography." In *The Uses of Air Photography*, edited by J.K.S. St Joseph, 12–34. London: John Baker, 1977.

Thomas, William L. *Man's Role in Changing the Face of the Earth.* Chicago: Wenner-Gren Foundation for Anthropological Research and the National Science Foundation in cooperation with University of Chicago Press, 1956.

Tietge, David J. *Flash Effect: Science and the Rhetorical Origins of Cold War America.* Athens: Ohio University Press, 2002.

Turner, Frederick Jackson. *The Frontier in American History.* New York: Henry Holt & Co., 1921. http://www.gutenberg.org/files/22994/22994-h/22994-h.htm.

Walker, Frank. *Geography From The Air.* London: Methuen/New York: Dutton, 1953.

Weems, Jason. "Aerial Views and Farm Security Administration Photography." *History of Photography* 28, no. 3 (Autumn 2004).

Worster, Donald. "Facing Limits: Abundance, Scarcity, and the American Way of Life." Talk at the Rachel Carson Center's lunchtime colloquium, May 5, 2012.

IMAGE COPYRIGHTS & SOURCES

Figure 1: © Photo courtesy of Skinner, Inc. www.skinnerinc.com.

Figure 2: Garden City, N. Y. / Doubleday: British Institute of Sociology. Fig. 366.1952.

Figure 3: Garden City, N. Y. / Doubleday: British Institute of Sociology., Fig 32. 1952.

Figure 4–5 a,b,c: © Getty Museum.

Picturing Climate Control: Visualizing the Unimaginable

JAMES RODGER FLEMING

As a historian of science and technology working on climate issues and engaged in high-level discussions about climate policy, I am continually amazed, but not surprised that the discourse is so narrowly focused on current science and current issues; moreover that it is poorly and somewhat predictably illustrated, mainly by technical diagrams and stock images. A visual approach to history has much to offer, and as I demonstrate here for the case of climate control proposals, can help make the unimaginable at least visible.

INTRODUCTION

Climate is a vast, but unframed canvas presenting extensive opportunities and considerable challenges to those who seek to depict it. How do people (artists, scientists, and the general public included) gain awareness and understanding of phenomena that cover the entire globe and change constantly on time scales ranging from geological eras and centuries to decades, years, and seasons?[1] Although we lack the ability to observe the climate system in its entirety (as an astronomer might view a planet through the eyepiece of a telescope, or an artist might depict a person or a landscape on a canvas), yet individuals immersed in and surrounded by the phenomena set out to picture climate and climatic change. This paper reviews various approaches by artists and scientists to depict the climate in landscapes, icescapes, and particularly skyscapes and introduces the aesthetic and politically powerful idea of anti-skyscapes. It concludes with a discussion of picture(e)ing climate control—or visualizations of active interventions in large-scale and global processes.

1 Fleming, *Historical Perspectives,* 1998.

With geoengineers proposing to "fix" the climate system to reduce global warming, the number of wild proposals in this field has been proliferating. Shade the planet by launching a solar shield into orbit. Shoot sulfates or engineered nanoparticles into the upper atmosphere, turning the blue sky milky white. Fertilize the oceans, turning the blue seas soupy green. All of the proposals and their associated graphic images are fantastic imaginaries. Some are idealistic, perhaps hopelessly so, depicting benign, humorous, but bumbling Heath Robinson- and Rube Goldberg-like devices to cool the climate. Others are much more dark and ominous involving Dr. Strangelove-like schemes for military control of the environment. All involve fantasies that can indeed be pictured. They are anti-skyscapes on the planetary level.

This article examines the visualizations of climate control from the late nineteenth century to recent and current practices. It is a contribution to cultural and media studies paying special attention to visual imaginaries—William Gibson's phantasmagoric semiotic ghosts; "Fragments of the Mass Dream" as mediated through the not-so-rosy lenses of climate angst and apprehension.[2]

DEPICTIONS OF WEATHER AND CLIMATE

Scientists attempt to render the invisible visible by depicting events and processes using maps, charts, and animations. Iconographic historical efforts include the 1688 map of the trade winds by Edmund Halley, a depiction of global isotherms in 1817 by Alexander von Humboldt, the synoptic storm maps published by Elias Loomis in 1846, an 1863 telegraphic weather map from the Paris Observatory, and a map depicting average pressures in France in 1864.[3] This process has continued ever since, with atmospheric scientists attempting to fill a broad and abstract canvas with graphic representations ranging from the micro-scale of cloud physics to the macro-scale of global climate change and from time scales ranging from microseconds to eons.[4] Most of these efforts involve scientific charts and graphs, or more recently, animations that communicate most directly within the scientific community, with only brief

2 Ibid.

3 Images reproduced in Hellmann, *Meteorologische Karten, 1897* and discussed, with other examples, in Anderson, "Mapping Meteorology," 2006, 69–91; on Halley and Humboldt, see article by Grevsmühl in this volume.

4 See, for example, Monmonier, *Air Apparent*, 1999, and Grevsmühl, "A la Recherche de l'Environnement Global," 2012.

excursions into the popular realm. The most familiar involve rising temperatures, rising carbon dioxide levels, falling stratospheric ozone levels, shrinking extent of polar sea ice, or earth as an "angry red planet" depicted by false color, deep red global maps of temperature projected 50 to 100 years into the future.[5]

In an attempt to visualize and popularize climate change, and perhaps galvanize opinion in advance of the United Nations' Climate Summit in Copenhagen in December 2009, the international photo agency NOOR launched its "Climate Change" project. Images depict Poland's coal industry, East Africa's "climate refugees," Greenland's shrinking ice, a Brazilian rain forest, Canadian oil sands, underground coal fires in India, Russia's Nenets reindeer herders, and the coastline of the low-lying Maldives. The project was well intentioned, but ultimately non-decisive, as the Copenhagen summit itself failed to reach any real agreement. Never mind that all of these pictures arc about something other than climate. A newer effort, "Solutions," contains photographs of wind farms and solar collectors, among other, less iconic objects.[6] But still, climate remains largely unframed and elusive.

In 2009, a prominent climate scientist and an award-winning photographer teamed up to produce *Climate Change: Picturing the Science,* a union of scientific analysis and stunning photography aimed at illustrating the effects of climate change on the global ecosystem.[7] Although largely successful (the book won the Sierra Club's Ansel Adams Award), the challenge of depicting both the human and the geological dimensions of climate change is apparent, even in the cover image (Figure 1). While the focus of the image is on the dry cracked mud in the foreground, a product of recent dry conditions, the landscape itself was carved some 11 to 5 million years ago by the uplift and erosion of ancient rock layers deposited in sea sediments some 300 million years ago. Thus the horizontal layers in the mountain ridge depict vast environmental and climatic changes of sea level over millions of years. Wetter times indeed!

---

5 *The Angry Red Planet,* a 1959 science fiction film.

6 "NOOR," see also article by Heine in this volume.

7 Schmidt and Wolfe, *Climate Change,* 2009. See also article by Heine in this volume.

Figure 1: Original caption: "Lake Powell is an artificial reservoir behind the Glen Canyon Dam on the Colorado River. After seven years of drought in the southwestern states, it currently holds only 50 percent of the water it did in 1999. The rock is visibly marked by a 'bathtub ring,' a reminder of wetter times." Schmidt and Wolfe, Climate Change, 2009, cover image and page 106. (→ color plate 18)

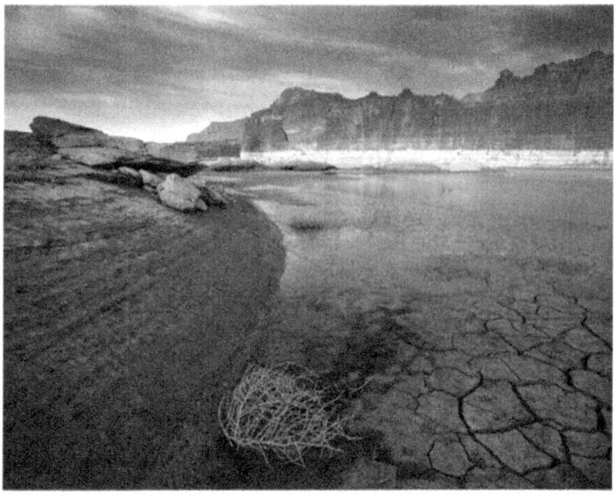

SKYSCAPES, AND ANTI-SKYSCAPES

Approximately half of every landscape painting, from the horizon up, is a skyscape, sometimes celebrated, often largely ignored, but unquestionably extremely difficult to render in most any media, given its transience and transcendence. A skyscape is a view of the sky or an artist's depiction of part of it. It is often strikingly beautiful, evanescent, and colorful, but it may also be an anti-skyscape—gloomy or menacing.[8] Undoubtedly, the genre has received less attention from scholars than its landed cousin. The vista of a natural skyscape, whether serene or ominous, is evanescent, rendered visible by reflected and refracted light. Blue (sky) and white (clouds) are only two of myriad colors that include the black of storm clouds and the near-infinite luminous pastels of sunrise and sunset, rainbow, aurora, and other meteors.

Anti-skyscapes can be infinitely more pernicious than even the most violent storm, which will surely pass. They were visible in the acidic skies of the Manchester factories, smog in London or LA, and dust storms over the prairies.

8 Fleming, "Skyscapes and Anti-skyscapes," 2013.

More recently, however, they are largely invisible. They are the polluted spaces of our planet, including now the planet itself, often rendered visible by scientists through idealized false color imagery and by artists in their own special ways. Anti-skyscapes such as miasmas, acid rain, the ozone hole, and carbon dioxide climate warming undergird narratives of environmental degradation, anxiety, despair, disease, dystopia, collapse, resilience, sustainability, collective action, inaction, heroic action, and even faith. No one has seen them, yet, in a way, everyone has. Today, highly politicized images of the atmosphere link scientific, environmental, technological, and aesthetic histories as they codify and symbolize fear, angst, urgency, control, and lack of control.

Cultural studies, science studies, and art history all provide perspectives on human-modified, but ubiquitous aerial spaces that are both ominous and threatening. The genre of skyscapes complements the terrestrial conversation and highlights emerging interests in environmental art, toxic airs, and climate engineering. Skyscapes are interdisciplinary, international, and intergenerational vehicles.[9] They are often overlooked. Anti-skyscapes are overlooked at our peril. Look up. And look out!

Skyscapes, depicting clear or threatening skies, can be found in most if not all art museums. Many are realistic, as in the work of Albert Bierstadt, John Constable, and George Innes; others, by George Catlin, William Turner, Claude Monet, and Alfred Stieglitz, are smoky and moody; recent works by contemporary artists Andrea Polli and HeHe make political statements. But what about the threatening sky we can't see? Invisible, potentially deadly miasmas afflicted the medieval body and its environment. The factory brought visible smoke, unhealthy particulates, and acid rain; the auto, too, transformed the blue skies over cities into brown clouds of photochemical smog, while emitting invisible carbon monoxide locally and invisible carbon dioxide globally.[10] Horrible poison gas clouds rolled over battlefields and choked both foe and ally, while clouds of tear gas clouds still accompany protests in city streets.

The atomic bombing of Hiroshima generated a new phenomenon—black rain, the remains of an incinerated city falling back to earth, but it also released the nuclear alchemy of invisible fallout on the world. Bomb blasts over the

9 For example, John Constable's work is interdisciplinary, linking art and science, Andrea Polli and HeHe convey global messages, and, as is the case for many artists, Alfred Steiglitz's photographs are interpreted differently by different generations. A recent article by Polli, "Who Owns the Air?" places her work in conversation with historians of chemistry and environmental historians.

10 See also Knebusch, "Art and Climate (change) Perception," 2008.

deserts and oceans soon gave way to bomb blasts in space designed to disrupt Earth's electrical and magnetic environment.[11] This was not the only alchemy, however, for the subtle but persistent cloud-seeding agent silver iodide, designed to "trick" a cloud into precipitating, soon became a hoped-for means for the military to control the sky.[12] As cool breezes of air-conditioning filled our homes and workplaces, seemingly inert and invisible chlorofluorocarbon molecules began their devastating ascent into the stratosphere. Now, because of stratospheric ozone depletion, children are taught to fear exposure to the clear blue sky. How many artists of earlier eras have depicted this emotional response?

CONTEMPORARY SKYSCAPES

There is a synergy between environmental activism and contemporary art as they co-evolve with technology, and as new modes of depicting the sky become possible. The above-mentioned contemporary artists Andrea Polli (a professor at the University of New Mexico) and HeHe (a collaborative French duo Helen Evans and Heiki Hansen) are expanding the boundary between science, technology, and environmental art. Polli has often collaborated with atmospheric scientists to depict the politics and reality of climate change and pollution through art.[13] Her work "Particle Falls," completed in collaboration with Chuck Varga, is an urban, environmentally motivated piece that renders the invisible visible. A blue waterfall projected by laser light onto the side of a large building in San Jose, California is linked to an interactive sensor that measures the amount of particulate matter in the air and can turn the beautiful waterfall into a yellow chaotic event with a puff of smoke from a cigarette or the fumes from a passing bus.[14] Observers at street level find it highly interactive.

HeHe have been using tactics similar to Polli in order to extend the dialogue on eco-consciousness to a larger scale. Their work, "Nuage Vert", or "Green Cloud" uses lasers to create a green outline of the pollution emitted from Helsinki's Salmisaari power plant. Although projected into the sky, the critique of "carbon emissions" is decentered from being solely an atmospheric malady,

11 Fleming, "Iowa Enters the Space Age," 2011.

12 Fleming, *Fixing the Sky,* 2010.

13 Polli, "Who Owns the Air?" 2004.

14 "Particle Falls" was installed in San Jose, California in 2010 and in Philadelphia, Pennsylvania, 2013.

to a complex and interconnected system mixing earth, wind, energy, and sky. HeHe notes:

"The physical dimension of the Salmisaari site, inhabits a special position in Helsinki; physically, visually and metaphorically. The vertical vectors of the architecture connect the underworld with the sky: the underground coal storage tunnels descend to 126m below sea level (the deepest point in Helsinki) whilst its chimney reaches 155m into the sky with its cloud disappearing into the lower atmosphere."[15]

The skyscape then becomes not a symbol of nature, industry, or progress, but the realization and template for the interconnectedness of both terrestrial and atmospheric pollution. It becomes a site for artistic and environmental activism. More than a naturalistic horizon, the mobility of air itself decenters it from a simple point source, and in this way, political projects can be seen to apply to individuals and cities, not just to the policy makers.

HeHe's work is ironic, in the sense that it attempts to render aesthetic, using laser light, a human-made cloud, yet it is also incredibly complex in blurring the line between art and activism, technology and aesthetics, and the public and private spheres. The project was collaborative in the sense that HeHe worked with owners and managers of the power plant to gain data on energy consumption, with scientists to configure the actual lasers, and with the Helsinki public through a poster and sticker campaign for an "unplug" event in which citizens reduced their power intake. As the levels of energy use were documented to decrease, HeHe increased the size of the laser cloud to show their progress. The sky then becomes the ultimate, unifying template to spread a public message, for as Helen Evans notes, "It shifts the discourse about climate change and carbon emissions from abstract immaterial models based on the individual, to the tangible reality of urban life."[16] The collaborative method also becomes a model for policy change, as artists create an equally aesthetic and social message, concentrating on the need for a community effort (the public, to reduce consumption, the policy makers, to push for greener bills, scientists, to create greener technology) in order to enact social change and promote environmental protection.

---

15 "Gallery: Nuage Vert, Green Cloud [...]," http://inhabitat.com/green-cloud-hehe-helsinki-environmental-art/green-cloud-2/.

16 Evans, "Nuage Vert," 2008.

CLIMATE CONTROL IN VISUAL CULTURE

It is time now to take a critical look at some iconic images of active interventions in large-scale and global processes and how they help visualize the absurdity of climate control through geo-engineering. A one-panel cartoon can often say as much, if not more, than a much longer technical discussion; and it can say it with visceral emotion. The general situation is clearly epitomized by the image of a white Western technocrat manipulating the levers of a weather control machine while looking to the sky (Figure 2).

Figure 2: Technocrat pulling the levers of weather control (Dust jacket from Fleming, Fixing the Sky, 2010. (→ color plate 19)

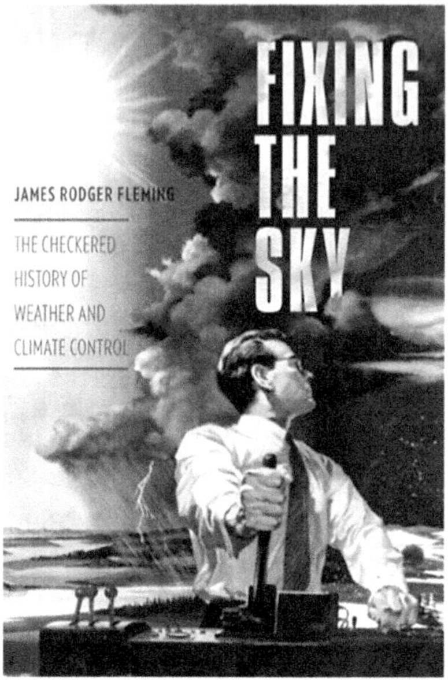

This image, by illustrator Fredrick Seibel, appeared on the cover of *Collier's* magazine in 1954. In the magazine Howard T. Orville, the chair of the presidential Advisory Committee on Weather Control, provided a vision of weather control being used as a weapon of war. Orville wrote: "We might deluge an enemy with rain to hamper a military movement or strike at his food supplies by withholding needed rain from his crops [...] But before we can hope to achieve all the *benefits* I have outlined, hundreds of meteorological unknowns must be solved at a cost of possibly billions of dollars."[17] Although speculative and wildly optimistic, such scenarios representing an official source helped fuel a weather race with the Soviet Union.

Images symbolizing both hubris and the "technological fix" applied to weather and climate control are rooted in antiquity. The classical story of Phaethon warns against human attempts to manage solar radiation. His blunder, disgrace, and ultimate demise involved trying to control his father Helios's Sun chariot—that which he could not control. The inexperienced youth was a poor driver and the chariot veered out of the zodiac. Phaethon, losing control completely, dropped the reins. The horses, when they felt the reins loose on their backs, dashed headlong, and unrestrained went off into unknown regions. Phaethon's recklessness caused extensive damage to the Earth before Zeus shot him out of the sky (Figure 3).

Figure 3: Phaeton, from the series The Four Disgracers (1588) by Hendrick Goltzius (Netherlandish, 1558–1617).

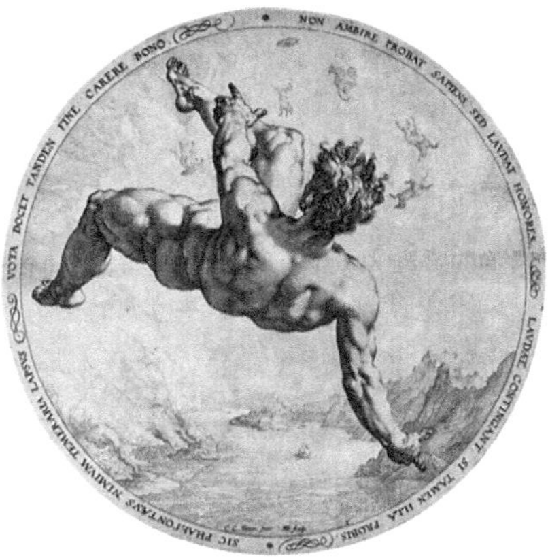

In 2007, the noted meteorologist Kerry Emanuel invoked the story of Phaethon to frame a short discussion of contemporary climate change science and politics. Emanuel, widely known and respected for his hurricane studies, called attention to a growing scientific consensus on climate change, yet admitted pointedly that "we are [...] conscious of our own collective ignorance of how the climate system works."[18] Emanuel ended his essay by returning to Phaethon, "Like it or not, we have been handed Phaethon's reins, and *we will have to learn how to*

17 Orville, "Weather Made to Order?," 1954, 25–26.

18 Emanuel, *What We Know About Climate Change,* 2008, 53.

control climate if we are to avoid his fate."[19] Unbelievably, he thus advocated repeating Phaethon's blunder. While doing nothing about climate change is out of the question, we could try to overreach and do too much.

Military images of shooting at the sky, like those of Orville from the 1950s, also have deep historical roots. Here, medieval archers have been ordered by the king to do something about a threatening hailstorm (Figure 4).

Figure 4: Medieval hail archers. (Olaus Magnus, Historia de gentibus septentrionalibus, 1555)

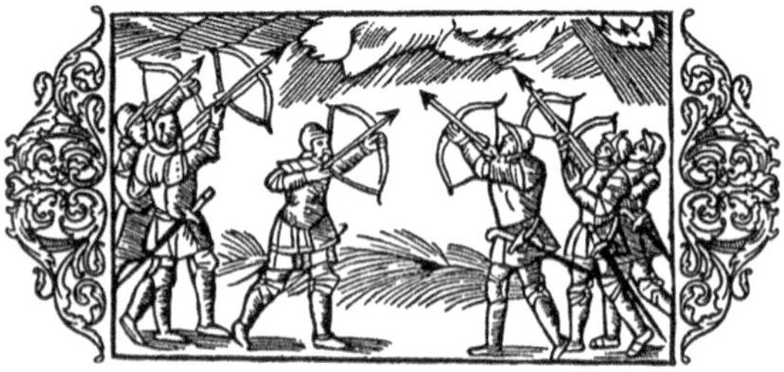

The theme continues through history: If something is wrong with the sky, shoot at it. This is true in the US National Academy of Sciences report, *Policy Implications of Greenhouse Warming: Mitigation, Adaptation, and the Science Base* (1992), which recommends using naval guns to shoot dust into the stratosphere to increase the reflection of sunlight. I am not kidding. Cartoonist John Ireland captured the pathetic and darkly pathological humor of the idea in his one-panel cartoon (Figure 5).

Washington insider Robert A. Frosch—a vice president of General Motors Research Labs, former deputy director of the Advanced Research Projects Agency, former assistant secretary of the navy for research and development, and former administrator of NASA—spearheaded the geoengineering aspects of the National Academy study. Frosch called his proposal to bombard the stratosphere using an array of naval guns "designer volcanic dust put up with Jules Verne methods."[20] He envisioned each $1 million, 16-inch gun being able to fire 1 ton of sulfate or aluminum oxide into the stratosphere about every ten minutes. Each barrel would need replacement after 1,000 to 1,500 shots. Thus a single cannon would have a useful life of less than two weeks, and a total of 300,000 cannons would be needed for a forty-year program!

19 Ibid., 69, [my emphasis].

20 Fagin, "Tinkering with the Environment," 1992, 7.

Figure 5: Shooting dust into the stratosphere to offset global warming, an eccentric proposal by the National Academy of Sciences in its report published in 1992. Nobel laureate Paul J. Crutzen revived the idea in 2006. (Cartoon by John Ireland, in Geographical Review, May 1992).

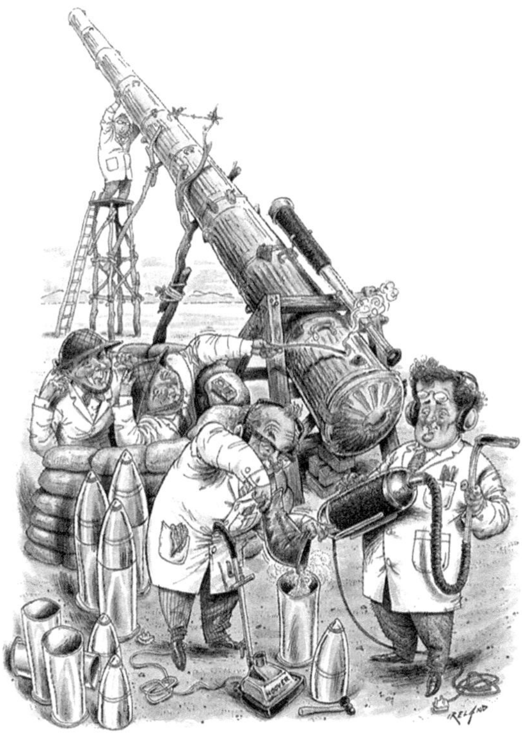

Frosch was aware that damaging side effects could result from such a program, such as stratospheric ozone destruction, widespread drought, low level pollution, or unacceptable atmospheric haze, but he did not emphasize that. He concluded that "the rifles could be deployed at sea or in military reservations where the noise of the shots and the fallback of expended shells could be managed."[21] Could such a long-term and violent bombardment be sustained without any accidents or other side effects? Is declaring war on the stratosphere the best mitigation strategy?

Naval artillery is only one of the many "muscular" ways to declare "war" on global warming by using military equipment. The cartoon depicted in figure 6 alludes to a 2006 proposal by the geoengineer Lowell Wood to attach a long

21 National Academy of Science, *Policy Implications*, 1992, 817–19.

hose to a nonexistent but futuristic military High Altitude Airship (a Lockheed-Martin—U.S. Defense Department stratospheric super blimp now on the drawing board with some twenty-five times the volume of the Goodyear blimp) to "pump" reflective particles into the stratosphere. According to Wood, "Pipe it up; spray it out!"[22] Wood has worked out many of the details—except for high winds, icing, and accidents, since an HAA is likely to wander as much as 100 miles from its assigned station. If the geoengineers cannot keep it up, however, imagine a 25-mile phallic "snake" filled with ten tons of sulfuric acid ripping loose, writhing wildly, and falling out of the sky. Carol Cohn said it best in her classic article "Sex and Death in the Rational World of Defense Intellectuals": "The dominant voice of militarized masculinity and decontextualized rationality speaks so loudly in our culture, it will remain difficult for any other voices to be heard [...] until that voice is delegitimated."[23] The geoengineers have been playing such games with the planet since computerized general circulation models were first developed. While this kind of research will undoubtedly continue, it should remain indoors between consenting adults. What have to be aired are the underlying assumptions.

Figure 6: "Screwing (with) the Planet," as interpreted by Henning Wagenbreth / New York Times, October 24, 2007.

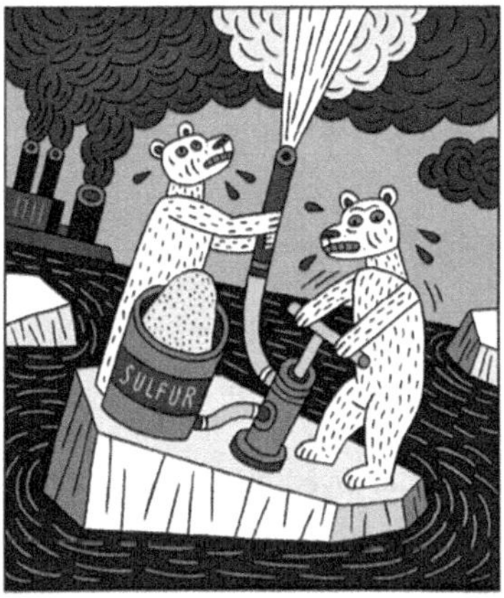

An editorial cartoonist for the *New York Times* captured the essence (and the absurdity) of this proposed technique (Figure 6). Two overheated polar bears are feverishly trying to pump sulfur into the air, but they seem to be having trouble

22 Wood, "Stabilizing Changing Climate," 2006.

23 Cohn, "Sex and Death," 1978, 717–18.

keeping their hose erect, especially if their ice floe shrinks any further. And whose warships are those in the distance? Do they carry Frosch/Crutzen sulfate cannons, or are they trying to stop the geoengineering? Russian opinion has long favored an open Arctic Ocean, and some scientists, including the noted Soviet geoscientist Mikhail Ivanovitch Budyko, believe that the beneficial effects of global warming might "pep up" cold regions and allow more grain and potatoes to be grown, making the country wealthier.[24] Better check with Vladimir Putin before we screw (with) the Arctic.

Another unsustainable climate change metaphor derives from scientists who claim to be planetary physicians.[25] Schmidt and Wolfe organized the entire text of their book, *Climate Change: Picturing the Science*, 2009 around just such medical metaphors:

- Symptoms (Taking the Temperature of the Planet)
- Diagnosis (The Prognosis for the Climate)
- Possible Cures (Preventative Planetary Care)

Nature magazine, reporting on the 1995 Second Assessment Report of the Intergovernmental Panel on Climate Change, printed just such a medical cartoon (Figure 7). In the image, three middle-aged white guys wearing ties (Bert Bolin, John Houghton, and Luiz Gylvan Meira Filho) are diagnosing Earth's CO_2-induced fever.[26]

Africa and India are clearly suffering, with the thermometer inserted somewhere near Madagascar in the Indian Ocean. La Jolla, California is another possible thermometer insertion point 180 degrees away. But diagnosis is not cure. If Earth needs treatment, what is the proper medicine and who signs the consent form for the planet? Or does Earth need invasive surgery? James Lovelock once likened geoengineering to crude planetary surgery, as practiced by the butcher/barber surgeons of old. According to Lovelock, "Our ignorance

24 Budyko, "Global Climate Warming," 1998.

25 Somerville, "Medical Metaphors," 2006.

26 Bolin was a Swedish meteorologist who served as the first chairman of the Intergovernmental Panel on Climate Change (IPCC); Welsh scientist Houghton served as the lead editor of the first three IPCC reports; and Filho, an aerospace engineer and meteorologist, was Secretary for Science and Technology Policy in the Ministry of Science and Technology in Brazil.

of the Earth system is overwhelming [...] the best option may simply be kind words and letting Nature take its course."[27]

Figure 7: Cartoon on diagnosing climate change, from Nature 455 (9 October 2008). (→ color plate 20)

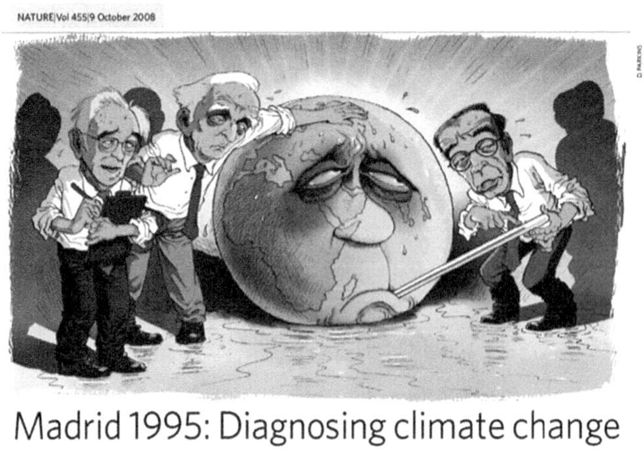

My final example of picturing climate control comes from the logo of a conference at the Massachusetts Institute of Technology in 2009, "Engineering a Cooler Earth: Can We Do It? Should We Try?" A male hand, god-like in scale, is on the thermostat (Figure 8). The thermostat is located precisely "nowhere," but perhaps in outer space. It indicates the temperature is 73 F, but the hand is turning the thermostat back to 54 F or five degrees cooler than Earth's long-term average.

If you look closely at the center of the dial, the land mass is centered over New Mexico—a symbolic choice due to the checkered history of the small towns located there. In the 1940s, Alamogordo was the site of the first atomic bomb test—an event that ushered in a new age of geoengineering fantasies, Roswell was ground zero for reports of alien spacecraft sightings, and overenthusiastic scientists in Socorro claimed a single silver iodide generator there had modified all the weather across the entire continent.

27 Lovelock, "A Geophysiologist's Thoughts," 2008, as quoted in "Medicine for a Feverish Planet: Kill or Cure?" Guardian, September 1, 2008.

Figure 8: Logo from the 2009 Massachusetts Institute of Technology conference, Engineering a Cooler Earth: Can We Do It? Should We Try?

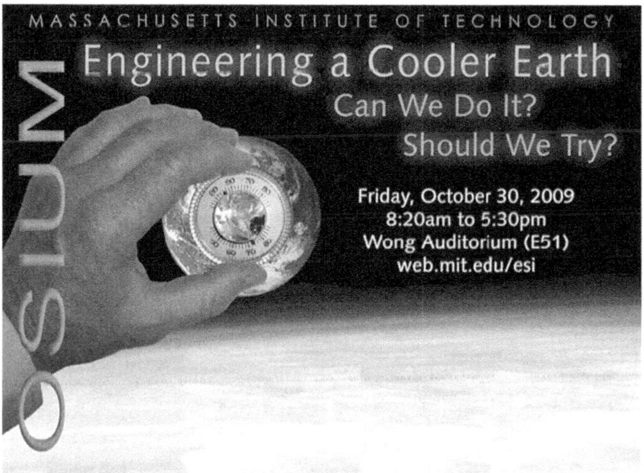

CONCLUSION

Climate is hard to depict, since it represents a set of diverse phenomena on a vast, but unframed canvas. Anti-skyscapes perhaps offer more hope, as they lie fully within the landscape tradition of fine art. This venerable tradition has recently been the beneficiary of new artistic talent which is both aesthetically and politically engaged. Climate control mechanisms, while easier to visualize, remain as yet in the realm of cartoonists' imaginations and are yet to attract serious artistic attention. Yet the quest to control or "fix" the sky is both powerful and perennial. Perhaps geoengineering art, more powerfully than cartoons, can help bridge the gap between knowledge and action, or more accurately between activism and opposition.

BIBLIOGRAPHY

Anderson, Katharine. "Mapping Meteorology." In *Intimate Universality: Local and Global Themes in the History of Weather and Climate,* edited by James Rodger Fleming, Vladimir Jankovic, and Deborah R. Coen. Sagamore Beach: Science History Publications, 2006.

Budyko, Mikhail Ivanovitch. "Global Climate Warming and Its Consequence." Blue Planet Prize 1998. Accessed July, 2012. www.ecology.or.jp/special /9902e.html.

Cohn, Carol. "Sex and Death in the Rational World of Defense Intellectuals." *Signs* 12 (1987): 687–718.

Emanuel, Kerry. *What We Know About Climate Change*. Cambridge: MIT Press, 2007.

Evans, Helen. "Nuage Vert." *Cluster Magazine* 7 (May 2008). hehe.org.free.fr/-hehe/texte/nv/index.html.

Fagin, Dan. "Tinkering with the Environment." *Newsday*, April 13, 1992.

Fleming, James Rodger. *Fixing the Sky: The Checkered History of Weather and Climate Control*. New York: Columbia University Press, 2010.

—. *Historical Perspectives on Climate Change*. New York: Oxford University Press, 1998.

—. "Iowa Enters the Space Age: James Van Allen, Earth's Radiation Belts, and Experiments to Disrupt Them." *Annals of Iowa* 70 (2011): 301–24.

—. "Skyscapes and Anti-skyscapes: Making the Invisible Visible." In *The Anti-Landscape*, edited by David Nye and Sarah Elkund, 27–42. Amsterdam: Rodopi, 2013.

Gibson, William. *The Gernsback Continuum*. Accessed December 11, 2013. lib.ru/GIBSON/r_contin.txt_with-big-pictures.html.

Grevsmühl, Sebastian Vincent. "A la Recherche de l'Environnement Global: de l'Antartique à l'Espace et Retour. Instrumentations, Images, Discours, et Metaphores." PhD diss., Ecole des Hautes Etudes en Sciences Sociales, 2012.

Hellmann, Gustav. *Meteorologische Karten 1688, 1817, 1846, 1863, 1864. Sechs Tafeln in Lichtdruck mit einer Einleitung*. Berlin: Asher, 1897.

Knebusch, Julian. "Art and Climate (change) Perception: Outline of a Phenomenology of Climate." In *Sustainability. A New Frontier for the Arts and Cultures,* edited by Sacha Kagan and Volker Kirchberg, 242–261. Frankfurt am Main: Verlag für Akademische Schriften, 2008.

Lovelock, James. "A Geophysiologist's Thoughts on Geoengineering." *Philosophical Transactions of the Royal Society,* A 366 (2008): 3883–3890.

Monmonier, Mark. *Air Apparent: How Meteorologists Learned to Map, Predict and Dramatize Weather*. Chicago: University of Chicago Press, 1999.

National Academy of Science. *Policy Implications of Greenhouse Warming: Mitigation, Adaptation, and the Science Base*. Washington, D.C.: National Academy Press, 1992.

Orville, Howard T. "Weather Made to Order?" *Collier's*, 28 May 1954: 25–29.

Polli, Andrea. "Who Owns the Air? Contemporary Art Addresses the Climate Crisis." In *Toxic Airs: Body, Place, Planet in Historical Perspective,* edited by James Rodger Fleming and Ann Johnson. Pittsburgh: University of Pittsburgh Press, in press.

Schmidt, Gavin, and Joshua Wolfe. *Climate Change: Picturing the Science*. New York and London: W.W. Norton, 2009.

Somerville, Richard. "Medical Metaphors for Climate Issues." *Climatic Change* 76 (2006): 1–6.

Wood, Lowell. "Stabilizing Changing Climate via Stratosphere-Based Albedo Modulation." PowerPoint presentation at NASA-Ames—CIW Workshop on Managing Solar Radiation, 2006.

IMAGE COPYRIGHTS

Images of Feasibility: On the Viscourse of Climate Engineering

ISABELL SCHRICKEL

This article investigates the esthetics of a certain type of image that has accompanied the climate engineering discourse for some years, lending it a degree of momentum on a visual level: a dynamic that should not be underestimated. The visualizations of the proposed large-scale interventions in the climate system in order to avoid climate change are discussed in their original context and subjected to a critical iconic analysis. This leads to the hypothesis that images function as agents of normalization of the discourse and of the familiarization of technical and natural processes. They represent a type of contemporary techno-scientific imagery that is characterized to a lesser extent by its epistemic functions than by how it opens up commercial and political scopes.

I

Within a few years, the landscapes in different European regions have changed dramatically. The recent transformations are not so much the result of the continuing growth of industrial infrastructures and the expansion of transportation and information networks but of the recent pervasion and occupation of the natural world for the purpose of extracting renewable energies: hydroelectric power stations in Iceland, solar fields in southern Spain or wind farms on the mountain ranges of the central highlands in Germany. Specific landscape features like mountains and coastal waters, fields and rivers are mobilized to serve the ever-mounting energy requirements of post-industrial society. These energy requirements of expanding infrastructures and virtual backstage architectures are needed to produce knowledge about the complex interactions between society, economy and environment, between present and future, and between nature and culture, and thereby determine our situation. It is predicted that by the year 2030, the Information and Communications Technologies (ICTs) will have a higher demand for electricity than the world's current total annual consumption.[1] The age of fossil fuels is characterized by selective and more local exploitation of energy, with energy requirements and land requirements kept distinct from each other, which is maybe a reason why we mistakenly perceive the technologies that emerged in that age, such as ICTs, as independent and decoupled from environmental resources. In the future, however, when peak oil and climate change enforce the priority of biofuels and renewable energies, land requirements and energy requirements will once again coincide—like in the pre-industrial Physiocracy.[2] The renaissance of land use for the extraction of energy becomes particularly visible in the monopoly of landscapes for oilseed rape and corn: agrarian monocultures for the production of biofuels. The demand for energy and other resources, and with this the material, political-economic disposition of the knowledge infrastructures and media technologies that contemporary societies and sciences are based on, is becoming more and more critical. This development might be an indication of the massive upheaval of the economic and political relation between technology and environment that human geographers, cultural historians and media theorists see occurring.[3] We have become accustomed to the sight of land use for energy extraction surprisingly quickly, although it poses an esthetic menace to various cultural landscapes both in the eyes of local citizens and for the UNESCO list of world heritage sites.[4] Today, the sites of the extraction of renewable energies seem to be a low-key aspect of contemporary logistics landscapes. In contrast to road and rail networks, harbor architectures and landscapes of consumption and disposal they even have a positive connotation, because something of ecologically sustainable value is being generated here. Instead of environmental degradation and pollution, here an economical revaluation and re-evaluation is taking place in a locally restricted area, for the benefit of the general public. Such eco-technological products symbolize the belief in green growth. They are to a large extent reversible and mobile and seem smart as they flexibly accumulate the forces of nature, exploiting an inexhaustible energy resource. And they do so almost noiselessly and apparently

1 Fettweis and Zimmermann, "ICT Energy Consumption," 2008.

2 Hornborg, "Accumulation," 2012, 19.

3 cf. Hörl, *Die Technologische Bedingung,* 2011; Parikka, *Medianatures,* 2011.

4 The UNESCO, for example, insists that wind turbines must not be visible at the horizon around the Wartburg Castle in Thuringia or Mont St. Michel in Brittany, otherwise the status as a World Heritage site may be revoked.

with no other harmful emissions: intelligent background technologies with a cool esthetic.

II

The representations of the futuristic visions that this article deals with call regenerative energy technologies to mind on a visual level, even though they are based on a completely different eco-technological paradigm. Whereas renewable energies aim to contribute to an eco-friendly economy, climate engineering technologies manipulate ecological conditions directly, thereby inverting the relationship between human and environment to an unheard of extent. Currently such technologies are being discussed to an increasing degree in the light of how anthropogenic climate change is coming to be perceived as highly threatening, and are advertised with a certain type of image. For several years now, the discussions on deliberate and large-scale technical interventions in the climate system to control climate change have been illustrated in scientific reports, political communication measures, popular science magazines, documentary channels,[5] etc. These images are the most visible witnesses of the bizarre techno-scientific earthing of much older futuristic projects like Hermann Oberth's space mirrors from the 1920s or Mikhail A. Gorodskiy's and Valentin Cherenkov's ideas for a high-altitude dust-ring similar to that encircling Saturn for the purpose of melting the Arctic glaciers and eliminating the seasonal climate differences from the late 1950s.[6] A fundamental shift in knowledge politics provides the precondition for the contemporary discussions: the climatic condition on planet earth has been more or less explicitly transformed from an externally preexistent and sustaining environmental reality into a matter of human activity and concern, not least due to the discovery of the anthropogenic influence. Climate policy became possible and absolutely essential because climate is no longer perceived as a given with an historically stable background—at any rate, a stability that could be projected as lying on the shaft of a hockey stick while its future lies on its blade[7]—but as an issue that is scientifically mapped and technologically explicated. In Peter Sloterdijk's words, concerns about the climate

5 cf. "Discovery Project Earth," a TV series of nine episodes on different climate engineering approaches.

6 Oberth, *Die Rakete zu den Planetenräumen*, 1923, 86–89; Rusin and Flit, *Man Versus Climate*, 1966, 60–67.

7 cf. Schneider, "Image Politics: Picturing Uncertainty," 2011, 191–209.

testify to "environment reversal."[8] Climate engineering appears to be the coherent answer. This was a premise for the atmospheric chemist and Nobel laureate Paul J. Crutzen asking in 2006 whether albedo enhancement through the injection of sulfur particles into the stratosphere might not solve the policy dilemma of climate change that is the result of the overly complex involvement of diverse political stances.[9] With the renewal of the term "anthropocene" in the year 2000, Crutzen also provided a discursive framework that even seems to legitimize research on technological interventions in order to avoid the "climate catastrophe."[10] Since that time, the number of scientific articles on the possibilities of climate engineering has increased tremendously. Scientists and engineers began to cooperate with private business joint ventures in order to work on a variety of technical approaches and to develop concepts and platforms for marketing. Additionally they provided with their expertise to think-tanks whose reportedly independent surveys and reports often attract interest in the political arena and thereby influence science policy. They expedite the discourse while developing a distinctive language, terms like "climate remediation" for example, that seem to anchor climate engineering action in a broader climate protection portfolio.[11] Some protagonists in this field actually occupy positions in all of these areas, and one cannot avoid the impression that a downright complex is being consolidated to prove that the world and its established political-economical order can be maintained even under changing environmental conditions.

An investigation of the network of these protagonists quickly reveals the conflicting interests and entanglements between different spheres. It shows to what extent climate change and the computer-simulated scenarios are already perceived as investment and marketing opportunities. Big corporations and super-rich individuals fund the initiatives and start-ups of entrepreneurial scientists to get some of their ideas and designs ready for the market. They employ engineers, have patents registered and generate publicity. The engagement of Bill Gates in this field and the formation of the company "Intellectual Ventures" is an instructive example of such constellations.[12] Beside the exploration of the bio-, geo- and energy technologies of the future, the company's business model

8 Sloterdijk, *Sphären,* 2004, 331.

9 Crutzen, *Enhancement by Stratospheric Sulfur Injections,* 2006.

10 Crutzen and Stoermer, "The Anthropocene," 2000, 17–18.

11 The Bipartisan Policy Center, "Task Force on Climate Remediation Research," 2011.

12 Gupta, "Geoengineering the Planet?" 2010, 31–37; Vidal, "Bill Gates Backs Climate Scientists," 2012.

seems to be the large-scale acquisition and licensing of patents and trading with them. It is called patent-trolling—a completely virtual business that causes enormous damage for national economies.[13] It is not uncommon for precisely these entrepreneurial climate researchers to work for the think-tanks that generally provide political protagonists with the first impression of the latest developments in the field to the political actors, and naturally they advocate the exploration of the climate engineering's potential. Ken Caldeira, who runs a lab at the Department of Global Ecology at the Carnegie Institution for Science in Stanford, is a scientist entangled in such a manner. He manages the Fund for Innovative Climate and Energy Research (FICER),[14] founded and funded by Gates, and thereby is able to codetermine if and what kind of climate engineering approaches are explored, as this engagement guarantees him a seat on a wide range of consultancy panels. He also worked for "Intellectual Ventures", he published some scientific articles together with its co-founder Nathan Myhrvold[15]—and holds a range of climate engineering patents. This complex of scientific expertise, entrepreneurial action and professional publicity work attempts to create some kind of mainstream legitimacy for the discourse, and to a certain extent, it does contribute to a normalization of discussions about the possible application of these large-scale technologies. In part due to the stimulus of these private initiatives, the American and European governments commission their technology assessment offices to evaluate the proposed schemes for solar radiation management (SRM) or carbon dioxide removal (CDR) and their feasibility in order to adjust their research and economic policy.[16] And although governments seem to be the most likely clients, some of the discussed climate engineering options

13 cf. Ewing and Feldman, "The Giants Among Us," 2012; Mirowski, *Science Mart,* 2011, 147–152; Bessen and Meurer, "The Direct Costs from the NPE Disputes," 2012.

14 Ken Caldeira: "Fund for Innovative Climate and Energy Research." Accessed December, 2013. http://dge.stanford.edu/labs/caldeiralab/FICER.html.

15 They wrote for example about the delayed benefits of a low-carbon energy system that reversely relativizes its attenuating impact on climate change: Myhrvold and Caldeira, "Greenhouse Gases," 2002, 014019.

16 Royal Society, "Geoengineering the Climate," 2009; United States Government Accountability Office / Center for Science, Technology, and Engineering, "Climate Engineering," 2011; Rickels et al., *Large-scale Intentional Interventions into the Climate System?,* 2011.

are cheap enough to be deployed by wealthy and capable individuals or corporations.[17]

III

The following reflections concern the "viscourse"[18] emerging from this field. More precisely, they analyze how special types of images beyond data visualizations and global climate maps, the icons of climate change and its caricatures (that are also presented in this book) yield a tempting scope for thought experiments for various actors from the fields of science, technology and politics. In May 2008, a press release on geoengineering was published by the Lawrence Livermore National Laboratory (LLNL). The release came along with a schematic overview of different climate engineering approaches that have been discussed in the last years within the techno-scientific community (Figure 1).

The LLNL was co-founded decades ago by Edward Teller, who was provided in the mid 1990s with the environment to draw up his late work on global climate manipulation—although it was at that time about the prevention of a new ice age that he saw coming. However, the mechanisms presented in the overview embody two different approaches for managing global warming scenarios. The first constantly compensates the consequences of the anthropogenic greenhouse effect by managing the incoming solar radiation (SRM) thus maintaining a stable temperature while disregarding the cause for its risings. The second approach relies on analogies to natural carbon dioxide sequestration by physical, chemical and biological processes in the earth system that regulated the atmospheric concentration of the heat-trapping gas in the geological past. One finds in the overview the visualization of the idea to inject aerosols (e.g. tiny sulfur particles) into the stratosphere to cause more reflexivity of solar radiation and to lower the temperature on the planet. This scheme was revived by Paul J. Crutzen, after it was originally launched by John von Neumann and others in the 1950s.[19] Furthermore one sees adjustable mirrors that are installed in the orbit, more precisely at the so-called Lagrange point between earth and sun, where the gravitational forces of each body are annihilated, allowing a static positioning. A

17 Victor et al., "The Geoengineering Option," 2009, 72.

18 Knorr-Cetina, "Viskurse in der Physik. Konsensbildung und Visuelle Darstellung," 2001.

19 von Neumann, "Can we Survive Technology?" 1963; Crutzen, "Enhancement by Stratospheric Sulfur Injections," 2006.

myriad of mirrors is adjusted according to the sun radiation requirements.[20] The application of mysterious chemicals to preserve the ozone layer is also represented, as well as the seeding of clouds, basically a short-term manipulation of the weather that has been explored and tested in experiments since the late 1940s.[21]

In the area of carbon dioxide removal (CDR), scientists emphasize the positive effects of reforestation, desert greening and the cultivation of genetically modified plants. Apart from this, more controversial technologies are suggested, such as carbon dioxide sequestration, where CO_2 is chemically filtered out of the air directly at the source of emission and dumped into deep geological formations, or ocean fertilization, which is expected to accelerate the growth of phytoplankton. Here, photosynthesis would transform atmospheric carbon dioxide into organic carbon and reduce the carbon dioxide concentration of the air. Similar visualizations—like the one that accompanied the Kiel Earth Institutes' scoping report "Large-Scale Intentional Interventions into the Climate System? Assessing the Climate Engineering Debate" from September 2011 (Figure 2) commissioned by the German Federal Ministry of Education and Research (BMBF)—also show airplanes that release reflecting aerosols and the sea-going hardware designed by Stephen Salter, Emeritus Professor of Engineering Design at the University of Edinburgh.

Figure 1: Lawrence Livermore National Laboratory: schematic representation of various geoengineering and carbon storage proposals, 2008, (original image is colored).

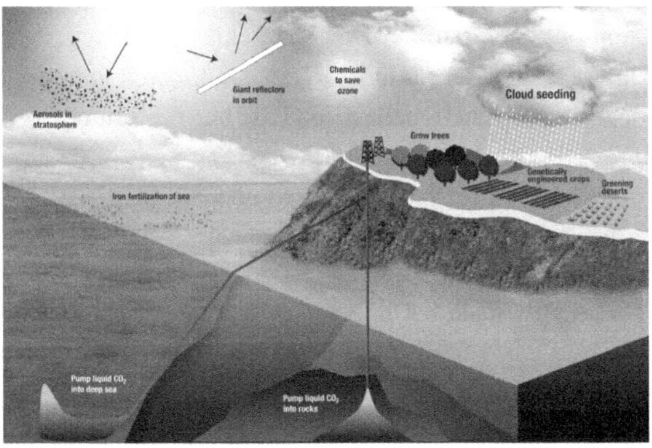

20 Oberth, "*Rakete zu den Planetenräumen*," 1923; Early, "Space-Based Solar Shield to Offset Greenhouse Effect," 1989; Angel, "Feasibility of Cooling the Earth," 2006.

21 Kwa, "The Rise and Fall of Weather Modification," 2001.

Large fleets of these wind-driven spray vessels would float on the oceans and distribute sea salt particles to boost marine stratocumulus clouds to increase the short wave albedo.[22] Furthermore, air capture devices are presented, so-called synthetic "trees" that would scrub carbon dioxide directly from the atmosphere as proposed by David Keith, Harvard physicist, TED-conferencier and founder of the start-up "Carbon Engineering," which was initially also funded by Bill Gates.[23] In contrast to temporary and spatially limited manipulation of the weather, climate engineering addresses the interactions of the atmospheric events in their entirety, even if not all of these technologies will be applied on a global scale.

The assumption that LLNL's schematic overview, produced by a graphic designer in 2008, was to be the prototype for a whole series of very similar visualizations turned out to be wrong, an enquiry to this effect at the LLNL was denied.[24] They claimed to have used older pictures found in other publications and fora. In fact it seems that this visiotype[25] first appeared in an article by David Keith in Nature in 2001.[26] Keith used the visualization from the climatologist Ben Matthews, who compiled already in 1996 a critical review of climate engineering proposals, their scientific and political context, and possible impacts for the Scientists For Global Responsibility (SGR).[27] However, since 2008, a multitude of similar images have been published in the reports of technology assessment institutes and engineering start-ups, in scientific articles and newspapers etc., and have found their way into the imaginations of scientists and politicians and—via the mass-media—also into those of a wider public. A simple Google image search with the term climate or geoengineering provides a good impression on the variation of this particular image type and its enormous range. An enquiry at the Kiel Earth Institute also showed that previous examples of similar images have functioned as models, but the sources could not be cit-

22 Latham et al., "Global Temperature Stabilization," 2008; Salter et al., "Sea-going Hardware," 2008.

23 Lackner, "Washing Carbon out of Air," 2010; Keith, "Climate Strategy with CO_2 Capture from the Air," 2005, Keith's TED-Talk from 2007 is documented here: http://www.ted.com/talks/david_keith_s_surprising_ideas_on_climate_change.html, posted November 2007, accessed December 2, 2013.

24 Anne M. Stark, e-mail message to author, January 14, 2013.

25 Pörksen, *Weltmarkt der Bilder*, 1997.

26 Keith, "Insight Feature Geoengineering," 2001.

27 Matthews, "Climate Engineering," 1996.

ed.[28] It is interesting though, that for the most part more or less external graphic designers not on the institutes' payroll were commissioned to produce these images. This implies that they are not the outcome of a reflected process of production within the institutes, by which the potential implications of specific designs could be reviewed. According to the statements of the images' publishers, they are made "merely" to give an overview on the discussed proposals and to enable comprehension. Thus their function is "only" documentary and communicative. The fact that such a purely synoptical character of images is highly unlikely to exist, as images tend to create realities beside that which they supposedly merely illustrate has been the long-standing position of studies in iconic criticism. Particularly images that are based on digital technologies, with their versatile dramaturgic capabilities, virtually demand active intervention in the design process. Scientific images abandoned long ago the realm of imagined objectivity, with its ascetic practices of image production. The interpretation of images does not follow their generation, instead, production and interpretation are processes that interpenetrate each other.[29] Therefore the images constitute a particular field of analysis in a way that transcends their purported substituting role—in this case, a mere neutral synopsis. They take on the role of models, and the prototypes that they refer to are both representations that deliver reference patterns and imaginary visualizations that only exist in anticipation.[30] This is why these visualizations must to be recognized as visual inventions that generate their own performative capacity beyond the intentions of their producers or, alternatively, conceal their underlying purpose. From this perspective, the images reveal an enlightening arena for contemporary image politics, which is to be examined in the following section.

IV

It is immediately conspicuous that the portrayed climate engineering technologies are in all three examples directly embedded into a natural landscape (Figures 1, 2, 3). Mostly, the images show a cross section, or a lateral cut through the different geological spheres: atmosphere, hydrosphere, biosphere,

28 Erven, telephone interview with author, January 7, 2013 and Ulrike Bernitt, e-mail message to author, January 7, 2013.

29 Heintz and Huber, "Der verführerische Blick," 2001, 22.

30 Ibid., 32.

Figures 2: Kiel Earth Institute: schematic representation of various geoengineering proposals, 2011. Figure 3: United States Government Accountability Office: schematic representation of various geoengineering proposals, 2011.

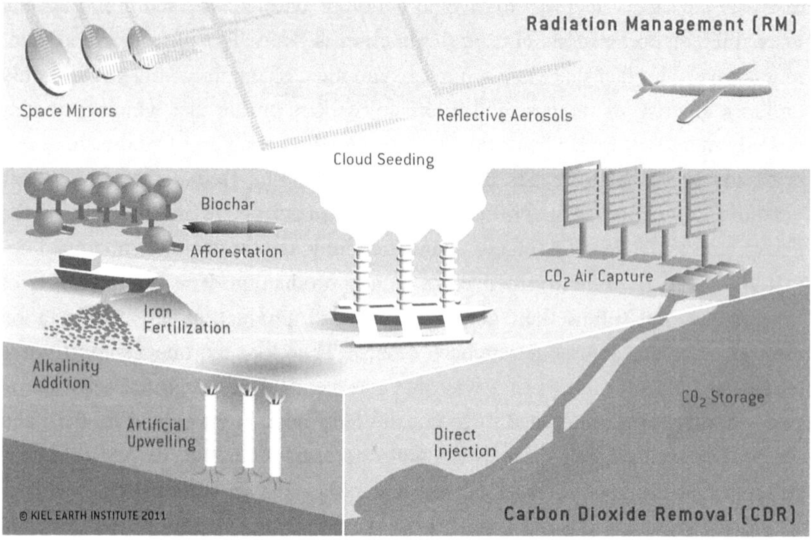

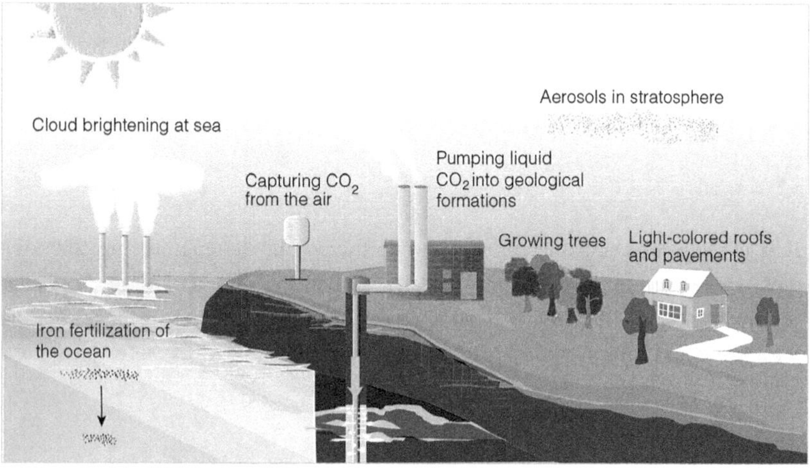

pedosphere and lithosphere—and the sun. Space technological systems and natural cycles are assimilated in a coherent image. Technology is presented as if it optimizes and regulates certain natural metabolisms and cycles and is thus naturalized. The oblique angle is typical for all of the images. This specific perspective enables an uninterrupted view, taking in thegreatest possible surrounding area, and has a certain tradition in landscape esthetics that was pointed out by the geographer and geopolitician Yves Lacoste: the elevated position generally reduces the covered parts of the surrounding area. Initially, such positionings were important for tactical, and therefore, military purposes. At the same time, they offered rare and privileged vistas, culturally encoding these elevated positions as pleasant views.[31] Today, these vantage spots are popular with tourists and hence commercially attractive. Lacoste also indicates that landscape esthetics has developed in a way that contributes to the obfuscation of certain problems: oblique angles from increasingly distant perspectives tend to alienate landscapes and present them to the observer as nothing but harmonically ordered and abstract arrangements. "What is beautiful is good. In light of the order that he discovers and admires due to its shape and colors, he feels nearly as majestic as a worldly sovereign, or as exalted as god, blithely contemplating his creation,"[32] according to Lacoste. This very perspective is used in all the images discussed here. They simulate the impression of objectivity, the "view from nowhere" (Thomas Nagel) that must be declared obsolete in the age of technoscience. Even though these landscapes are computer-generated, they are naturally still tied to a specific landscape esthetic powerful enough to suggest to the observer a sovereign overview, a view that is able to control the whole scenery. The landscape serves as an utterly empty abstract display for technology.[33]

The scale of the land that the climate engineering sites would actually cover and their scenic consequences remain unclear, though. The images do not contain information on that question, although the climate engineering proposals are basically designed to produce full-scale effects on the climate.

The objective and neutral synopsis invoked by the producers of the images is also problematic on another level. The images only show a rather small excerpt, indeed: the visionary, if not salvation-promising, part of the discourse. The recipes for slowing down climate change or saving carbon dioxide suggest the technological feasibility of controlling the climate. The balanced character of a large number of the reports—at least of the governmental reports—means the

31 Lacoste, "A quoi Sert le Paysage? Qu'est-ce un Beau Paysage?," 1995, 43.

32 Ibid., 70.

33 cf. Morse, "Nature Morte," 1996.

ethical considerations and critical evaluation of potential side-effects are completely omitted from the images. They basically show very simple cycles, cause-and-effect chains, functional mechanisms, thereby incidentally referring to other established climate research visualizations like those systemic images in which natural cycles are transferred into mechanical structures by the use of modularized boxes, arrows, figures and similar design elements (Figure 4). The intrinsically complex nature of the climate system and the effectively unknown side effects of such large-scale interventions are central to scientific discussions, but they obviously cannot be easily represented in images. In this respect, these visualizations of climate engineering technologies serve to reduce complexity where complexity might better not be reduced, as the greatest risk effectively lies in an oversimplification of the issue.[34] Lorraine Daston once described this as the aperspectival version of objectivity, in which science is designed according to the needs of communicability: "[...] narrowing the range of genuine knowledge to coincide with that of public knowledge. In the extreme case, aperspectival objectivity may even sacrifice deeper or more accurate knowledge to the demands of communicability."[35] Daston describes this development as a result of the reorganization of scientific practices in the 19th century "that multiplied professional contacts at every level, from the international commission to the well-staffed laboratory."[36] Contemporary globalized climate research is thus a paradigmatic case of regulatory science *par excellence*—a mutual construction between scientific research and policy needs[37]—and this obviously may call for more scientific communications than scientific thought. With this, one should be able to infer that images produced in the positivist milieu of engineering depend on a type of image politics that basically fails to depict the entangled processes of knowledge production in climate research. In an insightful article entitled "Visualization in Engineering" an Emeritus Professor of Mechanical and Process Engineering from ETH Zurich alluded to the fact that such images always function on behalf of product developers, their partners and the later users to eliminate doubts about the fundamental feasibility, functionality and practicability of the product, its recognizability, comprehensibility and

34 cf. Mahony and Hulme, "The Colour of Risk," 2012 or see the article by Mahony and Hulme in this volume.

35 Daston, "Objectivity and the Escape from Perspective," 1992, 600.

36 Ibid.

37 Shackley and Wynne, "Global Climate Change," 1995.

even doubts on the basic feasibility by representing it visually.[38] This is certainly the case for concrete devices like the spray vessels and the synthetic trees. It is indeed problematic when images that originate in an engineering context shift into the sphere of science, because both spheres do not rely on the same principles and values. It thus makes sense to specify the particular "viscourse" created by these images, as they in fact override the more balanced scientific discourse on climate engineering within science. They claim their own "realness" as the experimental proof of the implementability of these technologies, since they visualize them virtually, portraying them as functional mechanisms embedded in a natural and thus familiar environment. The "viscoursive" reduction of the climate engineering discourse creates the impression of a verisimilitude of such schemes, therefore they function primarily as an evident proof.

Figure 4: IPCC: natural and anthropogenic global carbon cycle, 2007.

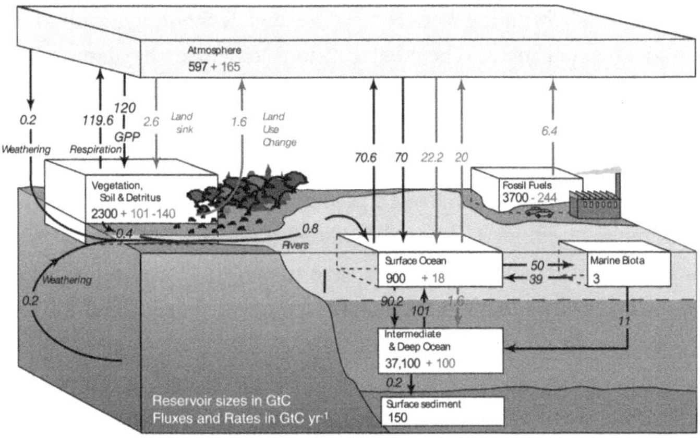

So far, real field experiments are limited to a small number of highly controversial examples,[39] so basically it is left completely to the images to constitute the

38 Breiing, "Visualisierung in den Ingenieurswissenschaften," 2001, 189; see also the articles in *Gegenworte,* "Visualisierung oder Vision?" 2008.

39 cf. 2009 / Russia: Putin's senior science advisor Yuri Izrael conducted an experiment to test the effects of aerosol spraying from a helicopter on solar radiation reaching the ground; January–March 2009 / South Atlantic: Indo-German cooperation project LOHAFEX, 6 tons of iron were deposited in the form of ferrous sulfate in an area of 300 square kilometers; July 2012 / Canada: American entrepreneur Russ George and

common identity of a discourse that is in fact hotly debated among the different parties involved. These images are employed in various contexts, in assessment reports, scientific articles, critical press reports and blogs, which largely demonstrate their adaptability. They are suitable for both the promotion and advertisement of climate engineering. In individual cases they do not necessarily reveal their original context. Therefore the images function as "boundary objects", as they are "both plastic enough to adapt to local needs and constraints of the several parties employing them, yet robust enough to maintain a common identity across sites. They are weakly structured in common use, and become strongly structured in individual-site use. [...] The creation and management of boundary objects is key in developing and maintaining coherence across intersecting social worlds."[40] This is how the sociologists Susan Leigh Star and James R. Griesemer define their concept, which is meant to help to understand the coordinating power and driving force of such objects and images within the processes of knowledge production. The bright and pastel coloring that is sometimes even simply derived from the institutes' corporate colors (Figure 2) conceals the risky bridging between nature and technology, natural cycles and technological experiment. The visualizations disguise the gaps of knowledge that relate to the application of such large-scale technologies, they almost caricaturize the intrinsic potential risk that goes hand in hand with the experimentalization of the climate system. This is how the images contribute to the normalization of the debates around potential applications for climate engineering to counteract climate change. Reducing the discourse to a viscourse of sensational technologies that claim to save the world causes a precarious simplification, which allows them to take on a more powerful role than the discourse.

The naïve and ubiquitous reproduction of the illustrations of climate engineering characterizes a kind of "visual defuturization" of those technologies which Andreas Lösch described as the replacement of futuristic images of the future with contemporary-seeming images of the future.[41] On the one hand this applies because they present them as being embedded into natural cycles and working in analogy to them, so that they are naturalized. On the other hand, on an esthetic level they recall eco-technologies like renewable energy technologies. Thus the images stage climate engineering not so much as futuristic visions

collaborators in the Haida Salmon Restoration Corp. conducted an iron fertilization experiment spreading 100 tonnes of iron sulfate into the Pacific Ocean.

40 Star and Griesemer, "Institutional Ecology, 'Translations' and Boundary Objects," 1989, 393.

41 Lösch, "Visual Dynamics," 2010, 91.

but more as the concrete potential of contemporary knowledge and pre-existing technologies. Furthermore, Lösch establishes the comparison of the visual representations of nanotechnologies. These forms of visual defuturization can be read as an indicator for the increased economization of such discourses. As indicated earlier in this text, this is already the case in the field of climate engineering. Since that shift, the message of nano images has no longer been "to recognize nano by looking into the future." The new mode is rather "to recognize the future by looking at the present market of nano-products." they are already everywhere.[42] This happens in the case of climate engineering in an analogous way with the familiarization of the technologies with processes that happen in nature, or vice versa, by naturalizing the technologies.[43] By implication, the climate itself is represented so to speak as a product of the engineering techniques of nature, now to be stabilized or optimized technologically. Lösch shows how the German Federal Ministry of Education and Research (BMBF), that also commissioned the scoping report at the Kiel Earth Institute, is regulating and economizing the discourses through such assessment publications and the subsequent identification of market potentials. Question of the economical sense and feasibility push social and cultural concerns into the background. Moreover, the visions of climate engineering seem to address the venture-capitalist who counts on the creativity of the free market to solve ecological problems. Like the images of nanotechnologies these futuristic images follow a speculative and techno-scientific rational that has the purpose of accumulating of capital through the virtual invention of investment objects to systematically track down new business opportunities.[44] In this context, the images of climate engineering function as visual models, as agents that render research and investment in climate engineering plausible. Insofar we are not dealing with classical scientific images. They are neither the results of elaborated and complex processes of scientific visualization nor are they essentially related to the process of knowledge production—therefore they have a very limited epistemic function. But these images are produced in the context of scientific expertise and they recur on this framework of legitimacy. They are made to transcend the sphere of science, to communicate and insofar they must allow critical questions.

42 Ibid.

43 cf. Matthews and Turner, "Of Mongooses and Mitigation," 2009.

44 cf. Lösch, "Visual Dynamics," 2010; Gupta, "Geoengineering the Planet?" 2010.

V

The images presented in this article are the products of the contemporary "visioneers"[45] of climate engineering, who will probably become the successors of the space and nano utopians of past decades. In this context, "visioneering" stands for a mode of image production that brings it close to engineering, as the term emphasizes the constructive character of the images, designed to implicate technological feasibility. They evoke a coherent world where urgent problems are solved by the application of technological fixes; with this they may influence the imagination of the observer even more than the expert graphics of climate researchers. Regardless whether climate engineering will ever be applied, one can also deduce from these images—by tracking their original context and perception—how and by whom some significant and some climate research is being orchestrated with very good publicity. It is symptomatic that institutes outsource the production of these images to external graphic designers who are certainly driven by different motivational factors. Their expertise for creating sleek, plausible and esthetically convincing visions supplies the "aperspectival objectivity" that renders them adaptable to different contexts and public spheres. Nonetheless, a critical perspective on these images reveals a planetary environment that is being made fully explicit by technological means, a post-human world where the natural environment serves as the milieu and the input for eco-technological control; as symbolized by the lateral cut through the whole geosphere, including the orbit, climate engineering addresses the climate system as a whole. Therefore, these proposals also support normalization of the planetary perspective, of all-embracing full-scale approaches to solve global problems.

45 McCray, *The Visioneers,* 2013.

BIBLIOGRAPHY

Angel, Roger. "Feasibility of Cooling the Earth with a Cloud of Small Spacecraft near the Inner Lagrange Point." *Proceedings of the National Academy of Sciences* 103, no. 46 (2006): 17184–89.

Berlin-Brandenburgischen Akademie der Wissenschaften. *Visualisierung oder Vision? Gegenworte* 20, (2008).

Bessen, James and Michael J. Meurer. "The Direct Costs from the NPE Disputes." *Boston University School of Law Working Papers* 12, no. 34 (June 2012): without page numbers.

Breiing, Alois. "Visualisierung in den Ingenieurswissenschaften." In *Mit dem Auge Denken. Strategien der Sichtbarmachung in Wissenschaftlichen und Virtuellen Welten*, edited by Bettina Heintz and Jörg Huber, 187–199. Zürich: Springer, 2001.

Crutzen, Paul J. "Enhancement by Stratospheric Sulfur Injections: a Contribution to Resolve a Policy Dilemma?" *Climate Change* 77, no. 3–4 (2006): 211–220.

Crutzen, Paul J. and Eugene F. Stoermer. "The Anthropocene." *IGBP Newsletter*, no. 41 (2000): 17–18.

Daston, Lorraine. "Objectivity and the Escape from Perspective." *Social Studies of Science* 22, no. 4 (1992): 597–618.

Early, James T. "Space-Based Solar Shield to Offset Greenhouse Effect." *Journal of the British Interplanetary Society* 42, (1989): 567–569.

Ewing, Tom and Robin Feldman. "The Giants Among Us." *Stanford Technology Law Review* 1, (2012).

Fettweis, Gerhard P. and Ernesto Zimmermann. "ICT energy consumption—trends and challenges." Proceedings of the 11th International Symposium on Wireless Personal Multimedia Communications (WPMC), Lapland, Finland, September 2008.

Gupta, Arun. "Geoengineering the Planet? Corporate Remaking of the Earth's Atmosphere," *Z-Communications* 23, no. 6 (2010): 31–37.

Heintz, Bettina and Jörg Huber. "Der Verführerische Blick. Formen und Folgen Wissenschaftlicher Visualisierungsstrategien." In *Mit dem Auge Denken. Strategien der Sichtbarmachung in Wissenschaftlichen und Virtuellen Welten*, edited by Bettina Heintz and Jörg Huber, 9–40. Zürich: Springer, 2001.

Hörl, Erich. *Die Technologische Bedingung. Beiträge zur Beschreibung der Technischen Welt.* Berlin: Suhrkamp, 2011.

Hornborg, Alf. "Accumulation: Land as Medium of Domination." In *Ecology and Power. Struggles over Land and Material Resources in the Past, Pre-*

sent, and Future, edited by Alf Hornburg, Brett Clark and Kenneth Hermele, 13–22. London: Routledge, 2012.

Keith, David. "Climate Strategy with CO_2 Capture from the Air." *Climatic Change* 74, (2005): 17–45.

Keith, David. "Insight Feature Geoengineering." *Nature* 49, (2001): 420.

Knorr-Cetina, Karin. "Viskurse in der Physik. Konsensbildung und visuelle Darstellung." In *Mit dem Auge Denken. Strategien der Sichtbarmachung in Wissenschaftlichen und Virtuellen Welten*, edited by Bettina Heintz and Jörg Huber, 305–320. Zürich: Springer, 2001.

Kwa, Chunglin. "The Rise and Fall of Weather Modification: Changes in American Attitudes toward Technology, Nature, and Society." In *Changing the Atmosphere: Expert Knowledge and Environmental Governance*, edited by Clark A. Miller and Paul N. Edwards. Cambridge / Mass.: MIT Press, 2001, 135–165.

Lackner, Klaus S. "Washing Carbon Out Of Air." *Scientific American* 302, (2010): 66–71.

Lacoste, Yves. "A quoi Sert le Paysage? Qu'est ce un Beau Paysage?" In *La Théorie de Paysage en France 1974–1994,* edited by Alain Roger, 42–73. Seyssel: Champ Vallon, 1995.

Latham, John, et al. "Global Temperature Stabilization via Controlled Albedo Enhancement of Low-level Maritime Clouds." *Phil. Trans. R. Soc. A* 366, (2008): 3969–3987.

Lösch, Andreas. "Visual Dynamics: The Defuturization of the Popular 'Nano-Discourse' as an Effect of Increasing Economization." In *Governing Future Technologies: Nanotechnology and the Rise of an Assessment Regime,* edited by Mario Kaiser et al., 89–108. Dordrecht: Springer, 2010.

Mahony, Martin and Mike Hulme. "The Colour of Risk: An Exploration of the IPCC's "Burning Embers" Diagram." *Spontaneous Generations: A Journal for the History and Philosophy of Science* 6, no. 1 (2012): 75–89.

Matthews, Ben. "Climate Engineering. A Critical Review of Proposals, Their Scientific And Political Context, And Possible Impacts." Last modified November, 1996. Accessed Dec. 2013. http://www.chooseclimate.org/cleng/.

Matthews, H. Damon and Sarah E. Turner. "Of Mongooses and Mitigation: Ecological Analogues to Geoengineering." *Environmental Research Letters* 5, (2009): 1–9.

McCray, W. Patrick. *The Visioneers. How a Group of Elite Scientists Pursued Space Colonies, Nanotechnologies, and a Limitless Future.* Princeton / Oxford: Princeton UP, 2013.

Mirowski, Philip. *Science Mart. Privatizing American Science.* Cambridge, Mass.: Harvard UP, 2011.

Morse, Margaret. "Nature Morte: Landscape and Narrative in Virtual Environments." In *Immersed in Technology. Art and Virtual Environments,* edited by Mary Anne Moser, 195–232. Cambridge, Mass.: MIT Press, 1996.

Myhrvold, Nathan and Ken Caldeira, "Greenhouse gases, climate change and the transition from coal to low-carbon electricity." *Environmental Research Letters* 7, (2012): 014019.

Neumann, John von. "Can We Survive Technology?" In *Collected Works*, vol. VI: *Theory of Games, Astrophysics, Hydrodynamics and Meteorology,* edited by A.H. Taub, 504–19. Oxford: Pergamon Press, 1963.

Oberth, Hermann. *Die Rakete zu den Planetenräumen.* Munich / Berlin: R. Oldenburg, 1923.

Parikka, Jussi. *Medianatures. The Materiality of Information Technology and Electronic Waste, Living Books about Life.* Bletchley: Open Humanities Press, 2011.

Pörksen, Uwe. *Weltmarkt der Bilder. Eine Philosophie der Visiotype.* Stuttgart: Verlag, 1997.

Rickels, Wilfried, Gernot Klepper, and Jonas Dovern, editors. *Large-scale Intentional Interventions into the Climate System? Assessing the Climate Engineering Debate.* Kiel: Earth Institute, 2011.

Royal Society. "Geoengineering the Climate. Science, Governance and Uncertainty." Royal Society policy document 10, 09. September 2009.

Rusin, Nikolai P. and Liya A. Flit. *Man Versus Climate.* Moscow: Peace Publishers, 1966.

Salter, Stephen et al.: "Sea-going Hardware for the Cloud Albedo Method of Reversing Global Warming." *Phil. Trans. R. Soc. A* 366 (2008): 3989–4006.

Schneider, Birgit. "Image Politics: Picturing Uncertainty. The Role of Images in Climatology and Climate Policy." In *Climate Change and Policy. The Calculability of Climate Change and the Challenge of Uncertainty*, edited by Gabriele Gramelsberger and Johann Feichter, 191–209. Berlin / Heidelberg: Springer 2011.

Shackley, Simon and Brian Wynne. "Global Climate Change: The Mutual Construction of an Emergent Science-Policy Domain." *Science and Public Policy* 22, no. 4 (1995): 218–30.

Sloterdijk, Peter. *Sphären: Schäume,* vol. 3. Frankfurt am Main: Suhrkamp, 2004.

Star, Susan Leigh and James R. Griesemer. "Institutional Ecology, 'Translations' and Boundary Objects. Amateurs and Professionals in Berkeley's

Museum of Vertebrate Zoology, 1907–1939." *Social Studies of Science* 19, (1989): 387–420.

The Bipartisan Policy Center. "Task Force on Climate Remediation Research. Geoengineering: A National Strategic Plan for Research on the Potential Effectiveness, Feasibility, and Consequences of Climate Remediation Technologies." 2011. http://bipartisanpolicy.org/sites/default/files/BPC%20 Climate%20Remediation%20Final%20Report.pdf.

United States Government Accountability Office / Center for Science, Technology, and Engineering. *Climate Engineering: Technical Status, Future Directions, and Potential Responses*, July 2011. http://www.gao.gov/pro ducts/GAO-11-71.

Victor, David G. M. Granger Morgan, Jay Apt, and John Steinbrunner. "The Geoengineering Option. A Last Resort Against Global Warming?" *The Foreign Affairs* 88, no. 2 (2009).

Vidal, John. "Bill Gates Backs Climate Scientists Lobbying for Large-scale Geoengineering." *The Guardian*, February 6, 2012. http://www.guardian.co.uk/ environment/2012/feb/06/bill-gates-climatescientists-geoengineering.

IMAGE COPYRIGHTS & SOURCES

Figure 1: © Lawrence Livermore National Laboratory, Kathleen Smith, publ. 27.05.2008 https://www.llnl.gov/news/newsreleases/2008/NR-08-05-04. html.

Figure 2: © Kiel Earth Institute, Rita Erven, 2011 http://www.kiel-earth-institute.de/scoping-report-climate-engineering.html.

Figure 3: © United States Government Accountability Office GAO, Center for Science, Technology, and Engineering: Climate engineering: Technical status, future directions, and potential responses. GAO, July 2011, GAO-11-71. http://www.gao.gov/assets/330/322208.pdf.

Figure 4: © IPCC, Working Group I, Climate Change 2007, The Physical Science Base, S. 515. http://www.ipcc.ch/pdf/assessment-report/ar4/wg1/ar4-wg1-chapter7.pdf.

Contributing Authors

Doyle, Julie, is a Reader in Media in the Faculty of Arts at the University of Brighton, UK. She is the author of *Mediating Climate Change* (2011) and co-editor of *Citizen Voices: Performing Public Participation in Science and Environment Communication* (2012). She is on the Board of Directors of the International Environmental Communication Association (IECA) and is a co-founder and former vice-chair of the Science and Environment Communication Section of ECREA (European Communication Research and Education Association). From 2009-2011, she participated in a Leverhulme Trust Funded Artist in Residence project with the artist, David Harradine, to explore new ways of visualising climate change. In 2011 she co-organised the *Visualising Science and Environment* symposium at the University of Brighton which brought together academics, artists, educators and NGOs from across the globe with the aim of promoting collaborative research and practice.

Feulner, Georg, studied physics at the Ludwig-Maximilans-Universität München (LMU, Germany) and the University of Cambridge (UK). He obtained a PhD in 2004 with a dissertation on observational cosmology before moving to the Potsdam Institute for Climate Impact Research (PIK, Germany) in 2006. Since 2012 he has served as deputy chair of the department "Earth System Analysis" at PIK and leads two groups on climate model development and deep-time paleoclimatology, respectively. His research interests include the influence of solar activity changes on the climate, Earth's energy budget and various paleoclimate problems ranging from the faint young Sun paradox to snowball Earth episodes and mass-extinction events in Earth's history.

Fleming, James Rodger, Ph.D. history, Princeton University, is a professor of science, technology, and society at Colby College, Maine and a visiting lecturer and officer at Columbia University in the City of New York. He has written extensively on the history of weather and climate, including social, cultural, and intellectual aspects. He is a fellow of the American Association for the Advancement of Science and the American Meteorological Society, series editor of Palgrave Studies in the History of Science and Technology, and chair of the AAAS Section on Societal Impacts of Science and Engineering. Jim's books include *Meteorology in America, 1800–1870* (1990), *Historical Perspectives on Climate Change* (1998), *The Callendar Effect* (2007), *Fixing the Sky: The Checkered History of Weather and Climate Control* (2010), and *Toxic Airs: Body, Place, Planet in Historical Perspective* (2014). He is currently writing a book called *Convergence: Atmospheric Science, Technology, and Society, 1900–1960.*

Grittmann, Elke, studied art history, journalism and mass communication and political communication at the University of Hamburg and obtained her Ph.D. in 2006 with a dissertation on the representation of politics in visual news coverage, which was published in 2007 under the title *Das Politische Bild. Fotojournalismus und Pressefotografie in Theorie, Praxis und Empirie* (Cologne: von Halem). From 1998 to 2009, she has worked as lecturer and scientific assistant at the University of Hamburg and the Leuphana University of Lüneburg. She has been a visiting professor at the University of Augsburg, the Westfälische University of Münster. Currently she has a visiting professorship at the Leuphana University of Lüneburg. Her main research fields are visual communication, sociocultural theories and transcultural communi-cation, methods of visual analysis, memory culture and journalism, gender and media.

Heine, Ulrike, studied art history, cultural sciences, and Russian philology at the University of Leipzig and the State University of Saint-Petersburg, Russia. In 2010, she earned a Master's degree with a thesis on the perception of the "Blue Noses," a contemporary group of artists from Russia. She started working on her PhD project on photographic images of climate change in the same year. From 2010 to 2013, Heine was a doctoral fellow of the research training group "Transnational Media Events from Early Modern Times to the Present" at the University of Giessen.

Hulme, Mike, is Professor of Climate and Culture in the Department of Geography at King's College London. He is interested in the numerous ways in which climate change is deployed in public and political discourse around the world, drawing on insights from human geography, environmental history, political philosophy and science and technology studies. He joined King's College after 25 years at the University of East Anglia, where he researched and taught the physical sciences of climate change, climate policy, and the cultural politics of environmental change. He is the author of *Why We Disagree about Climate Change* and *Exploring Climate Change through Science and In Society.*

Mahony, Martin, is a research associate in the Department of Geography at King's College London. He recently completed a PhD on the 'epistemic geographies' of climate change at the University of East Anglia, which investigated the role of various boundary objects in the science and politics of climate change, such as scientific visualizations, simulation models, and global warming targets. He is currently developing new projects on the history of climate modelling in the UK, and on visual and narrative representations of the Anthropocene and environmental migration.

Nocke, Thomas, studied computer science and computer graphics at the University of Rostock and obtained his Ph.D. in 2007 with a dissertation on data visualization for climate research which was published with the title *Visuelles Data Mining und Visualisierungsdesign für die Klimaforschung* (in German). Since 2006 he has been investigating the interdisciplinary topic of climate data visualization—considering both visual data analysis and visual climate knowledge communication aspects—at the Potsdam Institute for Climate Impact Research, Potsdam, Germany. Since 2012, he has headed the development activities related to the ClimateImpactsOnline.com web portal.

Parak, Gisela, studied art at the Academy of Fine Arts in Munich and art history at the Ludwig Maximilian University Munich and obtained her Ph.D. in 2008 with a dissertation on American photography, which was published in 2009 under the title *The American Social Landscape. Amerikanische Dokumentarfotografie im Wandel des 20. Jahrhunderts*. From 2010 to 2011, she conducted research on her second book as a Terra Foundation Postdoctoral Fellow at the Smithsonian American Art Museum, from 2011 to 2012 as a Volkswagen Visiting Fellow at Washington University St. Louis. In 2013, she edited the volume *Eco-Images: Historical Views and Political Strategies* in the Rachel Carson Center Perspective series. Currently, she is director of the Museum for Photography in Braunschweig.

Reusser, Dominik, studied chemistry and applied system science at Oregon State University and the University of Osnabrück. He obtained his Ph.D. in 2011 with a dissertation in hydrology at University of Potsdam with the title *Combining Smart Model Diagnostics and Effective Data Collection for Snow*

Catchments. Since 2010 he has been working at the Potsdam Institute for Climate Impact Research on the projects ci:grasp and IMACC, which are both related to the presentation of climate impact and adaptation research on web-based platforms. He is speaker of the PIK flagship project on sustainable transition pathways and coordinates the FP7 project TESS: Towards European Societal Sustainablity.

Sayler, Susannah and Edward Morris work with photography, video, writing and installation. Of primary concern are contemporary efforts to develop ecological consciousness and the possibilities for art within a social activist praxis. In 2006, they co-founded *The Canary Project*—a collaborative that produces visual media and artworks that deepen public understanding of climate change (www.canary-project.org). Works from The Canary Project have shown in diverse venues, including: art museums such as The Museum of Contemporary Art/Denver and the Everson Museum of Art (Syracuse, NY); science museums such as the Cleveland Museum of Natural History and the Museum of Science and Industry (Chicago, IL); universities; public art projects; magazines; city halls; etc. The Nevada Museum of Art's Center for Art + Environment is now home to the Canary Project archive. And the ZKM Center for Art and Media in Karlsruhe, Germany will exhibit part of the archive in 2012. In 2008–2009, Sayler and Morris were Loeb Fellows at Harvard University's Graduate School of Design. They are currently teaching in the Transmedia Department at Syracuse University.

Schneider, Birgit, studied art history and media studies, philosophy, and media art at the Karlsruhe University of Arts and Design, Goldsmiths College, London, and the Humboldt University of Berlin. From 1998 to 2003, she worked as a graphic designer. From 2000 to 2007, she was a research associate in the project "Das Technische Bild" at the Hermann von Helmholtz-Zentrum für Kulturtechnik, Humboldt University, where she wrote her dissertation on the history of punched-card weaving. Since 2008, she has been Dilthey Fellow of the Fritz Thyssen Foundation at the Institute for Arts and Media, University of Potsdam with her research project "Images of the Climate: A Typology of Climate Visualization and Its Changes Since 1800." In 2009, she was a substitute professor at the Bauhaus-Universität, Weimar. During 2010–11 she realized the artistic research project *Austertraum* in cooperation with the artist Stefan Saffer within the Initiative for Art and Sustainability ÜBER LEBENSKUNST funded by the German Federal Cultural Foundation. She is author of the book *Textiles*

Prozessieren (2007) and coeditor of the book *The Technical Image. A History of Styles in Scientific Imagery* (2014).

Schrickel, Isabell, studied media and communication studies and art history at the Universities of Berlin and Basel. In 2010, she received her master's degree with a thesis on the media history of weather forecasting. Between May 2011 and April 2013, she worked as a research associate at the DFG-project "Zeit·Bild·Raum: Das Projektionsplanetarium zwischen Medienästhetik und Wissensrepräsentation" at the Technische Universität Berlin. In the years 2009 and 2011, she taught at Humboldt University, in the summer term 2013 at Leuphana University. Since 2013 she has been a PhD candidate at mecs | Media Cultures of Computer Simulation, an Institute for Advanced Study funded by the DFG German Research Foundation. Her research focuses on the influence of computer simulations on the knowledge cultures of climate research.

Tollmann, Vera, studied Cultural Studies at the University of Hildesheim and at Liverpool John Moores University. Her work focuses on the practice and theory of the Internet, the discourse of climate change, and China's reception in the West. She has co-curated exhibitions that questioned the representation of climate change issues in the media (NGBK, Berlin, 2008) and reflected on the notion of sustainability (Kunsthalle Exnergasse, Vienna, 2007). In 2013, she co-curated the 9th edition of Video Vortex conference at the Moving Image Lab of Leuphana University Luneburg and presented her research at 59. Kurzfilmtage Oberhausen, School of Fine Arts of the University of Porto, Multimedija Institute Zagreb and transmediale's reSource 006 event in Berlin. Vera Tollmann's essays and reviews appear most regularly, among other publications, in springerin. Most recent publications include contributions to Tomislav Medak, Petar Milat (ed.): *Idea of Radical Media,* Multimedijalni institut, Zagreb (2013) and Johannes Ebert, Andrea Zell (ed.): *klima kunst kultur*, Steidl (2013).

Walsh, Lynda, is Associate Professor of English at the University of Nevada, Reno. Her research generally treats public constructions of the authority of science—most recently, in environmental debates. Her book *Scientists as Prophets: A Rhetorical Genealogy* (Oxford, 2013), examines the influence of prophetic rhetoric on the ethos or public character of scientists who become embroiled in climate change debates. It also considers the role visual rhetoric plays in these debates. Walsh's first book, *Sins Against Science: The Scientific Hoaxes of Poe, Twain, and Others* (SUNY, 2006), scrutinized the expanding cultural authority of scientists during the 19th century through the lens of mass media hoaxes.

Wrobel, Markus, has a background in both computer science and in information science and holds a PhD from Freie Universität Berlin. His focus is on the challenges posed by developing appropriate and usable information systems in the context of climate change, with a special interest in Human-Computer interaction, user interface design and usability. He works at the Potsdam Institute for Climate Impact Research (PIK) and acts as speaker of the cross cutting activity Impact Services of PIK-CCD. Markus has been responsible for design and development of various interactive, web- or desktop-based applications in the context of climate change communication, including the ADAM Digital Compendium (EC FP6), the MEDIATION Adaptation Platform (EC FP7), and the new version of the Climate Impacts: Global and Regional Adaptation Support Platform (ci:grasp). Currently, he is developing the user interface for the Global Calculator model with the Climate Media Factory.